NUREG-1934

I0482629

EPRI 1023259
Final Report

Nuclear Power Plant Fire Modeling Analysis Guidelines (NPP FIRE MAG)

Final Report

U.S. Nuclear Regulatory Commission
Office of Nuclear Regulatory Research
Washington, D.C. 20555-0001

Electric Power Research Institute
3412 Hillview Avenue
Palo Alto, CA 94303

United States Nuclear Regulatory Commission

Protecting People and the Environment

ELECTRIC POWER
RESEARCH INSTITUTE

Nuclear Power Plant Fire Modeling Analysis Guidelines (NPP FIRE MAG)

NUREG-1934   EPRI 1023259

Final Report
November 2012

U.S. Nuclear Regulatory Commission   Electric Power Research Institute (EPRI)
Office of Nuclear Regulatory Research (RES)   3420 Hillview Avenue
Washington, D.C. 20555-0001   Palo Alto, CA 94304-1338

U.S. NRC-RES Project Manager   EPRI Project Manager
M.H. Salley   R. Wachowiak

DISCLAIMER OF WARRANTIES AND LIMITATION OF LIABILITIES

THIS DOCUMENT WAS PREPARED BY THE ORGANIZATION(S) NAMED BELOW AS AN ACCOUNT OF WORK SPONSORED OR COSPONSORED BY THE ELECTRIC POWER RESEARCH INSTITUTE, INC. (EPRI). NEITHER EPRI, ANY MEMBER OF EPRI, ANY COSPONSOR, THE ORGANIZATION(S) BELOW, NOR ANY PERSON ACTING ON BEHALF OF ANY OF THEM:

(A) MAKES ANY WARRANTY OR REPRESENTATION WHATSOEVER, EXPRESS OR IMPLIED, (I) WITH RESPECT TO THE USE OF ANY INFORMATION, APPARATUS, METHOD, PROCESS, OR SIMILAR ITEM DISCLOSED IN THIS DOCUMENT, INCLUDING MERCHANTABILITY AND FITNESS FOR A PARTICULAR PURPOSE, OR (II) THAT SUCH USE DOES NOT INFRINGE ON OR INTERFERE WITH PRIVATELY OWNED RIGHTS, INCLUDING ANY PARTY'S INTELLECTUAL PROPERTY, OR (III) THAT THIS DOCUMENT IS SUITABLE TO ANY PARTICULAR USER'S CIRCUMSTANCE; OR

(B) ASSUMES RESPONSIBILITY FOR ANY DAMAGES OR OTHER LIABILITY WHATSOEVER (INCLUDING ANY CONSEQUENTIAL DAMAGES, EVEN IF EPRI OR ANY EPRI REPRESENTATIVE HAS BEEN ADVISED OF THE POSSIBILITY OF SUCH DAMAGES) RESULTING FROM YOUR SELECTION OR USE OF THIS DOCUMENT OR ANY INFORMATION, APPARATUS, METHOD, PROCESS, OR SIMILAR ITEM DISCLOSED IN THIS DOCUMENT.

REFERENCE HEREIN TO ANY SPECIFIC COMMERCIAL PRODUCT, PROCESS, OR SERVICE BY ITS TRADE NAME, TRADEMARK, MANUFACTURER, OR OTHERWISE, DOES NOT NECESSARILY CONSTITUTE OR IMPLY ITS ENDORSEMENT, RECOMMENDATION, OR FAVORING BY EPRI.

THE FOLLOWING ORGANIZATIONS PREPARED THIS REPORT:

U.S. Nuclear Regulatory Commission, Office of Nuclear Regulatory Research

Electric Power Research Institute

Hughes Associates, Inc.

National Institute of Standards and Technology

California Polytechnic State University

Westinghouse Electric Company

University of Maryland

Science Applications International Corporation

ERIN Engineering

NOTE

For further information about EPRI, call the EPRI Customer Assistance Center at 800.313.3774 or e-mail askepri@epri.com.

Electric Power Research Institute, EPRI, and TOGETHER...SHAPING THE FUTURE OF ELECTRICITY are registered service marks of the Electric Power Research Institute, Inc.

ABSTRACT

There is a movement to introduce risk-informed and performance-based (RI/PB) analyses into fire protection engineering practice, both domestically and worldwide. This movement exists in both the general fire protection and the nuclear power plant (NPP) fire protection communities. The U.S. Nuclear Regulatory Commission (NRC) has used risk-informed insights as a part of its regulatory decision making since the 1990s.

In 2001, the National Fire Protection Association (NFPA) issued NFPA 805, *Performance-Based Standard for Fire Protection for Light-Water Reactor Electric Generating Plants, 2001 Edition.* In July 2004, the NRC amended its fire protection requirements in Title 10, Section 50.48 of the *Code of Federal Regulations* (10 CFR 50.48) to permit existing reactor licensees to voluntarily adopt fire protection requirements contained in NFPA 805 as an alternative to the existing deterministic fire protection requirements. In addition, the NPP fire protection community has been using RI/PB approaches and insights to support fire hazard analysis in general.

One key element in RI/PB fire protection is the availability of verified and validated (V&V) fire models that can reliably estimate the effects of fires. The U.S. NRC, together with the Electric Power Research Institute (EPRI) and the National Institute of Standards and Technology (NIST), conducted a research project to verify and validate five fire models that have been used for NPP applications. The results of this effort are documented in a seven-volume NUREG report, NUREG-1824 (EPRI 1011999), *Verification & Validation of Selected Fire Models for Nuclear Power Plant Applications.*

This report describes the implications of the V&V results for fire model users. The features and limitations of the five fire models documented in NUREG-1824 (EPRI 1011999) are discussed relative to NPP fire hazard analysis (FHA). Finally, the report provides information on the use of fire models in support of various commercial NPP fire hazard analysis applications.

CONTENTS

FIGURES

TABLES

REPORT SUMMARY

Background

Beginning in the 1990s, when the U.S. Nuclear Regulatory Commission (NRC) adopted the policy of using risk-informed methods to make regulatory decisions whenever possible, the nuclear power industry has been moving from prescriptive rules and practices toward the use of risk information to supplement decision making. Several initiatives have furthered this transition within the fire protection field, including risk-informed, performance-based fire protection programs (FPPs) compliant with Title 10, Section 50.48(c) of the *Code of Federal Regulations* (10 CFR 50.48(c)) and FPP change evaluation under the existing Title 10 Section 50.48 and Regulatory Guide 1.189, *Fire Protection for Nuclear Power Plants*, October 2009. RI/PB fire protection often relies on fire modeling to estimate the effects of fires.

Objectives

- To provide guidance on the proper application of fire models to NPP fire scenarios

- To fulfill the need as a teaching tool and support the NRC/EPRI Fire PRA training

Approach

There were five different fire models evaluated in NUREG-1824 (EPRI 1011999), *Verification and Validation of Selected Fire Models for Nuclear Power Plant Applications*: (1) the NRC's Fire Dynamics Tools (FDT^s), (2) the Electric Power Research Institute's (EPRI) Fire-Induced Vulnerability Evaluation Revision 1 (FIVE-Rev1), (3) the National Institute of Standards and Technology's (NIST) Consolidated Model of Fire Growth and Smoke Transport (CFAST), (4) Electricité de France's (EdF) MAGIC, and (5) NIST's Fire Dynamics Simulator (FDS). To support consistency, the same fire models were used in the development of this report. The project team developed guidance on the selection and application of each model and treatment of uncertainty and/or sensitivity as part of the fire modeling analysis. Based on this guidance, the project team selected appropriate models and conducted fire modeling analyses of eight different fire scenarios of interest in nuclear power plants (NPPs). The results of each analysis were to be documented in a format appropriate for fire model calculation submittals.

Results

This report presents a step-by-step process for using fire modeling in NPP applications. The recommended methodology consists of a six step process: (1) define fire modeling goals, (2) characterize fire scenarios, (3) select fire models, (4) calculate fire-generated conditions, (5) conduct sensitivity and uncertainty analyses, and (6) document the results.

This report is designed to assist fire model users in applying this technology in the NPP environment. There are a number of unique construction and fire hazard attributes associated with NPPs. It was the authors' goal to explore and demonstrate the use of different models for this application. In addition, a fifth module, "Advanced Fire Modeling", has been added to the annual NRC/EPRI Fire PRA training workshop in 2011. This report expands upon the information provided in NUREG/CR-6850 (EPRI 1011989), *EPRI/NRC-RES Fire PRA Methodology for Nuclear Power Facilities*, and will serve as the training material for the Advanced Fire Modeling module.

The report contains a CD to assist the reader in reproducing the examples discussed in the appendices. The CD contains the installation files for CFAST and FDS. It also contains the latest versions (November 2012) of the NUREG-1805 spreadsheets. The CD also contains the input files for each model used to analyze the eight fire scenarios.

EPRI Perspective

The use of fire models requires a good understanding of their limitations and predictive capabilities, and also presents challenges that should be addressed if the fire protection community is to realize the full benefit of fire modeling and performance-based fire protection. EPRI, in partnership with NRC under a Memorandum of Understanding (MOU), will continue to provide training to the fire protection community, using this document to promote fire modeling and gain feedback on how the results of this work may affect known applications of fire modeling. In the long term, model improvement and additional experiments should be considered.

This report supersedes EPRI 10002981, *Fire Modeling Guide for Nuclear Power Plant Applications*, August 2002, as guidance for fire modeling practitioners in NPPs. The report has benefited from the insights gained since 2002 on the predictive capability of selected fire models in improving confidence in the use of fire modeling in NPP applications.

Note: Due to EPRI limitations on distribution of software, the FDT[s] spreadsheets, CFAST, and FDS are not included on the EPRI version of the CD. The FDTs can be obtained from the NRC web site (www.nrc.gov), and CFAST and FDS can be obtained from the NIST web site (www.nist.gov).

Keywords

Fire
Verification and Validation (V&V)
Risk-Informed Regulation
Fire Safety
Nuclear Power Plant (NPP)
Fire Modeling

Performance-Based
Fire Hazard Analysis (FHA)
Fire Protection
Probabilistic Risk Assessment (PRA)

PREFACE

This report describes research sponsored jointly by EPRI and U.S. Nuclear Regulatory Commission, Office of Nuclear Regulatory Research (RES). The main purpose of this report is to provide guidance on the application of fire models to nuclear power plant (NPP) fire scenarios and to serve as a teaching tool and support the Advanced Fire Modeling module of the NRC-RES/EPRI Fire PRA Course. The fire modeling analyses presented in this report represent the combined efforts of individuals from RES and EPRI. Both organizations provided specialists in the use of fire models/fire hazard analysis tools to support this work. These results are intended to provide technical analysis of the predictive capabilities of five fire modeling calculation tools. This report is the fifth in a series designed to assist those responsible for performing fire modeling in NPP applications.

In August 2002, EPRI published EPRI 1002981, *Fire Modeling Guide for Nuclear Power Plant Applications*. This report offered step-by-step guidance that analysts could follow when using fire modeling to support nuclear power plant fire protection applications. It also included FIVE-Rev1, an Excel-based library of fire models previously documented by EPRI, and additional models from fire protection literature.

In December 2004, the NRC published NUREG-1805, *Fire Dynamics Tools (FDT^s) Quantitative Fire Hazard Analysis Methods for the U.S. Nuclear Regulatory Commission Fire Protection Inspection Program*. This report provided an introduction to the principles of fire dynamics, and included an Excel-based library of fire models comparable to EPRI FIVE-Rev1.

In a follow-up effort as a part of the NRC/RES-EPRI Memorandum of Understanding (MOU), NRC/RES and EPRI jointly conducted a verification and validation of selected fire models for use in nuclear power plant fire modeling to gain insight into the predictive capabilities of these models. The results of this work were published in NUREG-1824 (EPRI 1011999), *Verification and Validation of Selected Fire Models for Nuclear Power Plant Applications*, May 2007. Using, in part, the findings of this work, the NRC conducted a Phenomena Identification and Ranking Table (PIRT) study to evaluate the current state of knowledge for fire modeling for NPP applications. The results of this work were published in NUREG/CR-6978, *A Phenomena Identification and Ranking Table (PIRT) Exercise for Nuclear Power Plant Fire Modeling Applications*, November 2008.

This document does not constitute regulatory requirements. RES participation in this study does not constitute or imply regulatory approval of applications based upon this methodology.

CITATIONS

This report was prepared by:

U.S. Nuclear Regulatory Commission,
Office of Nuclear Regulatory Research (RES)
Washington, D.C. 20555-0001
Principal Investigator:
D. Stroup

National Institute of Standards and Technology
Principal Investigators:
K. McGrattan
R. Peacock

University of Maryland
Principal Investigator:
J. Milke

Electric Power Research Institute (EPRI)
3412 Hillview Avenue
Palo Alto, CA 94303
Principal Investigator:
R. Wachowiak

Kleinsorg Group Risk Services
Division of Hughes Associates, Inc.
Principal Investigator:
F. Joglar
S. LeStrange
B. Najafi

California Polytechnic State University
Principal Investigator:
F. Mowrer

Hughes Associates, Inc.
Principal Investigator:
S. Hunt

Westinghouse Electric Company
Principal Investigator:
C. Worrell

Science Applications International Corp (SAIC)
Principal Investigators:
D. Birk

ERIN Engineering
Principal Investigator:
K. Zee

This report describes research sponsored jointly by the U.S. Nuclear Regulatory Commission's (NRC) Office of Nuclear Regulatory Research (RES) and the Electric Power Research Institute (EPRI).

The report is a corporate document that should be cited in the literature in the following manner:

Nuclear Power Plant Fire Modeling Analysis Guidelines (NPP FIRE MAG), U.S. Nuclear Regulatory Commission, Office of Nuclear Regulatory Research (RES), Washington, D.C., 2012, and Electric Power Research Institute (EPRI), Palo Alto, CA, NUREG-1934 and EPRI 1023259.

ACKNOWLEDGEMENTS

This report is the result of a multi-year effort by the authors, during which it has been significantly rewritten in response to comments from various reviewers. The authors wish to thank those reviewers for taking the time to provide their thoughts on this document. The authors express appreciation to the members of the peer review panel who provided comments on the original draft of this document in 2009: Professor Jose L. Torero and his students at BRE Centre for Fire Safety Engineering, The University of Edinburgh, Scotland; Professor Frederick Mowrer, formerly of the Department of Fire Protection Engineering, University of Maryland, and currently with the Department of Fire Protection Engineering, California Polytechnic State University; Mr. Patrick Finney, NRC Resident Inspector at Susquehanna Nuclear Plant; Mr. Naeem Iqbal, Fire Protection Engineer, Office of Nuclear Reactor Regulation (NRR), NRC; and Mr. Thomas Gorman, Project Manager, Pennsylvania Power and Light, Susquehanna Nuclear Plant.

Drafts of this report were Noticed twice in the <u>Federal Register</u> for public comment: first on December 29, 2009 (74 FR 68873), and second on August 2, 2011 (76 FR 46331). The authors thank those members of the public who provided comments during the two public comment periods, specifically Robert M. Brady, Schirmer Engineering Corporation; Patricia L. Campbell, GE Hitachi Nuclear Energy; Jason E. Floyd, Hughes Associates, Inc.; Pablo Guardado, Entergy; Michael D. Jesse, Excelon Nuclear; Daeil Kang, Korea Atomic Energy Research Institute; Nancy McNabb, National Fire Protection Association; Daniel Orr, NRC, Region 1; Mark Schairer, Engineering Planning and Management (EPM); and Robert Webster, AREVA, all of whom provided insightful comments on the two drafts of the document.

This report was used as a textbook for pilot offerings of an advanced fire modeling course held as part of the annual NRC/EPRI Fire PRA training in 2011 and 2012. The authors would like to thank the students in those two classes for their many constructive comments that significantly improved the final report.

Meetings were held before the Reliability and PRA Subcommittee of the Advisory Committee on Reactor Safeguards (ACRS) and before the full ACRS committee on March 21, 2012 and July 11, 2012, respectively. We would like to thank the members of the ACRS for their time and efforts in supporting the meetings and reviewing and providing comments and suggestions for improvements to the final report.

Finally, the authors express their thanks to Laurent Gay and Eric Wizenne, Electricité de France (EdF), for reviewing the MAGIC calculations. The authors would also like to acknowledge Mr. Bryan Klein of Thunderhead Engineering and Drs. Nathan Siu and Raymond Gallucci of the NRC for their valuable contributions to this report. The authors also express appreciation to Ms. Aixa Belen, Mr. Ken Canavan, Mr. Stuart Lewis, Mr. Nicholas Melly, Ms. Carolyn Siu, and Mr. Robert Vettori for their reviews of and comments on various drafts of this document. Publication of this document would not have been possible without assistance from the NRC's Office of Administration. Specifically, the authors thank Tojuana Fortune-Grasty and Guy Beltz for their efforts in publishing this report.

LIST OF ACRONYMS

ACH	Air Changes per Hour
ACRS	Advisory Committee on Reactor Safeguards
AGA	American Gas Association
AHJ	Authority Having Jurisdiction
ANS	American Nuclear Society
ASET	Advanced Science and Engineering Technologies
ASME	American Society of Mechanical Engineers
ASTM	American Society for Testing and Materials
ATF	Bureau of Alcohol, Tobacco, and Firearms
BE	Benchmark Exercise
BFRL	Building and Fire Research Laboratory
BRE	Building Research Establishment
BWR	Boiling Water Reactor
CAROLFIRE	Cable Response to Live Fire
CDF	Core Damage Frequency
CFAST	Consolidated Fire Growth and Smoke Transport Model
CFD	Computational Fluid Dynamics
CFR	*Code of Federal Regulations*
CHRISTIFIRE	Cable Heat Release, Ignition, and Spread in Tray Installations during Fire
COR	Code of Record
CSR	Cable Spreading Room
DBD	Design Basis Document
ECCS	Emergency Core Cooling Systems
EdF	Electricité de France
EPRI	Electric Power Research Institute
ERFBS	Electrical Raceway Fire Barrier System
FDS	Fire Dynamics Simulator
FDT^s	Fire Dynamics Tools (NUREG-1805)
FHA	Fire Hazard Analysis
FIVE-Rev1	Fire-Induced Vulnerability Evaluation, Revision 1
FFT	Fast Fourier Transform
FLASH-CAT	Flame Spread over Horizontal Cable Trays
FM/SNL	Factory Mutual & Sandia National Laboratories
FPA	Foote, Pagni, and Alvares
FPRA	Fire Probabilistic Risk Assessment
FRA	Fire Risk Analysis
GRS	Gesellschaft für Anlagen- und Reaktorsicherheit (Germany)
HGL	Hot Gas Layer
HRR	Heat Release Rate
HRRPUA	Heat Release Rate Per Unit Area
HVAC	Heating, Ventilation, and Air Conditioning
IAFSS	International Association of Fire Safety Science
iBMB	Institut für Baustoffe, Massivbau und Brandschutz
ICFMP	International Collaborative Fire Model Project
ID	Identification
IEEE	Institute of Electrical and Electronics Engineers

IPEEE	Individual Plant Examination of External Events
LERF	Large Early Release Frequency
LES	Large Eddy Simulation
LFS	Limiting Fire Scenario
LLNL	Lawrence Livermore National Laboratory
LOL	Low Oxygen Limit
MCC	Motor Control Center
MCR	Main Control Room
MEFS	Maximum Expected Fire Scenario
MOVs	Motor-Operated Valves
MQH	McCaffrey, Quintiere, and Harkleroad
MOU	Memorandum of Understanding
NBS	National Bureau of Standards (now NIST)
NEI	Nuclear Energy Institute
NFPA	National Fire Protection Association
NIST	National Institute of Standards and Technology
NPP	Nuclear Power Plant
NRC	U.S. Nuclear Regulatory Commission
NRR	Office of Nuclear Reactor Regulation (NRC)
PE	Polyethylene
PMMA	Polymethyl-Methacrylate
PRA	Probabilistic Risk Assessment
PVC	Polyvinyl Chloride
PWR	Pressurized Water Reactor
RCP	Reactor Coolant Pump
RES	Office of Nuclear Regulatory Research (NRC)
RG	Regulatory Guide
RI/PB	Risk-Informed, Performance-Based
RIS	Regulatory Issue Summary
RTE	Radiation Transport Equation
RTI	Response Time Index
SBDG	Stand-By Diesel Generator
SDP	Significance Determination Process
SFPE	Society of Fire Protection Engineers
SNL	Sandia National Laboratory
SWGR	Switchgear Room
THIEF	Thermally-Induced Electrical Failure
TP	Thermoplastic
TS	Thermoset
UL	Underwriters Laboratory
V&V	Verification & Validation
WTC	World Trade Center
XPE	Cross-Linked Polyethylene and Neoprene
ZOI	Zone of Influence

NOMENCLATURE

A	area (m^2)
A_{fl}	floor area (m^2)
A_o	opening area (m^2)
A_T	surface area of enclosure boundary (m^2)
c	specific heat, solid material (kJ/kg/K)
c_p	specific heat, gas, constant pressure (kJ/kg/K)
D	fire diameter (m) or optical density (m^{-1})
D^*	characteristic diameter
E	experimental measurement
g	acceleration of gravity (m/s^2)
h_k	heat transfer coefficient ($kW/m^2/K$)
H_c	ceiling height (m)
H_f	height of fire base above floor (m)
H_o	opening height (m)
k	thermal conductivity (kW/m/K)
k_m	constant
k_v	volumetric entrainment coefficient
K	light extinction coefficient (m^{-1})
K_m	mass-specific extinction coefficient (m^2/kg)
L	compartment length (m)
L_f	flame height (m)
m'	total mass per unit length of a single cable (kg/m)
m_c''	combustible mass per unit area of tray (kg/m^2)
$\dot{m}$	mass loss or flow rate (kg/s)
$\dot{m}''$	mass loss rate per unit area ($kg/s/m^2$)
M	model prediction
n	number of cables per tray
P	probability, wall perimeter (m)
$\dot{q}''$	heat flux (kW/m^2)
$\dot{q}_c''$	critical heat flux (kW/m^2)
$\dot{Q}$	heat release rate (kW)
$\dot{Q}^*$	fire Froude number
r	radial distance (m)
r_{cj}	ceiling jet distance (m)
s	specific area (m^2/g)
Δt	burnout time (s)
t	time (s)
T	temperature (°C)
V	volume (m^3)
$\dot{V}$	volume flow rate (m^3/s)
W	compartment width (m), width of tray (m)
x	cell dimension (m)
x_c	critical value
y	product yield (kg/kg)
Y	mass fraction (kg/kg)

z_i	smoke layer interface position above base of fire (m)

Greek:

α	fire growth coefficient (kW/s^2)
δ	model bias factor, difference operator
ΔH	heat of combustion (kJ/kg)
Δp	pressure difference (Pa)
φ	equivalence ratio
Ψ	stoichiometric oxygen to fuel ratio
μ	mean
ρ	density (kg/m^3)
σ	standard deviation
$\tilde{\sigma}_E$	relative standard deviation, experiment
$\tilde{\sigma}_M$	relative standard deviation, model
υ	stoichiometric coefficient, residue yield (kg/kg)

Subscripts:

∞, a	ambient
avg	average
dps	distributed point source
e	effective
f	fuel or fire
fl	floor
i	i^{th} element in set
L	leakage
n	total number of elements in a set
o	opening
O_2	relating to oxygen
p	plastic
ps	point source
req	required
s	soot or smoke
ten	tenability limit
tot	total

1
INTRODUCTION

1.1 Background

In 2001, the National Fire Protection Association (NFPA) issued the first edition of NFPA 805, *Performance-Based Standard for Fire Protection for Light-Water Reactor Electric Generating Plants*, 2001 Edition[1]. Effective July 16, 2004, the U.S. Nuclear Regulatory Commission (NRC) amended its fire protection requirements in Title 10, Section 50.48(c) of the *Code of Federal Regulations* (10 CFR 50.48(c)) to permit existing reactor licensees to voluntarily adopt fire protection requirements contained in NFPA 805 following a performance-based approach as an alternative to the existing deterministic[2] fire protection requirements. One important element in a performance-based approach is the estimation of fire hazard using mathematical fire models. Fire modeling is often used in constructing Fire PRAs to determine the effects of fire hazard so that the associated risk can be quantified.

As part of its fire modeling requirements, NFPA 805 states that "fire models shall be verified and validated" (section 2.4.1.2.3) and that "only fire models that are acceptable to the authority having jurisdiction (AHJ) shall be used in fire modeling calculations" (section 2.4.1.2.1). This is an important requirement because the verification and validation (V&V) of fire models is intended to ensure the correctness, suitability, and overall quality of the method. Specifically, verification is the process used to determine whether a model correctly represents the developer's conceptual description (i.e., whether it was "built" correctly), while validation is used to determine whether a model is a suitable representation of the real world and is capable of reproducing phenomena of interest (i.e., whether the correct model was "built").

In 2007, the NRC's Office of Nuclear Regulatory Research (RES) and the Electric Power Research Institute (EPRI) completed a collaborative project for the V&V of five select fire modeling tools. The results of this study, which was performed under the NRC/RES-EPRI Memorandum of Understanding (MOU), are documented in NUREG-1824 (EPRI 1011999), *Verification and Validation of Selected Fire Models for Nuclear Power Plant Applications*. The National Institute of Standards and Technology (NIST) was also an important partner in developing this publication, providing extensive fire modeling and experimentation expertise. The V&V effort is intended to support the use of fire modeling for various NPP fire hazard analysis applications.

This report builds on the V&V research described earlier by incorporating the results into a set of guidelines and recommendations for conducting fire modeling studies in support of

---

[1] All references in this report to NFPA 805 are specific to the 2001 edition of the standard, which is the code of record (COR) required by 10 CFR 50.48(c).
[2] In nuclear fire protection, the term "deterministic" is typically used to refer to prescriptive requirements while "deterministic" is often used as an adjective with "fire model" in the general fire protection field. Within this report, the usage should be clear from the context in which the word is used.

commercial nuclear industry applications. When the NRC's Advisory Committee on Reactor Safeguards (ACRS) issued a letter to Luis Reyes, Executive Director for Operations, recommending publication of NUREG-1824 (EPRI 1011999), they identified two major items to be included in the user's guide (Wallis, 2006). Specifically, the ACRS recommended that the user's guide include:

- Estimates of the ranges of normalized parameters to be expected in nuclear plant applications
- Quantitative estimates of the uncertainties associated with each model's predictions, preferably in the form of probability distributions

The ACRS indicated that quantitative estimates of the "intrinsic model uncertainty" would be a valuable input in risk-informed as well as non-risk-informed applications. Chapters 2 and 3 address the first ACRS recommendation. Chapter 4 specifically addresses the second ACRS recommendation, that is, the development of V&V results into quantitative estimates of model uncertainty. Finally, the appendices contain examples that illustrate the entire process for several nuclear power plant (NPP) scenarios.

1.2 Objective

The objective of this guide is to describe the process of conducting a fire modeling analysis, principally for commercial NPP applications. The process described in this guide addresses most of the technical elements relevant to fire modeling analysis, such as the selection and definition of fire scenarios and the determination and implementation of input values, sensitivity analysis, uncertainty quantification, and documentation. In addition, requirements associated with fire modeling analyses and analytical fire modeling tools are addressed through generic guidance, recommended best practices, and example applications.

1.3 Scope

1.3.1 User Capabilities

This guide should be used as a complement to, not a substitute for, "user's manuals" for specific fire modeling tools, fire dynamics textbooks, technical references, education, and training. This guide only compiles information and organizes it procedurally for NPP applications. Analysts are encouraged to review the references identified throughout the guide for in-depth coverage of the advantages and the range of applicability of specific models. Once a fire scenario has been selected, this guide will help the fire model user define the necessary modeling parameters, select an appropriate model, and properly interpret the fire modeling results. Since all models are merely approximations of reality, this guide also provides useful insights for translating real configurations into modeling scenarios. Due to the technical nature of this guide, users with the following areas of expertise will benefit the most from it:

- General knowledge of the behavior of compartment fires
- General knowledge of basic engineering principles, specifically thermodynamics, heat transfer, and fluid mechanics

- Ability to understanding the basis of mathematical models involving algebraic and differential equations

This guide focuses on the capabilities of the models selected for V&V. However, some generic guidance is also provided, and most of the discussion is applicable to any fire model of the respective type (algebraic model, zone model, or computational fluid dynamics (CFD) model). Five specific models are discussed in this guide:

(1) The NRC's Fire Dynamics Tools (FDT^S), NUREG-1805 and Supplements

(2) EPRI's Fire-Induced Vulnerability Evaluation, Revision 1 (FIVE-Rev1)

(3) NIST's Consolidated Model of Fire Growth and Smoke Transport (CFAST) Version (6)

(4) Electricité de France's (EdF) MAGIC code Version (4.1.1)

(5) NIST's Fire Dynamics Simulator (FDS) Version (5)

Finally, the user of this document would benefit from a familiarity with the use of fire modeling in a fire protection performance-based approach and NUREG/CR-6850 (EPRI 1011989), *EPRI/NRC-RES Fire PRA Methodology for Nuclear Power Facilities.*

1.3.2 Training Resources

For individuals seeking to enhance or update their expertise in the areas noted in section 1.3.1, there are several resources available, including academic courses, short courses, and written materials.

Information on academic institutions with degree programs or single classes in fire protection engineering can be found at the Society of Fire Protection Engineers (SFPE) web site:

http://www.careersinfireprotectionengineering.com/career_types.htm

A background in engineering fundamentals is essential for fire modelers, especially in the areas of fluid mechanics, thermodynamics, and heat transfer. These subjects are offered at virtually any academic institution with programs in fire protection and/or mechanical, aerospace, civil, and chemical engineering. While general courses provide basic background discussions, courses involving fire applications are preferable, and would be provided by the institutions offering courses or degree programs in fire protection engineering.

In addition to the academic programs, short courses in fire behavior and fire modeling are available through professional and industry associations, such as the SFPE (http://www.sfpe.org) and EPRI (http://www.epri.com).

Key written references on fire behavior and fire modeling include:

INTRODUCTION

INTRODUCTION

ASTM E1355–05a (2005), *Standard Guide for Evaluating the Predictive Capability of Deterministic Fire Models*, American Society for Testing and Materials, West Conshohocken, PA, 2005.

Buchanan, A. H. (2001), *Structural Design for Fire Safety*, John Wiley and Sons, LTD, Chichester, England, 2001.

Babrauskas, V., *Ignition Handbook*, Fire Science Publishers/Society of Fire Protection Engineers, Issaquah WA (2003).

Drysdale, D., *An Introduction to Fire Dynamics*, 3rd Ed., John Wiley, 2011.

Karlsson, B. and Quintiere, J.G., Enclosure Fire Dynamics, CRC Press, 2000

Quintiere, J.G., *Principles of Fire Behavior*, Delmar Publishers, 1998.

Quintiere, J.G., *Fundamentals of Fire Phenomena*, John Wiley, 2006.

NFPA, *Fire Protection Handbook*, National Fire Protection Association, 20th Ed., A.E. Cote, (Editor) 2008.

NUREG-1824 (EPRI 1011999), *Verification and Validation of Selected Fire Models for Nuclear Power Plant Applications*, Nuclear Regulatory Commission, 2007.

SFPE, *SFPE Engineering Guide to Assessing Flame Radiation to External Targets from Pool Fires*, SFPE Engineering Guide, Society of Fire Protection Engineers, Bethesda, MD, March, 1999.

SFPE, *SFPE Engineering Guide to Fire Exposures to Structural Elements*, SFPE Engineering Guide, Society of Fire Protection Engineers, Bethesda, MD, November, 2005.

SFPE, *SFPE Engineering Guide to Piloted Ignition of Solid Materials Under Radiant Exposure*, SFPE Engineering Guide, Society of Fire Protection Engineers, Bethesda, MD, January, 2002.

SFPE, *SFPE Engineering Guide to Predicting Room of Origin Fire Hazards*, SFPE Engineering Guide, Society of Fire Protection Engineers, Bethesda, MD, November, 2007.

SFPE, *SFPE Handbook of Fire Protection Engineering*, 4th Ed., P. DiNenno (Editor), National Fire Protection Association, 2008.

1.4 Fire Modeling Theory

Fire development in compartments is often divided into phases depending on the dominant processes at any given stage of development. Ignition is dictated by the characteristics of the fuel item being ignited (i.e., ignition temperature, geometry, orientation, and thermophysical properties[3]) and the strength of the ignition source. Once the flames are sustained on a burning fuel item, a smoke plume develops, transporting mass and heat vertically as a result of the buoyancy of the smoke (see Figure 1-1). The plume will entrain air as it rises, thereby causing the smoke to cool and become diluted; as a result, the quantity of smoke being transported will increase with increasing elevation. After a smoke plume strikes the ceiling, the smoke travels horizontally under the ceiling in a relatively thin layer, referred to as a ceiling jet. As the ceiling jet travels, the smoke cools with increasing distance from the plume impingement point, in part because of air entrainment into the ceiling jet as well as heat losses from the ceiling jet to the solid ceiling boundary.

---

[3] Thermophysical properties include thermal conductivity, specific heat, and density.

In an ideal situation, once the ceiling jet reaches the enclosing walls, a Hot Gas Layer[4] (HGL) develops. As a result of the continuing supply of smoke mass and heat via the plume, the HGL becomes deeper, and its temperature increases. Other properties of the smoke in the HGL also increase (including concentration of gas species and solid particulates).

Radiant heat from the HGL to other combustibles not involved in the fire increases their temperature. Similarly, the temperature of non-burning combustibles will also increase as a result of receiving thermal radiation from the burning item(s). As the other combustibles reach their respective ignition temperatures, they will also ignite. In some cases, the ignition of many other combustibles in the space caused by heating from the HGL occurs within a very short time span. This is commonly referred to as flashover.

Several aspects of fire behavior may be of interest when applying fire models, depending on the purpose of the modeling application. Analysts may seek to determine the effects associated with heating of targets submerged in smoke or receiving radiant heat from the flames, the response of ceiling-mounted detectors or sprinklers to the fire environment, or other phenomena.

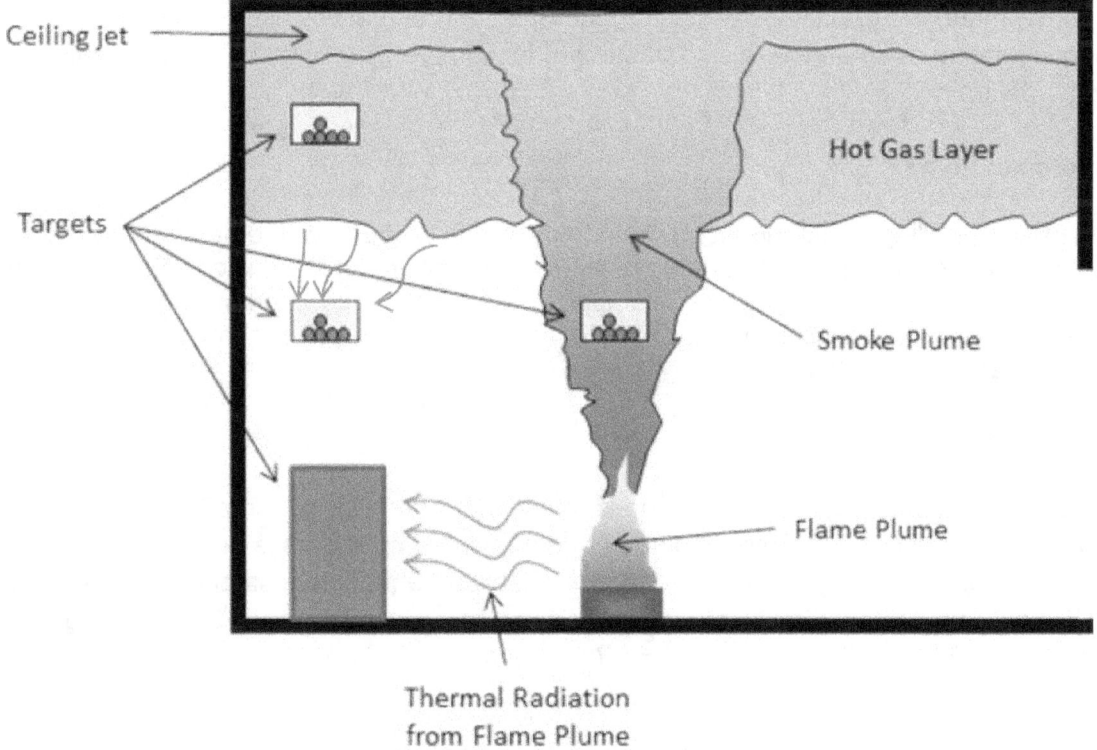

Figure 1-1. Characteristics of compartment fires.

---

[4] Hot Gas Layer or HGL is also called "smoke layer" or "hot upper layer" in other publications in fire protection engineering.

INTRODUCTION

The most common aspects of fire behavior that typically are of interest in such analyses include, but are not limited to:

- Rate of smoke production
- Rate of smoke filling
- Properties of the ceiling jet
- Properties of the HGL
- Target response to incident heat flux via either thermal radiation or convection

A detailed review of each of these aspects is provided in texts on fire dynamics. A brief review of each is provided here.

Rate of smoke production

Smoke is defined as a combination of the gaseous and solid particulates resulting from the combustion process, plus the air that is entrained into the flame and/or smoke plume. Consequently, the rate of smoke production at a particular height in the plume is the combination of the generation rate of combustion products and air entrainment rate into the flame and/or smoke plume between the top of the fuel and the height of interest. In most cases, the air entrainment rate greatly exceeds the generation rate of fuel volatiles. Thus, the correlations used to estimate the rate of smoke production are usually taken from experimental research on entrained air.

Rate of smoke filling

The rate of smoke filling is dependent on the rate of smoke production, the heat release rate (HRR), floor area, height and configuration of the space, and time from ignition. For a fire with a steady HRR, the rate of smoke filling in a compartment will decrease with time due to a decrease in the smoke production rate, which decreases as the height available to entrain air decreases when the HGL deepens.

Properties of the ceiling jet

The ceiling jet transports smoke and heat horizontally away from the plume after it impacts with the ceiling. The response of ceiling-mounted fire detectors or sprinklers is governed primarily by their interaction with a ceiling jet. The temperature and concentration of smoke in a ceiling jet is principally dependent on the height and configuration of the space, distance to the ceiling impact point of the smoke plume, and the HRR of the fire.

Properties of the HGL

As smoke and heat are transported to the HGL via the smoke and fire plumes, the properties of the HGL will change. The principal properties of interest include the depth, temperature, and gas concentrations in the HGL. The magnitude of the properties depends on the HRR of the fire, geometry of the space, ventilation openings (permitting material from the HGL to leave the space, providing air to the fire, and/or causing a stirring action), yields of combustion products, and the elapsed time after ignition. These changes can be tracked by considering the conservation of energy, mass, and species relative to the HGL.

Target response to incident heat flux via either thermal radiation or convection

The targets' temperature will increase as a result of receiving heat via either thermal radiation or convection. Radiation heat transfer is dependent on the intensity of thermal radiation emitted by a source, the size of the source, and the proximity of the target to the source. For this application, the flame height, the portion of heat released from the fire as radiation, and the distance separating the target from the flame are the dominant parameters. Convective heating occurs whenever the target is submerged in the smoke plume or HGL.

1.5 Fire Modeling Tools

1.5.1 Algebraic Models

Algebraic models may be standalone equations found in literature or may be contained within spreadsheets (such as the NRC's FDT^s), and can help give a general understanding of one of the fire environment phenomena. These equations are typically closed-form algebraic expressions, many of which were developed as correlations from empirical data. In some cases, they may take the form of a first-order ordinary differential equation, and, when used properly, can provide an estimate of fire variables, such as; HGL temperature, heat flux from flames or the HGL, smoke production rate, depth of the HGL, and the actuation time for detectors.

Algebraic models are helpful because they require minimal computational time and a limited number of input variables. When applying the results of the algebraic models, users need to be aware that the development of most equations involved approximations to simplify the analysis. Algebraic models are useful primarily as screening tools (i.e., to provide a rough approximation for an analysis, perhaps as a check of an aspect of the results of the computer-based models), and are also applicable when only one phenomenon can be treated in isolation: for instance, plume or ceiling jet correlations are not applicable if there is a significant HGL unless they are modified to account for this effect.

1.5.2 Zone Models

A zone model, such as the Consolidated Fire Growth and Smoke Transport Model (CFAST) or MAGIC, calculates fire environment variables using control volumes, or zones, of a space. The zones correspond to a cooler lower layer and a HGL, as depicted in Figure 1-2. The fundamental idea behind a zone model is that each zone is well-mixed and that all fire environment variables (temperature, smoke concentration, etc.) are therefore uniform throughout the zone. Conditions in each zone are calculated by applying conservation equations and the ideal gas law. The variables in each zone change as a function of time and rely on the initial conditions specified by the user. There is a well-defined boundary separating the two zones, though this boundary may move up or down throughout the simulation.

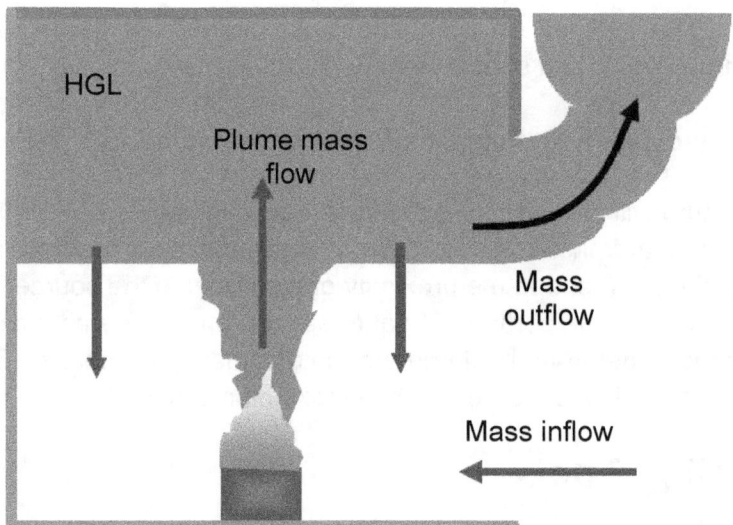

Figure 1-2. A two-zone enclosure fire with an HGL above and a cool lower layer below.

Zone models are most applicable in situations involving simple geometries or where spatial resolution within a compartment is not important. The preparation of input for a zone model, the computation time, and the amount of output data generated are slightly more extensive than a simple algebraic model; however, the overall computational time cost is still low.

Zone models can easily analyze conditions resulting from fires involving single compartments or compartments with adjacent spaces, and are often used to compute the HGL temperature, HGL composition, and target heat fluxes. They are also capable of modeling some effects of natural and mechanical ventilation in both horizontal and vertical directions. Some zone models allow the user to select a thermal plume model, which may assist in better characterization of a known fire scenario, while others use an axisymmetric smoke plume. Other features of a zone model may include a user-specified single zone or multiple fire plumes.

Simulations of spaces with complex ceilings or numerous compartments can be challenging with a zone model. Because zone models specify uniform conditions in the HGL and lower layer, results cannot be distinguished between locations at different distances from the fire. Due to the zone approach, smoke transport time lags are not considered in the simulation, which is an acceptable approximation in relatively small spaces but may lead to significant error in large-volume spaces or spaces with large aspect ratios.

Smoke production, fire plume dynamics, ceiling jet characteristics, heat transfer, and ventilation flows are all algebraic models embedded within zone models. Other parameters that can be calculated with a zone model include thermal behavior, detection response, and suppression response. The output of a zone model is typically simple to understand and is generally presented through an automatic user interface.

Most zone models have default values that must be recognized and adjusted as necessary to obtain an accurate solution. The model user must understand and justify the relevance of the default values used in any application. Fire model users are expected to assess the appropriateness of default values provided in the fire models and make changes or adjust values as necessary. User manuals and technical references for each zone model outline such values and may provide recommended ranges for the parameters.

1.5.3 CFD Models

A computational fluid dynamics (CFD) model is often useful when trying to determine fire variables at a specific location or when there are geometric features that are expected to play a significant role in the results beyond what can be calculated in a zone model approximation. A typical CFD model consists of a preprocessor, a solver, and a postprocessor. CFD models can provide a detailed analysis in both simple and complex geometries.

CFD models essentially apply a series of conservation and state equations across multiple cell boundaries in a space. The number of cell boundaries depends on the mesh size, which breaks the geometry into three-dimensional subvolumes called cells. Solutions to the conservation equations of mass, momentum, and energy are updated as a function of time within each numerical grid cell, with the solutions in all cells collectively describing the fire environment within the geometry at the cell resolution.

The number of grid cells defines the type of mesh. A fine mesh is made up of numerous grid cells. Since the equations are applied at each cell's boundaries, a more detailed distribution of fire parameters is characterized. A coarse mesh is made up of fewer grid cells and can result in less accurate results. The type of mesh and number of grid cells should be based on the geometry and the desired results. If a more detailed simulation is needed, then a finer mesh should be used. Be aware that a finer mesh significantly increases the computational running time of the model as well as the quantity of output data.

CFD models have much better spatial fidelity than zone models, being able to distinguish conditions in one part of the space from another. Because of the appreciable amount of time and effort required to apply CFD models as compared to zone models or algebraic models, CFD models are generally applied when:

- Spatial resolution is important, relative to either the locations of fuel packages or targets.
- Large compartments relative to the fire size are involved.
- Compartments have complex geometries, flow connections, or numerous obstructions in the upper part of the compartment.
- Large numbers of compartments are within the area of interest and the presence of each compartment is expected to affect the fire environment in the area of interest.

An example of a CFD simulation of a fire experiment is shown in Figure 1-3. The purpose of the calculation was to simulate an experiment that was part of the validation study described in NUREG-1824 (EPRI 1011999). In the experiment, a pan fire was placed in a relatively small compartment, and temperatures and heat fluxes were measured at various locations. The CFD simulation is able to describe the changing behavior of the fire as it interacts with its surroundings.

Time: 670.0

Figure 1-3. A Smokeview visualization of a CFD model of a compartment fire experiment.

While CFD models provide a detailed analysis of a space, they are costly to create, simulate, and maintain. The input files created in the preprocessing stage require a significant effort to create. The user must understand the code syntax and the implications and approximations embedded in the model. A firm understanding of fire dynamics is important in providing input data that is relevant to the application. Most CFD models have default values that must be recognized and adjusted as necessary to obtain an accurate solution. The model user must understand and justify the relevance of the default values used in any application. Fire model users are expected to assess the appropriateness of default values provided in the fire models and make changes or adjust values as necessary. User manuals and technical references for each CFD model outline such values and may provide recommended ranges for the parameters.

Depending on the complexity of the scenario and the computer's computational power, the solver within the model can take anywhere from a few hours to weeks to complete all the calculations. This time cost depends on the measured parameters, the size of the geometry, and the mesh size of the calculations. Outputs of CFD models are visualized through a post-processing program. The CFD model developed at NIST, Fire Dynamics Simulator (FDS), employs the program "Smokeview" to represent distributions of temperature, mass, heat flux, burning rate, etc. throughout the geometry. These parameters can be described through point locations, isocontours, or vector diagrams. Output data may also be stored in a comma-separated value file format that can be read by a standard spreadsheet program.

1.5.4 Fire Model Verification and Validation (V&V)

The use of fire models requires a good understanding of their limitations and predictive capabilities. For example, NFPA 805 states that fire models shall only be applied within the limitations of that fire model (section 2.4.1.2.2). ASTM E 1355, *Standard Guide for Evaluating the Predictive Capability of Deterministic Fire Models*, provides definitions of the terms *model verification* and *model validation*.

Model Verification is the process of determining that the implementation of a calculation method accurately represents the developer's conceptual description of the calculation method and the solution to the calculation method. The fundamental strategy of verification of computational models is the identification and quantification of error in the computational model and its solution.

Model Validation is the process of determining the degree to which a calculation method is an accurate representation of the real world from the perspective of the intended uses of the calculation method. The fundamental strategy of validation is the identification and quantification of error and uncertainty in the conceptual and computational models with respect to intended uses.

As noted in Section 1.1, NRC/RES and EPRI conducted a collaborative project for V&V of five fire models. The results of this project were documented in NUREG-1824 (EPRI 1011999), *Verification and Validation of Selected Fire Models for Nuclear Power Plant Applications*.

1.6 Fire Modeling Applications

Fire modeling is used in a variety of NPP applications. Examples include license amendments or exemption requests, fire induced circuit failure analyses, NFPA 805 performance based applications, and fire PRA support. This section provides a brief overview of common fire modeling applications.

1.6.1 License Amendments and Exemptions

Fire modeling has been used to justify requests for changes from the deterministic requirements contained in 10 CFR 50.48 and Appendix R of 10 CFR 50. License amendments and exemption requests are evaluated by NRC staff on a case by case basis.

1.6.2 Fire Induced Circuit Failures

The disposition of certain fire-induced circuit failures, such as multiple spurious operations (MSOs), is another accepted fire modeling application used in commercial NPPs. MSOs typically involve one or more fire-induced component failures that include spurious operation due to hot shorts, shorts to ground, or open circuits as a result of fire damage to electrical cables.

1.6.2.1 Deterministic Application

When one of the redundant safe-shutdown trains in a fire area is maintained free of fire damage by one of the specified means in Regulatory Position 5.3.1.1 of Regulatory Guide (RG) 1.189, Revision 2, then fire modeling may be used to demonstrate that components important to safe shutdown, including systems, structures, and components (SSCs) that are not part of the success path, are protected from fire damage. When fire modeling is used to demonstrate that components important to safe shutdown are protected from fire damage, the analysis would consider all in situ and transient fire sources in the area and all targets that involve components important to safe shutdown. Variations in fire compartment configuration (e.g., doors open or closed, ventilation system on or off, etc.) are to be included in this fire modeling analysis. The analysis needs to clearly demonstrate that the largest expected fire will not damage components important to safe shutdown. The phrase "largest expected" defines a fire that is consistent with the characteristics of the ignition sources and other combustibles in the area where the fire can propagate.

1.6.2.2 Risk-Informed, Performance-Based Application

RG 1.205 provides regulatory guidance on the evaluation of fire-induced circuit failures when used in applications that support risk-informed, performance-based decision making. Within a risk-informed, performance-based NFPA-805 license basis, fire-induced circuit failure[5] scenarios can be evaluated using a fire PRA. When applicable, the fire-induced circuit failure scenarios are included in the plant response model. Chapter 5 of NUREG/CR-6850 provides information on the definition of and methods for developing plant response models, such that the fire-induced circuit failure contribution is included in the quantification of core damage and large early release frequencies. The model needs to address both the possible plant impact caused by spurious operation(s) and the inability to restore equipment operability as a result of fire damage. With the applicable fire-induced circuit failure scenarios modeled in the risk assessment, quantitative risk calculations could be used to show that the scenarios present acceptable risk contributions within the overall plant fire risk profile, as discussed in RG 1.205.

Fire modeling can be used within the framework of the fire PRA to characterize the risk associated with fire scenarios. Determining the risk associated with fire scenarios may include such evaluations as developing a range of fire sizes that may damage cables at varying distances from an ignition source, with the corresponding times that it would take a fire to grow to those damaging sizes. In this context, the fire modeling objective is to predict the fire conditions that could induce cable damage and thus trigger fire-induced circuit failure scenarios, so that the corresponding risk contribution is captured in the analysis. If the fire-induced circuit failure is successfully resolved using fire modeling in a probabilistic approach, it must demonstrate that the damage necessary to cause the fire-induced circuit failure would not occur for the postulated fire scenario (e.g., a specific ignition source is screened from the analysis), or that the risk associated with the fire-induced circuit failure condition is acceptable, given the specific characteristics of the fire necessary to produce the postulated condition.

1.6.3 NFPA 805 Performance-Based Applications

NFPA 805 allows the use of either fire modeling or fire risk evaluations to demonstrate that performance based requirements are satisfied.

---

[5] The term fire-induced circuit failure is intended to address loss of function due to fire damage to cables and equipment as well as spurious operation(s) of that equipment either singly or in multiples.

1.6.3.1 Fire Modeling

The NFPA 805 requirements associated with fire modeling are organized into two sections, Section 2.4.1 and Section 4.2.4.1. Section 2.4.1 describes the requirements associated with the fire modeling calculations and analysis. Section 4.2.4.1 describes the requirements for the implementation of a performance-based fire modeling analysis.

Section 2.4.1.2 of NFPA 805 describes the requirements for the use of fire models, which include:

- The use of fire models acceptable to the authority having jurisdiction (AHJ), i.e., the U.S. Nuclear Regulatory Commission

- The application of fire models within their range and limitations

- Fire models shall be verified and validated

In the context of this application, the specific analytical capabilities within the fire model need to be verified and validated. Model capabilities not invoked in a specific calculation are outside the scope of this requirement. NUREG-1824 (EPRI 1011999) is an example of a verification and validation study for fire models specifically developed for NPP applications.

Section 4.2.4.1 is subdivided by process element as follows:

Identify Targets (NFPA 805 § 4.2.4.1.1): This subsection requires the description of the targets (e.g., equipment or cables) and target locations (specific locations of raceways/conduits containing the cables, electrical cabinets, or equipment) associated with them needed to achieve the nuclear safety performance criteria.

Establish Damage Thresholds (NFPA 805 § 4.2.4.1.2): This subsection requires the description of damage thresholds for the equipment and cables needed to achieve the nuclear safety performance criteria. The damage threshold (i.e., target vulnerability) for cables exposed to fire is expressed in most cases in the form of an incident heat flux on the cables or the cables' surface temperature shall be established in accordance with Section 2.5 of NFPA 805.

Determine Limiting Conditions (NFPA 805 § 4.2.4.1.3): This subsection requires the description of the combination of equipment or required cables with the highest susceptibility to any fire environment.

Establish Fire Scenarios (NFPA 805 § 4.2.4.1.4): This subsection requires the description of a given area's fire conditions resulting from the identified and analyzed fire scenarios. It should be noted that the scenario definition is consistent with the requirements listed under § 2.4.1.3 of NFPA 805. Appendix C of NFPA 805 provides two categories of fire scenarios used in the standard as follows:

Maximum Expected Fire Scenario: The *maximum expected fire scenario* (MEFS) is defined in NFPA 805 Section C.3.2, as the scenario that is used to determine by fire modeling whether the performance criteria are met in the fire area being analyzed. The input data for the fire modeling of the MEFS should be based on the following:

- Existing in-situ combustibles in the fire area

- Types and amounts of transient combustibles that industry experience and specific plant conditions indicate can reasonably be anticipated in the fire area

- Heat release and fire growth rates for actual in-situ and transient combustibles that are realistic and conservative based on available test data and applicable fire experience

- Ventilation within normal operating parameters with doors in the open or closed position

- Active and passive fire protection features operating as designed

Limiting Fire Scenario: The *limiting fire scenario* (LFS) is defined in NFPA 805 Section C.3.3 as fire scenario(s) in which one or more of the inputs to the fire modeling calculation (e.g., heat release rate, initiation location, or ventilation rate) are varied to the point that the performance criterion is not met. The intent of this scenario(s) is to determine that there is a reasonable margin between the expected fire scenario conditions and the point of failure ... [T]he LFS can be based on a maximum possible, though unlikely, value for one input variable or an unlikely combination of input variables. ... The values used for LFS input should remain within the range of possibility but can exceed that determined or judged to be likely or even possible.

For each fire scenario, the environmental conditions resulting from each MEFS are compared to the damage thresholds for the targets in the fire area. If damage thresholds are not exceeded, the targets in the fire area can be considered free of fire damage under the conditions of the postulated MEFS.

By definition, the effects of the LFS include damage to the targets in the fire area under consideration. Fire modeling parameters that have been varied to establish the LFS conditions are identified and described.

1.6.3.2 Fire Risk Evaluations

A fire risk evaluation for the NFPA 805 performance-based approach uses fire PRA methods, tools, and data to compare the risk associated with implementation of the proposed alternative. Sections 2.4.3 and 4.2.4.2 of NFPA 805 contain specific requirements for the use of fire risk evaluations. The fire PRA methods, tools, and data must be acceptable to the AHJ, and they should be appropriate for the nature and scope of the evaluation, be based on the as-built and as-operated and maintained plant, and reflect the operating experience at the plant. The evaluation uses core damage frequency (CDF) and large early release frequency (LERF) as measures for risk. The proposed alternative must ensure that the philosophies of defense in depth and sufficient safety margin are maintained. Section 1.6.4 of this report provides information on using fire modeling in support of fire PRA.

1.6.4 Fire Modeling in Support of Fire PRA

Fire PRA often applies fire modeling in the fire scenario development and analysis process. A fire scenario in a Fire PRA is often modeled as a progression of damage states over time, which is initiated by a postulated fire resulting from an ignition source. Each damage state is characterized by a time and a set of targets damaged within that time. Fire modeling is commonly used to determine the targets affected in each damage state and the associated time at which this occurs. The first damage state usually consists of damage only to the ignition source itself. Depending on the characteristics and configuration of the scenario, the last damage state may consist of an HGL formation that leads to full room damage by exceeding predetermined damage criteria. Damage states between the first and final states capture target sets compromised as the fire propagates through intervening combustibles.

Each scenario progression postulated in a Fire PRA is quantified to determine its contribution to fire risk. The most common fire risk metrics are the reactor Core Damage Frequency (CDF) and Large Early Release Frequency (LERF) of radioactive material. NUREG/CR-6850 provides guidance on the quantification of fire risk.

The Fire PRA standard (ASME/ANS RA-Sa-2008 and Addenda RA-Sa-2009) lists requirements for all the technical elements associated with a Fire PRA, includes specific requirements for the use of fire models. The standard addresses: (1) the selection of appropriate fire modeling tools for estimating fire growth and damage behavior, considering the physical behaviors relevant to the selected fire scenarios, and (2) the application of fire models that are sufficiently capable of modeling the conditions of interest within known limits of applicability. In the case of analytical fire models, the standard requires the use of appropriate fire modeling tools with the ability to model the conditions of interest within known limits of applicability.

1.7 Organization of the Guide

The guidance material provided in this document is divided into five chapters and a number of appendices, as outlined below.

- Chapter 2 presents a qualitative overview of the process for conducting fire modeling, including the basic principles of fire simulation, advantages and limitations of the technology, and brief descriptions of the five models
- Chapter 3 provides specific guidance on selecting models to address typical scenarios in commercial NPPs
- Chapter 4 contains information on determining the sensitivity and uncertainty associated with fire modeling calculations
- Chapter 5 contains the list of references identified throughout this document
- Chapter 6 contains general information on the eight scenarios documented in the appendices
- Appendices A through H provide detailed examples of fire modeling analyses of typical NPP scenarios:
  - Appendix A – Cabinet Fire in Main Control Room
  - Appendix B – Cabinet Fire in Switchgear Room
  - Appendix C – Lubricating Oil Fire in Pump Compartment
  - Appendix D – Motor Control Center Fire in Switchgear Room
  - Appendix E – Trash Fire in Cable Spreading Room
  - Appendix F – Lubricating Oil Fire in Turbine Room
  - Appendix G – Transient Fire in Multi-Compartment Corridor
  - Appendix H – Cable Tray Fire in Annulus
- CD containing the following files:
  - Appendix A Files
    - Algebraic models
    - CFAST and FDS input files
  - Appendix B Files

- Algebraic models
- CFAST and FDS input files
- Appendix C Files
  - FDS and MAGIC input files
- Appendix D Files
  - CFAST and FDS input files
- Appendix E Files
  - Algebraic models
  - CFAST and FDS input files
- Appendix F Files
  - Algebraic model files
  - FDS files
- Appendix G Files
  - Algebraic model files
  - MAGIC input files
- Appendix H Files
  - FDS input files
- Fire Modeling Software
  - CFAST Software Guides
  - FDS Software Guides
  - NUREG-1805 Spreadsheets
- Related Publications

2
THE FIRE MODELING PROCESS

This chapter provides a general step-by-step process for modeling fires in commercial nuclear power plants (NPPs). The recommended methodology comprises six steps: (1) define fire modeling goals and objectives, (2) characterize the fire scenarios, (3) select fire models, (4) calculate fire-generated conditions, (5) conduct sensitivity and uncertainty analyses, and (6) document the analysis. A simplified process involving the six steps is shown in Figure 2-1.

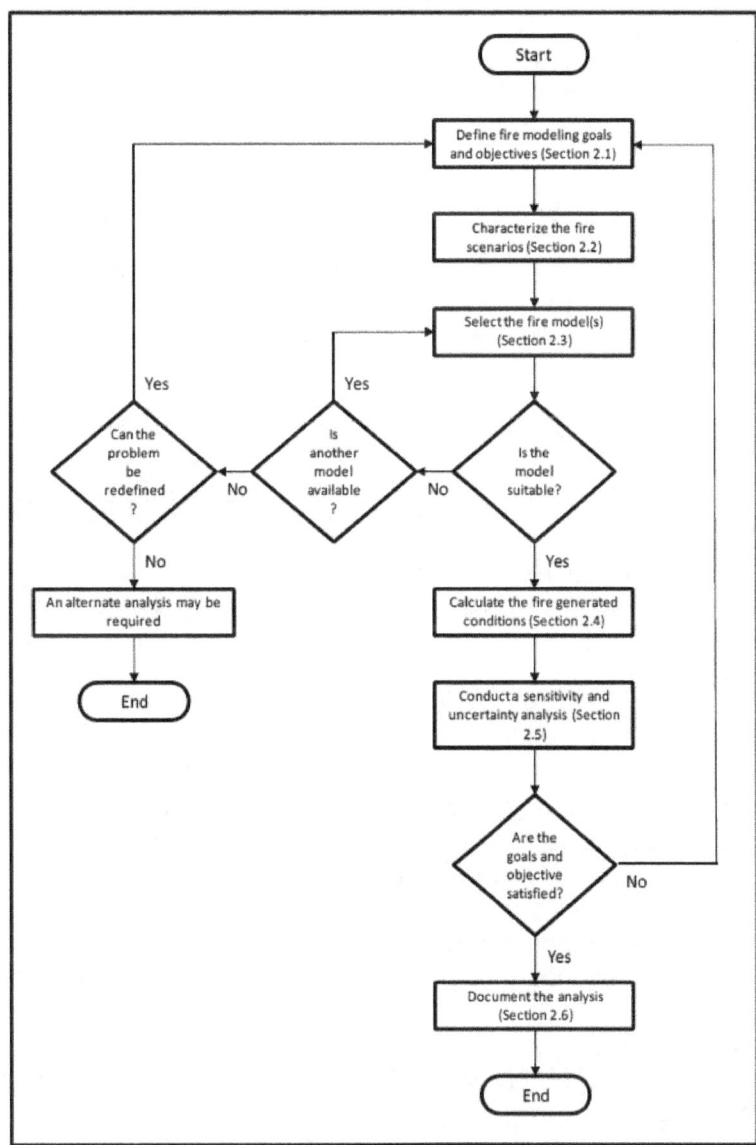

Figure 2-1. Fire modeling process.

2.1 Step 1: Define Fire Modeling Goals and Objectives

The first step in a fire model analysis is to identify and state the fire modeling goals and objectives. A fire modeling goal is a broad statement of what needs to be accomplished using fire modeling (SFPE, 2007). The goal(s) should also identify whether the analysis results are intended to help resolve a deterministic issue or are intended as input for a probabilistic risk assessment (PRA).

The objectives are more specific and describe in engineering terms how the goal(s) will be met (SFPE, 2007). Clearly defined objectives provide focus and are needed to correctly select the fire scenarios that will be evaluated and the fire modeling tools that will be used. In order to define the objectives, some understanding of the conditions by which success or failure are measured (i.e., the performance criteria) is necessary at the analysis outset. Any fire modeling objective may thus be viewed as a statement defining what needs to be accomplished, which criteria will be used to define success or failure, and which analysis process will be followed. The criteria should be stated in terms that can be achieved by the fire modeling analysis.

Some common situations in commercial NPPs where fire modeling may be used to determine which goals and objectives would be developed include, but are not limited to:

- Evaluating whether or when a fire could damage a single electrical cable or component

- Evaluating whether or when a fire could damage multiple electrical cables or components

- Evaluating whether conditions are habitable in an enclosure

- Evaluating the potential for fire propagation through or across a fire barrier

- Evaluating detection or sprinkler actuation

- Evaluating the potential for fire propagation between fire zones or fire areas, or to secondary combustibles

- Evaluating the structural response under fire conditions

The performance criteria are an important aspect for developing the fire modeling objectives and will be specific to the fire modeling application. The performance criteria will often include, but are not limited to one or more of the following:

- Maximum acceptable surface temperature or jacket-insulation interface temperature for a cable, component, secondary combustible, structural element, or fire-rated construction

- Maximum acceptable incident heat flux for a cable, component, structural element, or secondary combustible

- Maximum acceptable exposure temperature for a cable, component, structural element, or secondary combustible

- Maximum acceptable enclosure or key structural element temperatures

- Maximum smoke concentration or minimum visibility

- Maximum or minimum concentration of one or more gas constituents, such as carbon monoxide, oxygen, hydrogen cyanide, etc.

The performance criteria may also involve sequences of events, such as "detection or sprinkler actuation before cable damage, which occurs when the surface temperature exceeds a specified threshold value." NUREG/CR-6850 (EPRI 1011989) provides some performance thresholds for common NPP targets (see Appendix H) as well as for habitability (see Section 11.5.2.11).

A few simple examples will illustrate the various ways in which a fire modeling goal may be stated. In many NPP fire modeling applications, the motivation for a fire modeling analysis is a need to know whether or not an electrical cable or a component remains free of damage from a fire. This could be very specific (i.e., a particular exposure fire exposing a particular cable) or general (i.e., the maximum distance from which a particular type of fire could damage cables). In addition, it may only be necessary for a single fire to damage a single cable, or it may be necessary to simultaneously damage two particular cables with a known separation. The following are examples of goals that would lead to a fire modeling analysis:

- "Ensure that Panel 'X' and Cable 'Y' are not subject to failure given a fire in Room 'Z.'"

- "Tabulate the distance through which an electrical cabinet ignition source could adversely affect electrical targets for use as a screening tool in support of a PRA."

- "Ensure that Cable 'X' and Cable 'Y' both do not fail in Room 'Z' when exposed to a single transient fuel package fire."

- "Quantify the potential benefit of the sprinkler system in Room 'X' for preventing damage to Cable 'Y'."

Each of the goals explicitly states the purpose of the analysis and the means by which success is determined in terms that can be achieved by a fire modeling analysis. The objectives are more specific and may include specific steps that will be followed to satisfy the goal. Some examples of objectives that support the previously cited example goals are as follows:

- "Perform a calculation to determine if a fire in Fire Area 'X' involving Panel 'Y' could cause the surface temperature of Cable 'Z' to exceed a specified threshold value."

- "Evaluate the maximum distance from any surface of an electrical cabinet at which a 98th percentile heat release rate (HRR) fire in Fire Zone 'X' could cause the surface temperature of a cable to exceed a specified threshold value."

- "Perform a calculation to determine if a fire in Fire Area 'X' involving a transient fuel package could simultaneously cause the surface temperature of both Cables 'Y' and 'Z' to exceed a specified threshold value."

- "Determine if any ignition sources in Fire Zone 'X' could damage cables in Raceway 'Y' by causing the surface temperature of any cables in the raceway to exceed a specified threshold value before the sprinkler system actuates."

Thus, the objectives define in specific terms the means by which a goal will be fulfilled and the means by which success is determined.

2.2 Step 2: Characterize Fire Scenarios

The second step in the fire modeling process is to characterize the relevant fire scenarios that capture those technical elements necessary to address the goals. A fire scenario is defined within this guide as a set of elements needed to describe a fire event. These elements usually include the following:

- the enclosure details (i.e., compartment)

- the fire location within the enclosure

- the fire protection features that will be credited

- the ventilation conditions

- the target locations

- the secondary combustibles

- the fire, which is sometimes referred to as the "ignition source"

A number of the fire scenario elements may also be viewed as fire model input. This section provides a broad perspective on the considerations that apply when formulating the appropriate fire scenario, given a fire modeling goal. Chapter 3 provides additional guidance on specific fire scenario elements as they apply to various fire modeling goals and objectives evaluated with a particular fire model.

Note that when characterizing the fire scenarios, preliminary consideration should also be given to how many scenarios are needed to address a particular goal and which specific fire event characteristics each scenario should capture (i.e., which scenarios are needed). In general, at least one fire scenario would be necessary to assess the effects for a single ignition source-target set pair. The analyst should become familiar with the information necessary to develop input files for the fire modeling tools. In practice, this information should be collected during the process of selecting and describing fire scenarios to minimize the number of walkdowns and document/drawing reviews.

2.2.1 General Considerations

Various documents provide guidance for describing fire scenarios from a technical and regulatory perspective. Most of these documents are "application"-specific; for instance, NFPA 805 defines two general categories of fire scenarios, limiting fire scenarios (LFSs) and maximum expected fire scenarios (MEFSs). According to NFPA 805, the input values necessary to determine the MEFS are those that lead to the "most challenging fire that could reasonably be anticipated for the occupancy type and conditions in the space." For LFSs, "one or more of the inputs to the fire modeling calculation (e.g., heat release rate, initiation location, or ventilation rate) are varied to the point that the performance criterion is not met." If there is a "sufficiently large" margin between the MEFS and the LFS, then the fire scenario can be screened from the risk assessment (Gallucci, 2011).

In a Fire PRA, the goal is to quantify the risk contribution from individual scenarios and to identify potential risk-contributing scenarios (e.g., fires impacting important targets in the compartment). Although specific elements in the scenario selection process are "standardized" for guidance and completeness purposes, a certain degree of fire protection engineering judgment is also necessary. NUREG/CR-6850 (EPRI 1011989) contains information on fire frequency, cable (target) selection, HRR, damage criteria, and other information that is useful in developing fire scenarios.

Selected scenarios should represent a complete set of fire conditions that are important to the fire modeling goal. For example, if the goal of the fire modeling analysis is to estimate whether specific cable(s) will remain free of fire damage, the analyst should consider exposures that are close to the cables as well as exposures that are farther away. A small, localized fire exposure could be a greater challenge than a larger fire that is farther away, or vice versa. It may not always be appropriate to select, or at times even possible to define, the worst case fire scenario prior to conducting the analysis, due to the different exposure mechanisms associated with various ignition sources. In large enclosures with a limited number of targets to protect, such as a turbine building in an NPP when the protection of a safety-related circuit is the fire modeling goal, it is easier to locate the targets of interest and then identify those fire sources capable of affecting that target.

When attempting to characterize the fire scenario, plant walkdowns should be an essential aspect of the scenario selection. Many key decisions relevant to fire modeling, including those related to model selection and input parameters, are influenced by observations made during walkdowns. The occupants, the access level to a particular area, and the fire brigade/fire department access should be observed during the walkdowns, as applicable.

It should also be noted that not all the elements associated with a commercial NPP fire scenario can be directly modeled using the tools within the scope of this guide (e.g., the effect of suppression activities by the fire brigade or the conditions in a space after a sprinkler system has actuated). It is important, however, not to limit the scenario selection and description to those elements that can be modeled.

2.2.2 Enclosure Details

The enclosure details include the identity of the enclosures that belong in the fire model analysis, the physical dimensions of the enclosures included in the fire model, and the boundary materials of each enclosure. The enclosure(s) that belong in the fire model may depend on the fire modeling goal, the complexity and connectivity of the spaces in the general area of interest, the type of analysis conducted, and the type of fire model selected. It is possible that no enclosure may be involved, as would be the case for an exterior transformer fire. As a minimum, the space containing the fire would normally be included in the fire model, though treatments involving algebraic plume temperature or flame height correlations would not model the enclosure effects per se. Multiple enclosures might be necessary if there are flow connections (natural or forced) to adjacent areas and if the conditions in both areas could affect the analysis results or are of interest. Care should be taken to consider the potential effects of fires in adjacent areas on the targets of interest. In some cases, a heating, ventilation, and air conditioning (HVAC) recirculation system may involve areas that are fairly remote from the area of interest. Depending on the type of analysis conducted, the conditions within either or both areas may be of interest, and the fire model would thus include both spaces.

The physical dimensions of the enclosure and the boundary materials are model input and should be determined once the fire model has been selected, since the level of detail varies considerably among the fire models. One-zone models may only require a volume and boundary area; two-zone models will typically require the length, width, and height; and computational fluid dynamics (CFD)-type models will require details commensurate with the model grid resolution. The determination of the correct physical dimensions and boundary materials are described in Chapter 3 for various types of NPP fire scenarios.

2.2.3 Fire Location

The location of the fire will depend strongly on the fire modeling goal, the target location, and the fire modeling tool selected. For example, when evaluating the performance of a fire barrier system, fire scenarios challenging the barriers are of interest; when conducting a risk analysis, fire scenarios impacting safety-related circuits may be of primary interest. The selected scenarios for these two applications may not be the same.

When selecting the fire location, the fire scenario should challenge the conditions being estimated. For example, if the goal is to evaluate flame radiation to a target, locating the ignition source relatively far from the target may not provide the best representation of the fire hazards. If the goal is to determine whether a fire can cause two circuits in different raceways to fail, it may be appropriate to locate the fire between the two raceways. There will be situations in which the target location is fixed within the plan area of a space, but there is some flexibility in the vertical placement. A good example of this is an electrical cabinet fire. For a given electrical cabinet, the floor position is fixed; however, the base of the fire is not. Depending on the type of cabinet, it may be appropriate to locate the fire base at the cabinet floor (e.g., open back and containing thermoplastic (TP) cables), at the top (open top and no side vents), or somewhere in between. NUREG/CR-6850 Supplement 1 (EPRI 1019259), *Fire Probabilistic Risk Assessment Methods Enhancements,* recommends a fire elevation equal to the top of the upper vent or 0.3 m (1 ft) below the cabinet top for cabinets meeting certain physical constraints.

In the case of transient fuel package fires or other types of fires that are not fixed, some consideration of the effects of the wall or corner on the upper gas layer temperature is necessary. If the primary exposure mechanism is the Hot Gas Layer (HGL), then locating the fire in a corner or near a vertical boundary will produce higher HGL temperatures. However, the type of analysis may dictate that multiple locations be postulated not necessarily yielding the worst fire effects. Other features that affect the fire location could include the presence or slope of a floor, particularly when a melting plastic or liquid hydrocarbon fuel is considered and transient fuel packages may be staged on mezzanine levels, scaffolding, or platforms. The October 19, 1989, turbine lubricating oil fire at the Vandellos NPP in Spain illustrated this problem. Additional information can be found in NUREG/CR-6738, *Risk Methods Insights Gained from Fire Incidents.*

The following general guidelines and considerations for locating the fire for different fire exposure mechanisms may be followed as applicable:

- Targets in the fire plume or ceiling jet. Locating a source on top of a cabinet ignition source usually results in the most severe fire conditions, since the cabinet walls will not affect fire-generated conditions. Furthermore, since the fire is located in the highest possible position, flames are expected to be higher, and temperatures in the plume and ceiling jet will also be high. The user should judge whether this is conservative based on the goal of the analysis. For example, this would not necessarily be conservative if detection of the fire was a critical aspect of the analysis.

- Targets affected by flame radiation. Combustible materials that are not fixed, such as transient fuel packages and unconfined liquid spills, should be located so that there is an unobstructed (when no passive fire protection system is credited) view between the source and the target. A horizontal path between flame and target provides the highest heat flux to the target.

- Targets engulfed in flames. Flame height calculations should be performed to determine whether the selected location will result in targets engulfed in flames. Proper justification should be provided as to the location of the fire to ensure that the target is out of the flames. For example, consider the case where the analyst locates the fire on top of an enclosed cabinet, resulting in a cable tray engulfed in flames. This would represent the most severe exposure for the cable tray since the fire is expected to start somewhere inside the cabinet. The analyst may choose to lower the fire's position and ignore the cabinet walls after a visual examination identifies the actual location of the combustibles.

- Targets immersed in the HGL. The fire's elevation may influence how far down the HGL will develop as predicted by some fire models, although other important scenario characteristics will also be influential.

2.2.4 Credited Fire Protection

The fire protection features that will be credited in a fire modeling analysis usually require a fire protection engineering evaluation of the system's effectiveness in performing its design objectives. This may include both an assessment of the system compliance with applicable codes, including maintenance and inspection, and an assessment of the system performance against the particular fire scenario considered. The evaluation should determine whether the detection, suppression, and/or passive systems can protect the selected target from fire-generated conditions. Once the decision to credit a fire protection system is made, the analyst should specify the type of system selected for the scenario.

There are several common fire protection features that may be present in a typical NPP area:

- Fire detection systems. These include smoke detectors, heat detectors, or high sensitivity detection systems

- Fire suppression systems. These include automatic or manually activated fixed systems, fire extinguishers, and fire brigades

- Passive fire protection systems. These include structural fire barriers, fire doors, Electrical Raceway Fire Barrier Systems (ERFBSs), radiant shields, and fire stops

- Administrative controls. These include combustible or transient-free zones, combustible fuel load limits, and hotwork procedures

When assessing the performance of a system against the postulated fire hazard, it is necessary to consider the conditions under which the system is designed (fire size, fuel load, exposure temperature, plant operation mode, etc.). For example, an ordinary hazard sprinkler system may not have a sufficient water spray delivery to protect against a large hydrocarbon pool fire. Another example would be passive fire protection systems that are rated against an ASTM

E119 (2008) fire exposure. Under some exposure conditions, such as prolonged flame impingement or large hydrocarbon pool fire scenarios, such systems may not provide sufficient fire resistance. There may be circumstances in which the thermal material properties are well defined, such that the response of the passive systems can be analytically determined. In other situations, it may be necessary to specify at the start of the fire modeling analysis that the passive systems do not protect the target or prevent fire propagation because the exposure exceeds the design capability of the passive fire protection feature.

When considering the effect of active fire protection features, a valid set of response characteristics of the system is needed. For manually actuated fixed suppression systems or manual intervention in the fire modeling analysis, additional information relating to the occupants, the fire brigade, and the fire department are usually necessary. This may include the means of access to the area considered, the presence of a fire watch, the potential for plant personnel to be in the area, etc. Notice that the fire modeling tools within the scope of this guide may not be capable of modeling the impact of some of the fire protection features that may be credited in a given scenario. Credit for active fire suppression may thus be limited to preventing further increases in the HRR or alteration of the HRR profile using suppression curves such as those described in Appendix P of NUREG/CR-6850 (EPRI 1011989) and Chapter 14 of NUREG/CR-6850 Supplement 1 (EPRI 1019259). Nevertheless, fire protection features are designed to impact the outcome of a scenario, so their effects should be included in the analysis.

2.2.5 Ventilation Conditions

Ventilation conditions collectively refer to the operation of the mechanical ventilation system (e.g., the system will continue in normal operational mode, the system will transfer to smoke purge mode, the system will transfer off with closed dampers, etc.) and the position of doors or other openings during the fire event (e.g., doors closed, doors open, doors opening at fire brigade arrival, etc.). Typically, both normal and off-normal ventilation conditions are considered. Spaces in which doors are normally closed may have the doors propped open or opened during the fire by plant personnel, or damaged during the fire. HVAC flows that are normally present in a space may change during the fire due to dampers closing, activation of purge modes, filter plugging, or fan damage by the hot gases. Characterization of the flow field from mechanical devices may be important in some scenarios, especially if the inlet or outlet of the mechanical system is in close proximity to the fire or target.

2.2.6 Target Locations

The target location refers to the physical dimensions of the target relative to the source fire or the fire model coordinate system. These could include its horizontal and vertical distances from the ignition source or source fire, or its spatial position within the room itself. It may be necessary to further specify the location of a vulnerable portion of a target, such as the junction box on a service water pump motor. The orientation of the target with respect to the exposure fire may be of interest as well. An elevated target that is exposed only in the vertically upward direction may be susceptible to thermal radiation from an HGL, but possibly shielded from thermal radiation from the source fire itself. Note that in some types of analyses (e.g., a control room abandonment calculation), occupants may be a target.

The fire exposure mechanisms should also be assessed when quantifying the target location. Fire exposure mechanisms, such as flame impingement, fire plume, ceiling jets, HGLs, and flame radiation, should be considered based on the relative location of the ignition source, intervening combustibles, and the targets. The subsequent fire model analysis should quantify relevant fire conditions and include a discussion of the proper disposition of those that are not expected to affect the target.

2.2.7 Secondary Combustibles

Secondary combustibles include any combustible materials that, if ignited, could affect the exposure conditions of the target set considered. Intervening combustibles, which are typically those combustibles that are located between the ignition source or source fire and the target, are examples of secondary combustibles. However, secondary combustibles would also include combustible materials that are not between the fire and the target but are exposed to the fire effects. In this case, if the secondary combustibles were to ignite, the total HRR in the enclosure (if applicable) would increase, resulting in a hotter gas layer, and the radiant energy from the burning secondary combustible would augment the exposure from the initial source fire, regardless of its location.

Secondary combustibles would include both fixed and transient materials. Typical fixed combustibles include exposed cable jackets or cable insulation, combustible thermal insulation, and combustible wall lining materials. Transient combustibles vary considerably from plant to plant and plant area to plant area, but they may include trash containers, waste accumulations, hoses, hand tools, cleaners and solvents, protective clothing, plastic containers, and so on. It is essential to perform a visual survey of an area to obtain an understanding of the types of combustibles present and the activities in the space, which can provide insight into the types of combustibles that may be present from time to time. The combustible load calculations, fire protection procedures, and fire hazards analysis could provide additional details on the nature of fixed and transient fuel packages in a particular plant area.

Combustible materials in sealed or rated containers may be excluded from consideration if the container is capable of resisting the effects of the fire. Some examples include cabinets containing flammable liquids, solid bottom cable trays with fixed top covers, and bus ducts.

Secondary combustibles will take on the characteristics of a target prior to their ignition (see Section 2.2.6). In this regard, the physical location and orientation of the secondary combustibles with respect to both the ignition source and the target set are determined. The performance criterion for a secondary combustible target is the ignition condition, which will usually be a critical radiant heat flux or exposure temperature or an integrated heat flux. Unlike a true target, once the performance criterion has been met, the secondary combustible is ignited and then takes on the characteristics of a second source fire (see Section 2.2.8).

2.2.8 Source Fire

The source fire is the forcing function for the fire scenario. As all fire effects are directly related to the characterization of the source fire, great care must be taken in characterizing it. A source fire is often described as the "ignition source," which introduces the concept of having both a fuel package and a credible ignition mechanism. There are many ignition mechanisms in a nuclear power plant; however, ignition sources are typically grouped by component or item such as into electrical cabinets, transient fuel packages, self-ignited cable trays, hotwork-ignited cable fires, and overheated motors.

THE FIRE MODELING PROCESS

The source fire is typically characterized by a HRR, though other important aspects include the physical dimensions of the burning object, its composition, and its behavior when burning. The HRR may be specified as a continuous function of time (e.g., proportional to the square of time or a t^2 fire), or it may be an array of HRR and time data. Algebraic models may only permit a constant HRR. There may be situations in which the HRR is a function of the ventilation rather than the object burning. Burning behaviors that may need consideration include whether the material can melt and form a liquid pool, whether it can spread by dripping, and where a liquid could pool. The HRR for many common ignition sources may be developed using the guidance provided in both NUREG/CR-6850 (EPRI 1011989) and NUREG/CR-6850 Supplement 1 (EPRI 1019259). A summary of the information contained in these documents that may be used to develop HRR profiles is as follows:

- Appendix E of NUREG/CR-6850 (EPRI 1011989) provides distributions of peak HRR values as a function of the types (qualified vs. unqualified) and amounts (single vs. multiple bundles) of cables inside vertical electrical cabinets

- Appendix G of NUREG/CR-6850 (EPRI 1011989) provides guidance on defining the transient HRR profiles for electrical cabinets, transient combustible materials, and flammable/combustible liquids

- Appendix R of NUREG/CR-6850 (EPRI 1011989) provides guidance on calculating the flame spread and HRR from cable fires

- Appendix S of NUREG/CR-6850 (EPRI 1011989) provides guidance on estimating fire propagation times between electrical cabinets, which would affect the total HRR profile for a given area

- Chapter 8 of NUREG/CR-6850 Supplement 1 (EPRI 1019259) provides further guidance on fire propagation among groups of electrical cabinets

- Chapter 11 of NUREG/CR-6850 Supplement 1 (EPRI 1019259) provides further guidance on cable tray fires, including applicability of the Appendix R of NUREG/CR-6850 cable fire parameters

- Chapter 17 of NUREG/CR-6850 Supplement 1 (EPRI 1019259) provides further guidance on the fire growth rate for transient combustible materials

When fire modeling is used to support a Fire PRA, the HRR for a source fire may be represented as a conditional probability distribution. In this case, depending on the type of analysis, a screening value may be selected (e.g., a 98th percentile peak HRR as recommended in NUREG/CR-6850), or the effects may be represented using multiple points on the conditional probability distribution.

As was the case with secondary combustibles, combustible materials in sealed or rated containers may be excluded from consideration if the container is capable of resisting the effects of the fire. In addition, self-ignited cable fires are generally postulated only for non-IEEE-383 qualified power cables (NUREG/CR-6850/EPRI 1011989).

2.3 Step 3: Select Fire Models

A number of models are available for performing fire simulations. These models range from algebraic models to sophisticated CFD computer codes that require days to set up a scenario and perform the associated calculations. Given the availability of different models, the analyst is responsible for understanding the advantages and limitations of a particular model in a specific situation in order to achieve the established goals. In general, fire models can be classified into three groups: (1) algebraic models, (2) zone models, and (3) CFD models. The level of effort required to describe a scenario and the computational time consumed by each group increase in the order in which they are listed. Table 2-1 provides a summary of the three groups of models, their advantages and disadvantages, and typical applications.

In practical fire modeling applications, it is likely that a combination of all three types of models would be useful for analyzing a specific problem. For example, algebraic models might be used to estimate the radiative flux to a target for determination of a zone of influence (ZOI) or minimum separation distance. A zone model would provide the temperature of the HGL and height as a function of time for evaluating cable temperatures. CFD model calculations could be used to provide more detailed information on fire-induced conditions in areas where the algebraic models and zone models are not conclusive. Complex models can also be used as a means of estimating the degree of conservatism in a simple model analysis.

The first step in selecting a model is to determine whether the scenario can be analyzed using algebraic models, zone models, or CFD models. This guide focuses on the following five fire models: Fire Dynamics Tools (FDT^s) (NUREG-1805, 2004), Fire Induced Vulnerability Evaluation (FIVE-Rev1) (EPRI 1002981, 2002), Consolidated Fire Growth and Smoke Transport (CFAST) model (Jones et al., 2004), MAGIC (Gay et al., 2005), and Fire Dynamics Simulator (FDS) (McGrattan et al., 2009). The FDT^s and FIVE-Rev1 are a set of relatively simple algebraic models codified in the form of electronic spreadsheets. CFAST and MAGIC represent the class of fire models commonly referred to as zone models, which divide a compartment of interest into two zones, an elevated temperature upper layer and a cool lower layer. FDS is an example of a CFD model, which divides each compartment into thousands or millions of cells. Temperatures and other quantities of interest are calculated for each cell.

Algebraic models can be performed by hand with relatively little computational effort. Karlsson and Quintiere (2000) classify algebraic models into three categories: (1) those that deal with combustion, (2) those that estimate resultant environmental conditions, and (3) those that address heat transfer. Algebraic models related to the combustion process estimate fire intensity based on the flammability characteristics of the fuel. Equations that estimate fire-generated conditions include plume, ceiling jet, and compartment temperatures. Heat transfer equations deal with target temperatures and heat fluxes in the plume, ceiling jet, and lower and upper layer regions.

Zone models are computer algorithms that solve conservation equations for energy and mass. The fundamental approximation associated with zone models is that the enclosure is divided into a limited number of distinct gas zones of uniform properties. In fire applications, the enclosure is usually divided into two zones. The HGL is the volume of smoke generated by the fire and accumulated below the ceiling of the enclosure. This layer is modeled as homogeneous and will have uniform density and temperature. Its temperature and depth are affected by the amount of mass and energy entering or leaving the volume in each time step during the simulation. The lower layer, which can also experience a temperature increase, is characterized by colder fresh air between the floor and the bottom of the HGL. This layer will also have uniform density and temperature.

CFD models are sophisticated algorithms that solve a simplified version of the Navier-Stokes equations. To run CFD codes, the enclosure must be divided into a large number of control volumes (perhaps several million), and the equations solved for each control volume. CFD models then provide a detailed estimate of temperature profiles because calculations are performed for each control volume specified in the enclosure. CFD models also handle turbulent gas flows. Another advantage of CFD models is their ability to simulate fire conditions in geometries other than rectangular floor compartments with flat ceilings. Some CFD models also attempt to estimate HRR values based on fuel flammability properties provided by the analyst. The drawback of CFD models is the computational time and the level of effort required to set up a scenario, as computational times are usually on the order of days. The time required to set up a problem usually depends on the complexity of the geometry.

Another consideration when selecting a CFD-type model is that the amount of detail supplied to the model is significantly greater than it is for the simpler empirical and zone models. Given the large amount of information required for input, there is an intrinsically higher likelihood of errors being introduced into the input, which is different from the model uncertainty and parameter uncertainty described in Chapter 4. Furthermore, the features that may be described in the input could include ductwork, cable trays, electrical cabinets, and other fixed contents that may later be modified, relocated, or removed. New cabinets, cable trays, or other fixed contents that would have been included in the fire model had they been present may be added to an area. Although these changes may be minor, at the very least they would require an assessment by a fire modeler as to whether the original analysis is still applicable or whether the model needs to be adapted for the change. In some situations, such as the determination of a sprinkler actuation time, such small modifications could have a significant effect on the model results.

Table 2-1. Summary of common fire model tools.

Fire Model Class	Examples	Typical Applications	Advantages	Disadvantages
Algebraic models	FDT^S FIVE-Rev1	Screening calculations; ZOI; target damage by thermal radiation, HGL, or thermal plume acting in isolation.	Simple to use; minimal inputs; quick results; ability to perform multiple parameter sensitivity studies.	Limited application range; treats phenomena in isolation; typically applicable only to steady state or simply defined transient fires (e.g., proportional to the square of time or t^2 fires).
Zone Model	CFAST MAGIC	Detailed fire modeling in simple geometries; often used to compute HGL temperatures and target heat fluxes.	Simple to use; couples HGL and localized effects; quick results; ability to perform multiple parameter sensitivity studies.	Error increases with increasing deviation from a rectangular enclosure; large horizontal flow paths not well treated.
Computation Fluid Dynamics Model	FDS	Detailed fire modeling in complex geometries, including computing time to target damage and habitability (main control room (MCR) abandonment or manual action feasibility).	Ability to simulate fire conditions in complex geometries and with complex vent conditions.	Significant effort to create input files and post-process the results; long simulation times; difficult to model curved geometry, smoke detector performance, and conditions after sprinkler actuation.

An important consideration in the fire model selection process is the type of analysis performed. Because of the large number of potential scenarios in a typical NPP, it is usually not practical to default to the most sophisticated tool available. The analysis typically begins with a series of screening analyses (NUREG/CR-6850/EPRI 1011989) performed to identify a subset of fire scenarios and targets that require further analysis with greater resolution. The screening process will typically use fairly simple fire modeling tools, such as algebraic models or generic solutions. When such screening is conducted, it is important to remain within the model limitations and the verification and validation (V&V) basis for the screening model. Section 2.3.6 and Chapter 4 provide additional guidance on the significance of the fire model V&V basis and steps that the user should take to ensure that the fire model is used within acceptable limits.

2.3.1 Fire Dynamics Tools (FDT^s)

Fire Dynamics Tools (FDT^S) is a set of algebraic models preprogrammed into Microsoft® Excel® spreadsheets. The FDT^S library is documented in NUREG-1805, *Fire Dynamics Tools (FDT^s): Quantitative Fire Hazard Analysis Methods for the U.S. Nuclear Regulatory Commission Fire Protection Inspection Program* (NUREG-1805, 2004) and Supplement 1 (NUREG-1805 Supplement 1, 2012). The primary objective of the FDT^S library and the accompanying documentation is to provide a methodology for U.S. Nuclear Regulatory Commission (NRC) fire protection inspectors to use in assessing potential fire hazards in NRC-licensed NPPs. The methodology uses simplified, quantitative fire hazard analysis techniques to evaluate the potential hazard associated with credible fire scenarios.

The FDT^S library includes a suite of spreadsheets that can be used to calculate various fire parameters under varying conditions. Documentation of the theoretical bases underlying the equations used in the FDT^S spreadsheets helps to ensure that users understand the significance of the inputs that each spreadsheet requires, and why a particular spreadsheet should (or should not) be selected for a specific analysis. The governing equations for FDT^S are well established within the fire science community and are documented in handbooks and scientific publications, such as the *NFPA Fire Protection Handbook* (NFPA Handbook, 2008), the *SFPE Handbook of Fire Protection Engineering* (SFPE Handbook, 2008), and other fire science literature.

The complete list of spreadsheets included in the FDT^S library is shown in Table 2-2. A number of the calculation methods included in the FDT^S were part of the V&V study conducted by the NRC, the Electric Power Research Institute (EPRI), and the National Institute of Standards and Technology (NIST) (NUREG-1824 Volume 3, EPRI 1011999, 2007). These spreadsheets are identified in Table 2-2. The NRC maintains a website at http://www.nrc.gov/reading-rm/doc-collections/nuregs/staff/sr1805/final-report/index.html, where both new and updated spreadsheets are posted.

Table 2-2. Routines included in the FDT^s.

FDT^S Spreadsheet Function Name	NUREG-1805 Chapter and Function Description	NUREG-1824 Verification and Validation Status
02.1_Temperature_NV.xls	**Chapter 2.** Predicting Hot Gas Layer Temperature and Smoke Layer Height in a Compartment Fire with Natural Ventilation (Compartment with Thermally Thick/Thin Boundaries): Method of McCaffrey, Quintiere, and Harkleroad (MQH)	V&V provided
02.2_Temperature_FV.xls	**Chapter 2.** Predicting Hot Gas Layer Temperature in a Compartment Fire with Forced Ventilation (Compartment with Thermally Thick/Thin Boundaries): Method of Foote, Pagni, and Alvares (FPA) and Method of Deal and Beyler	V&V provided
02.3_Temperature_CC.xls	**Chapter 2.** Predicting Hot Gas Layer Temperature in a Compartment Fire with Door Closed (Compartment has Sufficient Leaks to Prevent Pressure Buildup; leakage is Ignored): Method of Beyler	V&V provided
03_HRR_Flame_Height_Burning_Duration_Calculation.xls	**Chapter 3.** Estimating Burning Characteristics of Liquid Pool Fire, HRR, Burning Duration, and Flame Height	V&V provided for flame height only

FDT^S Spreadsheet Function Name	NUREG-1805 Chapter and Function Description	NUREG-1824 Verification and Validation Status
04_Flame_Height_Calculations.xls	**Chapter 4.** Estimating Wall Fire Flame Height, Line Fire Flame Height Against the Wall, and Corner Fire Flame Height	V&V not provided
05.1_Heat_Flux_Calculations_Wind_Free.xls	**Chapter 5.** Estimating Radiant Heat Flux from Fire to a Target Fuel (Wind-Free Condition): Point Source Radiation Model (Target at Ground Level); Solid Flame Radiation Model (Target at Ground Level); and Solid Flame Radiation Model (Target Above Ground Level)	V&V provided for the point source model and the solid flame radiation model (above ground) only
05.2_Heat_Flux_Calculations_Wind.xls	**Chapter 5.** Estimating Radiant Heat Flux from Fire to a Target Fuel (Presence of Wind): Solid Flame Radiation Model (Target at Ground Level); and Solid Flame Radiation Model (Target Above Ground Level)	V&V not provided
05.3_Thermal_Radiation_From_Hydrocarbon_Fireballs.xls	**Chapter 5.** Estimating Radiant Heat Flux from Fire to a Target Fuel: Estimating Thermal Radiation from Hydrocarbon Fireballs	V&V not provided
06_Ignition_Time_Calculations.xls	**Chapter 6.** Estimating the Ignition Time of a Target Fuel Exposed to a Constant Radiative Heat Flux Method of Estimating Piloted Ignition Time of Solid Materials Under Radiant Exposures Method of: (1) Mikkola and Wichman; (2) Quintiere and Harkleroad; (3) Janssens; (4) Method of Toal, Silcock, and Shields; and (5) Method of Tewarson	V&V not provided
07_Cable_HRR_Calculations.xls	**Chapter 7.** Estimating Full-Scale Heat Release Rate of a Cable Tray Fire	V&V not provided
08_Burning_Duration_Soild.xls	**Chapter 8.** Estimating Burning Duration of Solid Combustibles	V&V not provided
09_Plume_Temperature_Calculations.xls	**Chapter 9.** Estimating Centerline Temperature of a Buoyant Fire Plume	V&V provided
10_Detector_Activation_Time.xls	Estimating Detector Response Times: **Chapter 10.** Estimating Sprinkler Response Time **Chapter 11.** Estimating Smoke Detector Response Time **Chapter 12.** Estimating Heat Detector Response Time	V&V not provided
13_Compartment_Flashover_Calculations.xls	**Chapter 13.** Predicting Compartment Flashover Compartment Post-Flashover Temperature: Method of Law Minimum Heat Release Rate Required to Compartment Flashover: (1) Method of McCaffrey, Quintiere, and Harkleroad (MQH); (2) Method of Babrauskas; and (3) Method of Thomas	V&V not provided
14_Compartment_Over_Pressure_Calculations.xls	**Chapter 14.** Estimating Pressure Rise Attributable to a Fire in a Closed Compartment	V&V not provided
15_Explosion_Claculations.xls	**Chapter 15.** Estimating the Pressure Increase and Explosive Energy Release Associated with Explosions	V&V not provided

FDT[s] Spreadsheet Function Name	NUREG-1805 Chapter and Function Description	NUREG-1824 Verification and Validation Status
16_Battery_Compartment_ Flammable_Gas_Conc.xls	**Chapter 16.** Calculating the Rate of Hydrogen Gas Generation in Battery Compartments: Method of Estimating Hydrogen Gas Generation Rate in Battery Compartments; Method of Estimating Flammable Gas and Vapor Concentration Buildup in Enclosed Spaces; and Method of Estimating Flammable Gas and Vapor Concentration Buildup Time in Enclosed Spaces	V&V not provided
17.1_FR_Beams_Columns_ Substitution_Correlation.xls	**Chapter 17.** Calculating the Fire Resistance of Structural Steel Members (Algebraic Models): Beam Substitution Correlation (Spray-Applied Materials); and Column Substitution Correlation (Spray-Applied Materials)	V&V not provided
17.2_FR_Beams_Columns_ Quasi_Steady_State_Spray_ Insulated.xls	**Chapter 17.** Calculating the Fire Resistance of Structural Steel Members (Algebraic Models): Heat Transfer Analysis using Numerical Methods Protected; and Steel Beams and Columns (Spray-Applied)	V&V not provided
17.3_FR_Beams_Columns_ Quasi_Steady_State_Board_ Insulated.xls	**Chapter 17.** Calculating the Fire Resistance of Structural Steel Members: Heat Transfer Analysis using Numerical Methods Protected Steel Beams and Columns (Board Materials)	V&V not provided
17.4_FR_Beams_Columns_ Quasi_Steady_State_ Uninsulated.xls	**Chapter 17.** Calculating the Fire Resistance of Structural Steel Members: Heat Transfer Analysis using Numerical Methods Unprotected Steel Beams and Columns	V&V not provided
18_Visibility_Through_Smoke. xls	**Chapter 18.** Estimating Visibility Through Smoke	V&V not provided

2.3.2 Fire Induced Vulnerability Evaluation (FIVE-Rev1)[6]

In August 2002, EPRI published the *Fire Modeling Guide for Nuclear Power Plant Applications* (EPRI 1002981, 2002) for the first time. Since then, it has provided fire protection engineers in the commercial nuclear industry with a broad overview of fire modeling theory and applications, including representative calculations performed with various state-of-the-art fire models. With this guide, EPRI included a library of preprogrammed Microsoft® Excel® equations, which are used to estimate some aspects of fire-induced conditions. This collection of algebraic models is referred to as the Fire Induced Vulnerability Evaluation model (FIVE-Rev1). In general, the equations in the library are closed-form analytical expressions that can be solved by hand. The capabilities of the various equations in the library include predicting temperature and convective heat fluxes in the fire plume or ceiling jet, radiation heat flux, upper-layer temperature, time to detection, and target heating, among others. Some of the equations in FIVE-Rev1 were included in the V&V study (NUREG-1824 Volume 4, EPRI 1011999, 2007). Like the FDT[s], several of the equations used in the examples have not been subject to V&V. Subsequent efforts will be directed at the V&V of these equations and models. The calculations included in the FIVE-Rev1 are summarized in Table 2-3.

---

[6] FIVE-Rev1 is a proprietary software package available to EPRI members (www.epri.com).

Table 2-3. Routines included in FIVE-Rev1.

FIVE-Rev1 Function	Function Description	NUREG-1824 Verification and Validation Status
Qf	Heat release rate profile considering t^2 growth and four stages.	V&V not provided
Firr	Estimates flame irradiation at distance r from the fire source. Point source approximation for REMOTE targets.	V&V provided
FHeight	Flame height based on Heskestad flame height correlation.	V&V provided
TpAlpert	Plume temperature at a specific height based on Alpert plume temperature correlation.	V&V not provided
TpMcCaffrey	Plume temperature at a specific height based on McCaffrey plume temperature correlation.	V&V not provided
TpHeskestad	Plume temperature at a specific height based on Heskestad plume temperature correlation.	V&V not provided
Plcflux	Estimates convective heat flux in the fire plume.	V&V not provided
VpAlpert	Plume velocity at a specific height based on Alpert's plume temperature correlation.	V&V not provided
VpMcCaffrey	Plume velocity at a specific height based on McCaffrey plume temperature correlation.	V&V provided
VpHeskestad	Plume velocity at a specific height based on Heskestad plume temperature correlation.	V&V provided
EpZukoski	Air entrainment into plume based on Zukoski plume entrainment correlation.	V&V not provided
EpThomas	Air entrainment into plume based on Thomas plume entrainment correlation.	V&V not provided
EpHeskestad	Air entrainment into plume based on Heskestad plume entrainment correlation.	V&V not provided
PdHeskestad	Estimates plume diameter based on Heskestad plume correlation.	V&V not provided
TcjAlpert	Unconfined ceiling jet temperature based on Alpert ceiling jet correlation.	V&V provided
TcjDelichatsios	Confined ceiling jet temperature based on Delichatsios ceiling jet correlation.	V&V not provided
Cjcflux	Estimates convective heat flux in the ceiling jet.	V&V not provided
VcjAlpert	Unconfined ceiling jet velocity based on Alpert ceiling jet correlation.	V&V not provided
MQHTemperature	Compartment temperature after a specified time, given a steady HRR based on MQH approach.	V&V provided
MQHFlashover	Heat release rate required for flashover after a specified time based on MQH approach.	V&V not provided
FiveTemp	Estimates compartment temperature based on FIVE-Rev1.	V&V provided
Detact	Activation time of heat detection devices based on heat release rate profiles.	V&V not provided
Aset	Time required by Hot Gas Layer to reach a specific height based on heat release rate profiles and openings at the bottom of the enclosure.	V&V not provided
CThrr	Estimates heat release rate from cable trays. The correlation is based on 14 experiments with a stack of 12 horizontal cable trays and 2 experiments with a combination of 12 horizontal cable trays and 3 vertical trays.	V&V not provided

FIVE-Rev1 Function	Function Description	NUREG-1824 Verification and Validation Status
Visib	Estimates the length of a visible path in a smoke environment. The correlation applies to light-reflecting signs.	V&V not provided
Ttar	Estimates target temperature under constant heat flux.	V&V not provided
Ttdam	Time to target damage under constant heat flux.	V&V not provided

2.3.3 Consolidated Fire Growth and Smoke Transport (CFAST) Model

The Consolidated Fire Growth and Smoke Transport (CFAST) model is a two-zone computer fire model. For a given fire scenario, the model subdivides a compartment into two control volumes, which include a relatively hot upper layer (i.e., the HGL) and a relatively cool lower layer. Mass and energy are transported between the layers via the fire plume and mixing at the vents. Combustion products accumulate via the plume in the HGL. Each layer has its own energy and mass balances. The most important approximation for the model is that each zone has uniform properties, that is, that the temperature and gas concentrations are constant throughout the zone, only changing as a function of time. The CFAST model describes the conditions in each zone by solving equations for conservation of mass, species, and energy, along with the ideal gas law. The Technical Reference Guide for CFAST (Jones et al., 2004) provides a detailed discussion concerning the specific derivation of these conservation laws. Documentation for CFAST also includes a User's Guide (Peacock et al., 2008b), which details the use of the model, and a Model Development and Evaluation Guide (Peacock et al., 2008a), which presents the latest model V&V results.

For some applications, including long hallways or tall shafts, the two-zone approximation may not be appropriate. To address this, CFAST includes empirical algorithms to simulate smoke flow and filling in long corridors and for a single well-mixed volume in tall shafts. CFAST also includes several correlations (as sub-models), based on experimental data that are used to calculate various physical processes during a fire scenario: smoke production, fire plume dynamics, heat transfer by radiation, convection, conduction, natural flows through openings (vertical and horizontal), forced or natural ventilation, thermal behavior of targets, heat detectors, and water spray from sprinklers.

CFAST models horizontal flow through vertical vents (doors, windows, wall vents, etc.), vertical flow through horizontal vents (ceiling holes, hatches, roof vents, etc.), and mechanical ventilation through fans and ductwork. Natural flow is determined by the pressure difference across a vent, using Bernoulli's law for horizontal vent flow, and by algebraic models for vertical vent flow. Mechanical ventilation is based on an analogy to electrical current flow in series and parallel paths where flow is split in parallel paths proportional to the flow resistance in each path and resistance to flow is additive for paths in series.

CFAST includes algorithms to account for radiation, convection, and conduction within a modeled structure. Radiative transfer occurs among the fire(s), gas layers, and compartment surfaces (ceiling, walls, and floor). It is a function of the temperature differences and emissivity of the gas layers, as well as the compartment surfaces. Convective heat transfer between gas layers and compartment or target surfaces is based on typical correlations available in the literature. The V&V results for CFAST are documented in Volume 5 of NUREG-1824 (EPRI 1011999). Additional validation results, particularly for plume temperature predictions that were not included in the NUREG-1824 (EPRI 1011999) Volume 5 results, are included in the CFAST Model Development and Evaluation Guide (Peacock et al., 2008a).

2.3.4 MAGIC[7]

MAGIC is another two-zone computer fire model, developed and maintained by Electricité de France (EdF) specifically for use in NPP analysis. MAGIC is supported by three EdF publications, including: (1) the technical manual, which provides a mathematical description of the model (Gay et al., 2005b); (2) the user's manual, which details how to use the graphical interface (Gay et al., 2005a); and (3) the validation studies, which compare MAGIC's results to experimental measurements (Gay et al., 2005c). These three proprietary publications and the MAGIC software are available through EPRI to EPRI members. Additional V&V results for MAGIC are documented in Volume 6 of NUREG-1824 (EPRI 1011999).

MAGIC is fundamentally the same type of model as CFAST and thus solves the same basic set of differential equations. The combustion model and vent flow models are similar as well. Despite this, MAGIC differs from CFAST in that it does not have the corridor or shaft sub-models, and the ceiling jet and wall jet treatments are different. The user should consult the technical manual for a complete description of the MAGIC sub-models (Gay et al., 2005b).

Once a given simulation is completed, MAGIC generates an output file with all of the solution variables. Through a "post-processor" interface, the user selects the relevant output variables for the analysis. Typical outputs include the temperatures of hot and cold zones, concentrations of oxygen and unburned gases, smoke migration into each compartment, the mass flow rates of air and smoke through the openings and vents, the pressures at the floor level of each compartment, the temperatures at the surfaces of the walls, and the thermal fluxes (radiative and total) exchanged by the targets placed by the user.

2.3.5 Fire Dynamics Simulator (FDS)

Fire Dynamics Simulator (FDS) (McGrattan et al., 2007) is a CFD model of fire-driven fluid flow. The model numerically solves a form of the Navier-Stokes equations appropriate for low-speed, thermally driven flow, with an emphasis on smoke and heat transport from fires. The partial derivatives of the equations for conservation of mass, momentum, and energy are approximated as finite differences, and the solution is updated in time on a three-dimensional, rectilinear grid. Thermal radiation is computed using a finite volume technique on the same grid as the flow solver. Lagrangian particles are used to simulate smoke movement and sprinkler discharge. FDS computes the temperature, density, pressure, velocity, and chemical composition within each numerical grid cell at each discrete time step. There are typically hundreds of thousands to several million grid cells, and thousands to hundreds of thousands of time steps. In addition, FDS computes the temperature, heat flux, mass loss rate, and various other quantities at solid surfaces.

Time histories of various quantities at a single point in space, or global quantities, such as the fire's HRR, are saved in simple, comma-delimited text files that can be plotted in a spreadsheet program. Some output quantities typically used to assess the conditions in a space or to compare against measured data are not inherently calculated in a CFD-type model, such as the HGL temperature and interface height. FDS does provide an estimate of these types of parameters using the continuous vertical temperature distribution and an integrated averaging scheme, as described by McGrattan et al. (2007).

Most field or surface data are visualized with a program called Smokeview, a tool specifically designed to help analyze results generated by FDS. FDS and Smokeview are used in concert

---

[7] MAGIC is a software package available to EPRI members (www.epri.com).

to model and visualize fire phenomena. Smokeview does this by presenting animated tracer particle flows, animated contour slices of computed gas variables, and animated surface data, and also presents contours and vector plots of static data anywhere within a scene at a fixed time. The FDS User's Guide (McGrattan et al., 2007) provides a complete list of FDS output quantities and formats, while the Smokeview User's Guide (Forney, 2008) explains how to visualize the results of an FDS simulation. Volume 7 of NUREG-1824 (EPRI 1011999) contains the results of V&V efforts for FDS. Additional V&V results for FDS are contained in the FDS documentation series (McGrattan et al., 2007).

FDS solves conservation equations of mass, momentum, and energy for an expandable mixture of ideal gases in the low Mach number limit. This means that the equations do not permit acoustic waves, the result of which is that the time step for the numerical solution is bounded by the flow speed, rather than the sound speed. Situations in which this limitation may be encountered include jet fires, deflagrations, and detonations. This limitation also reduces the number of unknowns by one, as density and temperature can be related to a known background pressure. Flow turbulence is treated by large eddy simulation.

For most simulations, FDS uses a mixture fraction combustion model. The mixture fraction is a conserved scalar that represents, at a given point, the mass fraction of gases originating in the fuel stream. In short, the combustion is controlled by the rate at which fuel and oxygen mix. Unlike versions of FDS prior to version 5, the reaction of fuel and oxygen is not necessarily instantaneous and complete, and there are several optional schemes that are designed to estimate the extent of combustion in under-ventilated spaces. The mass fractions of all of the major reactants and products can be derived from the mixture fraction by means of "state relations," expressions achieved through a combination of simplified analysis and measurement. The combustion model used by FDS is an area of active development. Consequently, FDS users should consult the latest code documentation for a description of new features or sub-models.

Numerical parameters play a very important role in a CFD model like FDS. A numerical parameter is any input value that is needed for the mathematical solution of the equations, but has little or no physical meaning. For example, the time step with which the numerical solution of the HGL temperature is computed does have units of seconds, but it is not a value that has meaning outside of that particular algorithm; nevertheless, these numerical parameters can affect the solution, and their sensitivity should be assessed in some way. For the spreadsheet and zone models, this procedure is relatively straightforward because the calculations run in less than a minute. One simply varies the value and ensures that the solution does not change appreciably. Specifically, one should simply demonstrate that the solution converges towards a particular value as the parameter is varied; for instance, using a smaller and smaller time step ought to lead to convergence of any evolution equation.

In FDS, the numerical parameter with the greatest importance is cell size. CFD models solve an approximate form of the conservation equations of mass, momentum, and energy on a numerical grid. The error associated with the discretization of the partial derivatives is a function of the size of the grid cells and the type of differencing used. FDS uses second-order accurate approximations of both the temporal and spatial derivatives of the Navier-Stokes equations, meaning that the discretization error is proportional to the square of the time step or cell size. In theory, reducing the grid cell size by a factor of two reduces the discretization error by a factor of four; however, it also increases the computing time by a factor of at least sixteen (a factor of two for the temporal and each spatial dimension). Clearly, there is a point in diminishing returns as one refines the numerical mesh. Determining which size grid cell to use in any given calculation is known as a grid sensitivity study.

Determining an optimal grid size in FDS is usually a matter of assessing the size of the fire. The physical diameter of the fire is not always a well-defined property; a compartment fire does not have a well-defined diameter, whereas a circular pan filled with a burning liquid fuel has an obvious diameter. Regardless, it is not the physical diameter of the fire that matters when assessing the "size" of the fire, but rather its characteristic diameter, D^*:

$$D^* = \left(\frac{\dot{Q}}{\rho_\infty c_p T_\infty \sqrt{g}}\right)^{2/5}$$
(2-1)

where $\dot{Q}$ is the fire HRR (kW), ρ_∞ is the ambient density of air (kg/m^3), c_p is the specific heat of air (kJ/kg/K), T_∞ is the ambient air temperature (K), and g is the acceleration of gravity (m/s^2).

In many instances, D^* is comparable to the physical diameter of the fire. FDS employs a numerical technique known as large eddy simulation (LES) to model the unresolvable or "sub-grid" motion of the hot gases. The effectiveness of the technique is largely a function of the ratio of the fire's characteristic diameter, D^*, to the size of a grid cell, δx. In short, the greater the ratio $D^*/\delta x$, the more the fire dynamics are resolved directly, and the more accurate the simulation. Past experience has shown that a ratio of 5 to 10 usually produces favorable results at a moderate computational cost for problems where gross smoke movement is of interest.

As an example, suppose the HRR of the fire were 700 kW. The characteristic diameter may then be calculated as follows:

$$D^* = \left(\frac{700 \text{ kW}}{1.2 \text{ kg/m}^3 \times 1.012 \text{ kJ/kg/K} \times 293 \text{ K} \times \sqrt{9.81 \text{ m/s}^2}}\right)^{2/5} = 0.83 \text{ m}$$
(2-2)

To perform a grid sensitivity analysis, a good place to start might be a cell resolution of 15 cm (6 in), which means that $D^*/\delta x = 5.5$. Then choose a grid of 10 cm (4 in), and then 5 cm (2 in). At this point, the calculation time will have increased by a factor of roughly three hundred, making it potentially impractical to compute; however, if it can be shown that there is little difference between the 5 cm (2 in) and 10 cm (4 in) grids, then the objective has been achieved. The meaning of "little difference" can be interpreted in several ways. Given that NUREG-1824 (EPRI 1011999), the fire model V&V study, lists the relative error expected of the various models for the various quantities, it is reasonable to interpret the difference in results on different grids in light of what is expected of the model accuracy.

Although the fire size and dimensions often determine the optimum grid resolution, there are other factors that can influence the selection of the grid resolution. These include the number of cells used to resolve a flow path dimension, the number of cells used to describe the fire dimension, and the number of cells used to resolve the conditions in a partially isolated volume. These considerations are related in that it is generally advisable to include at least three cells across any flow path, such as a door or a window, and fire dimension, regardless of the minimum number of cells computed using the fire characteristic diameter. In some cases, partially isolated volumes are created by various obstructions; if the temperature and flow conditions are of interest in these areas, a minimum of three cells across any dimension should be provided. Another consideration that could influence the grid resolution is the dimension of the obstructions that are expected to influence the result. For example, if it is necessary to quantitatively assess the effect that various conduits and light fixtures may have on the actuation time of a nearby sprinkler, the maximum grid resolution would be comparable to the dimensions of the smallest distinct obstruction included in the model.

FDS input files are frequently created with the assistance of preprocessing software, which may include commercial software packages that can create input files for FDS or spreadsheet tools created by users to insert obstructions or create stair-step approximations to curved geometries. This type of software can reduce the tediousness of creating the geometric representation of a space, but is not part of the FDS model. Any input files created by such software should be carefully checked by the user to ensure that the geometry or boundary data are exactly as intended.

2.3.6 Verification and Validation (V&V)

The use of fire models requires a good understanding of their limitations and predictive capabilities. NFPA 805, for example, states that fire models shall only be applied within the limitations of the given model and shall be verified and validated. Previously, the NRC's Office of Nuclear Regulatory Research (RES) and EPRI conducted a collaborative project for the V&V of the five selected fire models described in Sections 2.3.1 through 2.3.5. NIST was also an important partner in this project. The results of this project were documented in the seven volumes of NUREG-1824 (EPRI 1011999), *Verification and Validation of Selected Fire Models for Nuclear Power Plant Applications*.

The parameters for which NUREG-1824 (EPRI 1011999) provides V&V information are shown in Table 2-4. Not all output parameters are available in all models. The information in Table 2-4 may be a useful element to consider when selecting the appropriate fire model tool; for example, it is clear that the libraries of algebraic models (FDT[s], FIVE-Rev1) have limited capabilities when compared to the zone and CFD models. These libraries do not have appropriate methods for estimating many of the fire scenario attributes evaluated in this study. The correlations that the libraries do contain are typically empirically deduced from a broad database of experiments; they are based on fundamental conservation laws, and have gained a considerable degree of acceptance in the fire protection engineering community. However, because of their empirical nature, they are subject to many limiting approximations. The user must be cautious when using these tools.

CFD model predictions can be more accurate in complex scenarios; however, the time it takes to obtain and understand a prediction may also be an important consideration in the decision to use a particular model for a specific scenario. FDS is computationally expensive in all respects (preprocessing, simulation, and post-processing), and, while the two-zone models produce answers in seconds to minutes, FDS provides comparable answers in days to weeks. In general, FDS is better suited to estimate fire environments within more complex configurations.

The fire experiments selected for inclusion in the V&V were limited to high-quality, real-scale experiments with direct applicability to NPP applications. As it was not possible to consider all possible NPP applications, a method for determining the applicability of validation results to other specific NPP fire scenarios has been described in NUREG-1824 Volume 1 (EPRI 1011999). The applicability of the validation results is determined using normalized parameters traditionally used in fire modeling applications. Normalized parameters allow users to compare results from scenarios of different scales by normalizing the physical characteristics of the scenarios.

Table 2-4. Fire modeling attributes included in NUREG 1824/EPRI 1011999 (2007).

Fire Modeling Attributes	Fire Model				
	FDT^s	FIVE-Rev1	CFAST	MAGIC	FDS
Hot Gas Layer (HGL) Temperature	YES	YES	YES	YES	YES
Hot Gas Layer (HGL) Height	NO	NO	YES	YES	YES
Ceiling Jet Temperature	NO	YES	YES	YES	YES
Plume Temperature	YES	YES	NO	YES	YES
Flame Height	YES	YES	YES	YES	YES
Radiated Heat Flux to Targets	YES	YES	YES	YES	YES
Total Heat Flux to Targets	NO	NO	YES	YES	YES
Total Heat Flux to Walls	NO	NO	YES	YES	YES
Wall Temperature	NO	NO	YES	YES	YES
Target Temperature	NO	NO	YES	YES	YES
Smoke Concentration	NO	NO	YES	YES	YES
Oxygen Concentration	NO	NO	YES	YES	YES
Room Pressure	NO	NO	YES	YES	YES

Table 2-5 identifies normalized parameters that may be used to compare NPP fire scenarios with validation experiments. A full description of each parameter follows Table 2-5. The validation range for the normalized parameters shown in Table 2-5 were derived from NUREG-1824 (EPRI 1011999), Table 2-4, and are intended to provide guidance on which groups of validation experiments to consider when evaluating a certain attribute based on the validation results. These parameters may not be the only ones appropriate for evaluating the applicability of a specific experiment; Table 2-5 of NUREG-1824 (EPRI 1011999) Volume 1 lists the ranges of values for different physical characteristics and normalized parameters based on the experiments considered in the validation study.

It is seen in Table 2-5 that the fire diameter plays a role in three of the normalized parameters: the fire Froude Number, the flame length ratio, and the radial distance ratio. The fire diameter may be estimated from the total plan area of the burning fuel using the following equation (SFPE, 2011):

$$D = \sqrt{\frac{4A}{\pi}} \tag{2-3}$$

where D is the effective fire diameter (m) and A is the plan area of the burning fuel (m^2) equal to the fuel package length dimension multiplied by the width dimension. Equation 2-3 produces a diameter of a circle having an equivalent area to that of the burning fuel and is therefore applicable to fires with a plan area that is nearly circular. When the aspect ratio of a source fire with a rectangular plan or footprint (i.e., the ratio of the long dimension to the short dimension) becomes large, the flame length and thermal plume take on characteristics of a line type fire (Grove et al., 2002). There is no clearly defined aspect ratio limit; however, a value of up to about five is shown to be reasonable for using the effective diameter (Grove et al., 2002). Note that the form of the Froude Number for a line type source fire is nearly the same as the axisymmetric source, with the exception that the source area is used in place of the diameter (Yuan et al., 1996). The fire tests used to V&V the various fire models in NUREG-1824 all had aspect ratios below two.

Table 2-5. Summary of selected normalized parameters for application of the validation results to NPP fire scenarios (NUREG-1824/EPRI 1011999, 2007).

Quantity	Normalized Parameter	General Guidance	NUREG-1824 Validation Range
Fire Froude Number	$$\dot{Q}^* = \frac{\dot{Q}}{\rho_\infty c_p T_\infty D^2 \sqrt{gD}}$$	Ratio of characteristic velocities. A typical accidental fire has a Froude number of order 1. Momentum-driven fire plumes, like jet flares, have relatively high values. Buoyancy-driven fire plumes have relatively low values.	0.4 – 2.4
Flame Length Ratio	$$\frac{H_f + L_f}{H_c}$$ $$\frac{L_f}{D} = 3.7\,\dot{Q}^{*2/5} - 1.02$$	A convenient parameter for expressing the "size" of the fire relative to the height of the compartment. A value of 1 means that the flames reach the ceiling.	0.2 – 1.0
Ceiling Jet Distance Ratio	$$\frac{r_{cj}}{H_c - H_f}$$	Ceiling jet temperature and velocity correlations use this ratio to express the horizontal distance from target to plume.	1.2 – 1.7
Equivalence Ratio	$$\varphi = \frac{\dot{Q}}{\Delta H_{O_2} \dot{m}_{O_2}}$$ $$\dot{m}_{O_2} = \begin{cases} 0.23 \times \frac{1}{2} A_0 \sqrt{H_0} & \text{(Natural)} \\ 0.23\,\rho_\infty \dot{V} & \text{(Mechanical)} \end{cases}$$	The equivalence ratio relates the energy release rate of the fire to the energy release that can be supported by the mass flow rate of oxygen into the compartment, $\dot{m}_{O_2}$. The fire is considered over- or under-ventilated based on whether φ is less than or greater than 1, respectively.	0.04 – 0.6
Compartment Aspect Ratio	L/H_c or W/H_c	This parameter indicates the general shape of the compartment.	0.6 – 5.7
Radial Distance Ratio	$$\frac{r}{D}$$	This ratio is the relative distance from a target to the fire. It is important when calculating the radiative heat flux.	2.2 – 5.7

Froude Number

The Froude Number is a measure of the buoyant strength of the fire plume. A large Froude Number indicates a strong source, and, given a sufficiently high Froude Number, the fire plume may take on characteristics of a jet fire. The Froude Number is determined through the following equation:

$$\dot{Q}^* = \frac{\dot{Q}}{\rho_\infty c_p T_\infty D^2 \sqrt{gD}} \tag{2-4}$$

where $\dot{Q}^*$ is the Fire Froude number (non-dimensional), $\dot{Q}$ is the fire heat release rate (kW), ρ_∞ is the density of ambient air (kg/m³), c_p is the heat capacity at constant pressure for ambient air (kJ/kg-°C), T_∞ is the ambient temperature (K), D is the fire diameter or the effective fire diameter (m²), and g is the acceleration of gravity (9.81 m/s²). The heat release rate that is used to generate the validation range is the peak heat release rate; thus, when comparing the model application range to these values, the peak heat release rate should be used. The effective fire diameter may be computed using Equation 2-3. The ambient density and temperature are typically coupled through the following equation:

$$\rho_\infty = \frac{352}{T_\infty} \tag{2-5}$$

where all terms have been defined. The heat capacity for air varies slightly with temperature, and, at 298 K, it is equal to about 1.012 kJ/kg-°C.

Flame Length Ratio

The flame length ratio is a measure of the flame height relative to the upper horizontal boundary (ceiling). The NUREG-1824 validation tests all involved fires with a flame height that was at or below the ceiling. In situations where the flame height is greater than the ceiling, the flames will extend radially outward from the impingement point. This configuration is not well characterized by simple algebraic models, or even by zone models that use plume and flame height correlations.

$$\text{Flame Length Ratio} = \frac{H_f + L_f}{H_c} \tag{2-6}$$

where H_f is the base height of the fire (m), L_f is the flame height (m), and H_c is the enclosure height (m). Note that the flame length ratio does not apply to fires modeled outside an enclosure. The flame height is computed using the following equation:

$$\frac{L_f}{D} = 3.7\,\dot{Q}^{*2/5} - 1.02 \tag{2-7}$$

where D is the fire diameter or the effective fire diameter (m²) and $\dot{Q}^*$ is the fire Froude Number calculated using Equation 2-4 (non-dimensional). The effective fire diameter may be computed using Equation 2-3.

THE FIRE MODELING PROCESS

Ceiling Jet Distance Ratio

The ceiling jet distance ratio is a measure of the ceiling jet position at which data is sought relative to the enclosure height and is applicable primarily when the temperature and velocity of the ceiling jet are quantities of interest. A low ceiling jet distance ratio indicates that the position is within the impingement zone and that the conditions would be dominated by the thermal plume. A high ceiling jet position ratio suggests that the position is approaching the edge of the ceiling jet, at least as idealized by algebraic correlations. A high ceiling jet ratio also suggests that a considerable portion of the ceiling would need to be free of obstructions in order to conform to the underlying approximations of the ceiling jet models. The ceiling jet ratio is given by the following equation:

$$\text{Ceiling Jet Ratio} = \frac{r_{cj}}{H_c - H_f} \tag{2-8}$$

where r_{cj} is the horizontal distance within the ceiling jet from the fire centerline (m), H_c is the enclosure height (m), and H_f is the base height of the fire (m). The ceiling jet ratio is applicable primarily when sprinkler or heat detector actuation is calculated. It is not applicable when the fire is modeled outside an enclosure.

Equivalence Ratio

This quantity is the ratio of the generation rate of fuel to the supply rate of oxygen. When the equivalence ratio is equal to one, the exact amount of oxygen required for complete combustion is available. When the ratio is greater than one, the environment is fuel rich and the fire is considered to be under-ventilated. The reverse is true when the ratio is less than one. The upper limit for the equivalence ratio in enclosure fires is about three (Gottuk et al., 2008). Modeling under-ventilated fires is challenging in part because the fuel mass loss rate is typically specified by the user and is not adjusted by the fire models to the conditions actually present. The equivalence ratio may be estimated for natural and forced ventilation by the expression:

$$\varphi = \frac{\dot{Q}}{\Delta H_{O_2} \dot{m}_{O_2}} \tag{2-10}$$

where φ is the equivalence ratio (non-dimensional), $\dot{Q}$ is the heat release rate of the fire (kW), ΔH_{O_2} is the "heat of combustion" for oxygen (kJ/kg), and $\dot{m}_{O_2}$ is the mass flow rate of oxygen into the enclosure (kg/s). The "heat of combustion" for oxygen is the energy released per unit mass of oxygen and varies by less than five percent among nearly all carbon-based fuels (Janssens, *SFPE Handbook, 4th edition*). The value for ΔH_{O_2} is typically taken to be 13,100 kJ/kg, an average value over a wide range of common fuels. A notable exception to this convention is hydrogen, whose heat of combustion based on oxygen consumption is approximately 18,000 kJ/kg. For other pure fuels, it is possible to calculate a more accurate value than 13,100 kJ/kg, in which case Eq. (2-10) can be modified accordingly.

The mass flow rate of oxygen into the enclosure is given by the following equation:

$$\dot{m}_{O_2} = \begin{cases} 0.23 \times \frac{1}{2} A_0 \sqrt{H_0} & \text{(Natural)} \\ 0.23\, \rho_\infty \dot{V} & \text{(Mechanical)} \end{cases} \tag{2-11}$$

where A_0 is the effective area of the openings (m²), H_0 is the effective height of the openings (m), ρ_∞ is the density of ambient air given by Equation 2-5 (kg/m³), and $\dot{V}$ is the volumetric flow rate of air into the enclosure (m³/s). In the case of natural ventilation, Equation 2-11 is applicable only to enclosures with one or more vertical openings. Note that the oxygen mass flow parameter is the oxygen flow into an enclosure; if the enclosure has forced exhaust only, the mass flow rate during the fire will likely change as the temperature of the enclosure increases. The mass flow that corresponds to the heated environment should be used when computing the equivalence ratio. An alternate means of estimating the mass flow rate of oxygen into the enclosure is necessary if there are horizontal openings, which could include using the model-predicted vent flow rates. This approach may also be used for enclosures in which there is only forced exhaust. If there are multiple vertical openings, the effective opening height (H_0) is given by the following equation (Buchanan, 2001):

$$H_o = \frac{\sum_{i=1}^{n} A_i H_i}{\sum_{i=1}^{n} A_i} \tag{2-12}$$

where A_i is the area of the ith opening (m²), H_i is the height of the ith opening, and n is the total number of openings. The effective opening area (A_0) is given by the following equation (Buchanan, 2001):

$$A_o = \sum_{i=1}^{n} A_i \tag{2-13}$$

where A_i is the area of the ith opening (m²).

In many compartments, both natural and forced ventilation is present. The recommended procedure to follow in this case is to select the dominant ventilation mode (i.e., the mode that produces the highest mass flow of oxygen). If the two modes are comparable, the oxygen masses may be added. As was the case for the ceiling jet ratio and the flame length ratio, the equivalence ratio applies only to fires that are modeled within an enclosure.

Compartment Aspect Ratio

The compartment aspect ratio is a measure of the deviation of the enclosure dimensions from a cube. When at least one of the compartment aspect ratios is large, the enclosure takes on the characteristics of a corridor. In such cases, the transport time of the combustion products and a non-uniform layer can both become significant parameters that require consideration. When at least one of the compartment aspect ratios is low, the enclosure takes on the characteristics of a shaft. In these cases, stratification of the combustion products, the interaction of the fire plume and the enclosure boundaries, or choked flow could become parameters that influence the results. These situations may lie outside the development basis for algebraic and zone fire models. The compartment aspect ratio is computed using the following equation:

$$\text{Compartment Aspect Ratio} = \begin{cases} \dfrac{L}{H_c} \\[2ex] \dfrac{W}{H_c} \end{cases} \tag{2-14}$$

where L is the compartment length (m), W is the compartment width (m), and H_c is the compartment height (m). The compartment aspect ratio is applicable to fires evaluated within an enclosure.

Radial Distance Ratio

The radial distance ratio is a measure of the distance from the center of the source fire at which a heat flux quantity is predicted. It is applicable only when the heat flux parameter is computed. A low radial distance ratio indicates that the target location is close to the fire and that near-field thermal radiation effects could be significant. Large radial distance ratios indicate that the target is located far from the fire. The radiant heat flux at high radial distance ratios will approach that predicted from a point heat source, barring other sources of external thermal radiation or geometric factors (flame deflections under a ceiling or boundary re-radiation).

The radial distance ratio is given by the following equation:

$$\text{Radial Distance Ratio } = \frac{r}{D} \tag{2-15}$$

where r is the actual distance between the target and the center of the fire base (m) and D is the fire diameter as computed using Equation 2-3.

For a given set of experiments and NPP fire scenarios, the user can calculate the relevant normalized parameters. If the fire scenario parameters fall within the ranges evaluated in the study, then the results of the study offer appropriate validation for the scenario. If they fall outside the range, then a validation determination cannot be made based on the results from the study. For any given fire scenario, more than one normalized parameter may be necessary for determining the applicability of the validation results.

The V&V study provides valuable insight into the predictive capability of the five fire models. This insight is ultimately characterized in terms of a bias and a standard deviation for a number of output parameters. The closer the bias is to unity, the more accurate the fire model tends to be in predicting the given parameter; the smaller the standard deviation, the smaller the expected scatter around the mean bias. Chapter 4 of this guide describes how the V&V uncertainty information can be used to assign a probability function to the output data.

NUREG-1824 (EPRI 1011999) provides V&V documentation for specific versions of fire models. Because the fire models considered are under active development, new releases are expected to and do occur. The user has the option of using the model version that has been verified and validated in NUREG-1824 (EPRI 1011999) or re-evaluating cases in NUREG-1824 (EPRI 1011999) to demonstrate that the predictive capability of the model has not decreased for the application at hand. It is expected that NUREG-1824 (EPRI 1011999) will be updated from time to time, as the need arises.

2.3.7 Fire Modeling Parameters Outside the Validation Range

The development of the sample problems documented in the appendices to this report suggests that many commercial NPP fire modeling applications can fall outside the range of applicability of the validation study documented in NUREG-1824 (EPRI 1011999). The primary reason for this is that the range of applicability, as defined by the dimensionless parameters, is governed by the experiments selected for the validation study. The selected experiments are representative of various types of spaces in commercial NPPs, but do not encompass all possible geometries or applications. Accordingly, the analyst will encounter many areas or applications that will fall outside this application range. The predictive capabilities of the fire models in specific scenarios can extend beyond the range of applicability defined in NUREG-1824 (EPRI 1011999). Additional analysis and justification is required by the analyst to address situations where some or all of the analysis parameters fall outside the range of applicability defined in NUREG-1824 (EPRI 1011999). The additional analysis and justification should address the applicability of fire modeling results generated by input parameters outside the validation range to support the conclusions of the study. This section describes the recommended strategies for addressing this situation.

2.3.7.1 Sensitivity Analysis

In the context of applicability of validation results, sensitivity analysis refers to varying selected input parameters in the "conservative" direction so that they fall within the applicability range. If the fire modeling conclusions are not affected by the variations in the parameters, the analyst may use the sensitivity analysis results to further justify the conclusions. Based on the dimensionless terms listed above, the following sensitivities could be evaluated:

- Froude number: The two parameters that can be practically varied are the fire diameter and the HRR. For fire sizes (i.e., HRR) that are small for the postulated diameter, the resulting Froude number can fall under the low end of the applicability range. Similarly, for fires that are relatively large for the postulated diameter, the Froude number can fall above the applicability range. In the former situation, the analysts may consider reducing the fire diameter and keeping the HRR profile unchanged. In most fire modeling tools, the fire diameter is simply used to determine HRRs or to calculate the fire plume conditions, such as the flame height or plume temperature. Considering that the HRR is "fixed" in this sensitivity study, the fire diameter may not be a relevant parameter in the analysis, with the important exception of scenarios where the fire plume conditions are relevant. A similar approach could be used for the latter situation. Increasing the fire diameter can "force" the dimensionless term into range. It should be stressed that fire diameter is often a parameter that influences predicted flame height and fire plume conditions, and that the effects of diameter variations should be explicitly addressed in the analysis. This includes other dimensionless terms where the fire diameter is a key input (e.g., target distance to diameter (r/D), etc.).

- Flame length relative to ceiling height: This is a convenient parameter for expressing the "size" of the fire relative to the height of the compartment. A value of 1 means that the flames reach the ceiling. The validation range extends up to a value of 1.0, which should cover most of the scenarios of interest in commercial NPPs. Scenarios that are expected to fall out of the range are:

  o Those associated with relatively short flames. Typical ceiling heights in NPP scenarios range from about 3 to 6.1 m (10 to 20 ft), excluding the containment and turbine buildings, which have relatively large openings between elevations. Consequently, flame lengths shorter than 0.6 to 1.2 m (2 to 4 ft) will be considered outside of validation range. A sensitivity analysis increasing the HRR values should provide a conservative estimate of fire conditions within the validation range. In cases where the conclusion of the analysis does not change given the increased fire intensity (e.g., no damage within the flame length of fire plume), the suggested sensitivity analysis can be used as the justification for the evaluation of a compartment that falls outside the validation range.

- o Flame extensions under ceilings. In this particular case, not only are such flame lengths out of the range of validation, but also the models for predicting this phenomenon have not been verified or validated with a process similar to the one documented in NUREG-1824 (EPRI 1011999).

- Ceiling jet radial distance relative to the ceiling height: Ceiling jet temperature and velocity correlations use this ratio to express the horizontal distance from target to plume. Ceiling jet applications in commercial NPPs should be carefully evaluated due to the numerous obstructions near the ceiling (e.g., cable trays, HVAC ducts, piping, etc.). Most of its applications include determination of time to detection and sprinkler activation, in which the ceiling jet velocity is a sub-model in the analysis. An alternative option is a sensitivity analysis consisting of moving the fire location to distances that would fall within the validation range; it is recognized, however, that in many situations the fire location cannot be altered, particularly in the case of fixed ignition sources or transient fires postulated near areas where redundant targets are in close proximity (pinch-points). In general, longer horizontal distances will result in longer activation time results; by contrast, shorter horizontal distances would result in "conservative" time-to-damage results. In situations where the ceiling jet geometry deviates significantly from the idealized flat horizontal surface, as may be the case when there are large numbers of obstructions or bays, a CFD model may be the better choice for calculating detector response times.

- Equivalence ratio as an indicator of the ventilation rate: The validation available is for well-ventilated fires: that is, no model validation information is available for under-ventilated compartment fires, including fire extinction due to lack of oxygen. In general, fires that are considered well ventilated in the enclosure should result in bounding conditions as long as the HRR profile is appropriate. Conditions in the enclosure are not expected to be worse in a fire where the combustion process is affected by lack of oxygen than they would be under fire conditions where the combustion process is unaffected. However, under-ventilated fire conditions should be considered carefully as sudden air inflows into compartments with under-ventilated fire conditions could produce relatively severe fire conditions.

- Compartment aspect ratio: It is expected that some compartments in commercial NPPs would have geometric characteristics outside the validation range (e.g., relatively long, narrow corridors with high ceilings, etc.). These parameters are important in fire scenarios involving HGL calculations, as the size and configuration of the compartment are important input parameters. Clearly, these parameters should not be applicable in scenarios where the enclosure conditions are not considered, such as flame radiation calculations using the point source model and plume temperature calculations using algebraic models where it has been determined that enclosure conditions are not a factor. As part of the sensitivity analysis, the analyst may consider "shortening" the length, width, or height of the compartment to values that fall within the validation range, with the expectation that this will result in an elevated level of hazardous fire-generated conditions, as predicted by the model (i.e., a conservative calculation). In cases where the conclusion of the analysis does not change given the "smaller" compartment (e.g., the HGL temperature does not exceed damage threshold of cables in either case), the suggested sensitivity analysis can be used as the justification for the evaluation of a compartment that falls outside the validation range.

- Radial distance relative to the fire diameter: This ratio is the relative distance from a target to the fire, and is important when calculating the radiative heat flux. Note that the validation range starts at a distance approximately twice the fire diameter. In practice, targets at a very close distance to the fire (approximately two fire diameters or less) should be expected to fail, given the relatively low damage threshold levels for cables. An alternative option is a sensitivity analysis, which consists of moving the fire location to distances that would fall within the validation range; it is recognized, however, that in many situations the fire location cannot be altered, particularly in the case of fixed ignition sources or transients fires postulated near "pinch-points." In general, shorter horizontal distances will result in higher heat flux levels.

2.3.7.2 Additional Validation Studies

There are other fire model validation studies besides NUREG-1824 (EPRI 1011999) that can serve as a basis for establishing the applicability of fire modeling results. In developing the examples documented in the appendices of this report, the research team identified relevant validation studies outside of NUREG-1824 (EPRI 1011999), as summarized below:

- Scenarios involving targets within the fire plumes: A useful discussion of fire plumes is contained in Gunnar Heskestad's chapter in the *SFPE Handbook of Fire Protection Engineering, 4th ed.*, "Fire Plumes, Flame Height, and Air Entrainment." The plume correlations used in the empirical and zone models are described, as well as their range of applicability. NUREG-1824 (EPRI 1011999) contains experimental measurements of fire plumes, but the range is somewhat limited. The plume correlations used by the models have a much wider range of applicability than that exercised in NUREG-1824 (EPRI 1011999).

- Scenarios involving targets within the ceiling jet: Similarly, Ronald Alpert's chapter "Ceiling Jet Flows" in the *SFPE Handbook* contains a description of the various correlations used to estimate the temperature and gas velocity of ceiling jets. There are extensive references to the original experimental test reports from which the correlations were derived.

- Scenarios involving targets exposed to flame radiation: A useful collection of techniques and validation data for thermal radiation calculations is found in the *SFPE Engineering Guide for Assessing Flame Radiation to External Targets from Pool Fires,* written by the SFPE Task Group on Engineering Practices, 1999.

- Scenarios involving flashover/post-flashover conditions: A series of experiments was conducted at NIST as part of an investigation of the collapse of the World Trade Center towers. Validation calculations with FDS are described in the report NIST NCSTAR 1-5F, *Federal Building and Fire Safety Investigation of the World Trade Center Disaster: Computer Simulation of the Fires in the WTC Towers*, September 2005.

- Scenarios involving electrical failure of cables: The Cable Response to Live FIRE (CAROLFIRE) program led to the development and validation of the Thermally-Induced

Electrical Failure (THIEF) model (NUREG/CR-6931, Volume 3). This model can be used to estimate the temperature within an electrical cable that is exposed to an elevated temperature or heat flux.

- Scenarios involving cable burning: The Cable Heat Release, Ignition, and Spread in Tray Installations in Fire (CHRISTIFIRE) program led to the development and validation of the Flame Spread in Horizontal Cable Trays (FLASH-CAT) model (NUREG/CR-7010, Volume 1). This model addresses the growth and spread of fire within vertical stacks of horizontal, open-back cable trays.

In addition to NUREG-1824 (EPRI 1011999) and the various documents cited above, the individual model developers typically maintain a collection of validation cases that are included as part of the model documentation. The algebraic spreadsheet models, FDT^s and FIVE-Rev1, are based directly on experimental correlations. Validation of these models is typically not part of the model documentation; rather, there are references to source material like the *SFPE Handbook* or the original test reports. Validation studies by the CFAST and FDS developers are contained within:

NIST Special Publication 1086, *CFAST – Consolidated Model of Fire Growth and Smoke Transport, Software Development and Model Evaluation Guide*, 2008.

NIST Special Publication 1018, *Fire Dynamics Simulator, Technical Reference Guide, Volume 3, Validation*, 2007.

In summary, the purpose of the sensitivity analysis is to "re-shape" the scenario with parameters that fall within the V&V range and result in more severe fire generated conditions (e.g., higher HGL or plume temperature, higher incident heat flux, etc.). Depending on the application, one or more parameters may need to be varied affecting multiple dimensionless parameters. It is recommended that the results from the sensitivity calculations always be compared to those resulting from the base case to ensure that the input parameter manipulation produces more severe fire generated conditions.

2.4 Step 4: Calculate Fire-Generated Conditions

This step involves running the model(s) and interpreting the results. When running a computer model, the following general steps are recommended:

1. Determine the output parameters of interest. If the goal of the simulation is to estimate wall temperatures, for example, the analyst should be interested in internal and external wall temperatures. The analyst should ensure that the model will provide the output of interest, or at least the fire conditions that can help achieve the objectives of the analysis. The output file should be labeled with a distinctive file name.

2. Prepare the input file. In this step, the analyst enters the input parameters into the model. The best way to enter input parameters is to follow the same guidelines described in the scenario description section. Each model has a user's manual with instructions on creating the respective input file. These files are created either through user-friendly menus and screens or through a text editor. If a text editor is used, it is strongly recommended that the analyst start with an example case prepared by code developers, and make appropriate changes to that file.

3. Run the computer model. The running time for zone models is on the order of minutes, depending on the complexity of the scenario and the speed of the computer. Calculations using a CFD model may take up to days or weeks in complex scenarios, including multiple compartments, multiple fires, and mechanical ventilation systems.

4. Interpret the model results. Check that the results are intuitively consistent with the input and expectations. Check that the output results accurately reflect the desired input; common verifications would include the fire size and location, the location and status of any doors or boundary openings, and the forced ventilation flow rate and location. The model output should be checked for indications of a solution error. For example, the pressure and the HGL should fall within the ranges observed in test data; the HGL temperature should be greater than the lower gas layer temperature; and there should not be anomalous areas of flow acceleration or temperature change. Determine whether or not the fire scenario resulted in conditions that exceed the performance criteria, as applicable.

5. Organize the results in a form that answers the question(s) and allows the calculations to be reproduced. If the results are used in a PRA screening analysis, this may take the form of a ZOI dimension or a maximum HGL temperature. If the results are part of a deterministic analysis, the output form may be a conclusion with regard to the performance of some component and an associated safety margin if the component is predicted to be free of damage.

For the FDT^s and FIVE-Rev1, the input data is entered directly into a spreadsheet, and the results are presented in the spreadsheet. Some of the FDT^s spreadsheets include graphical and tabular results. FIVE-Rev1 typically provides a single result for a given set of input data; however, many of the calculations in FIVE-Rev1 are implemented as Microsoft Excel functions. These functions can be called from any cell in the spreadsheet, and can be used to easily calculate the plume temperature at a specific location above the fire as a function of time for a fire with a time-varying HRR.

CFAST, MAGIC, and FDS can handle user-specified transient heat release rates, as they calculate the results for each zone or cell at each time step. The time step required to maintain stable calculations is typically determined by the model. The interval at which results are presented is a user-specified value. CFAST, MAGIC, and FDS can output results as text files, which can be read or plotted using commercially available spreadsheet programs; CFAST and FDS can also output their results in a form appropriate for SMOKEVIEW (Forney, 2010). SMOKEVIEW is a software tool that visualizes smoke and other attributes of the fire using traditional scientific methods, such as displaying tracer particle flow, two- or three-dimensional shaded contours of gas flow data (e.g., temperature), and flow vectors showing flow direction and magnitude. MAGIC includes its own post-processor for visually analyzing the results of a simulation. Post-processing may also be performed using other graphical or graphical animation software. If using a software package that is not designed for viewing the particular fire model results, the user should check that the output parameters are interpreted and displayed as intended.

2.5 Step 5: Conduct Sensitivity and Uncertainty Analyses

This document recommends a comprehensive treatment of uncertainty and/or sensitivity analysis as part of a fire modeling analysis for the following reasons:

- Models are developed based on idealizations of the physical phenomena and simplifying approximations, which unavoidably introduces the concept of model uncertainty (i.e., model error) into the analysis.

- A number of input parameters are based on available or generic data or on fire protection engineering judgment, which introduces the concept of parameter uncertainty into the analysis.

The concepts of model and parameter uncertainty have traditionally been addressed in fire modeling using uncertainty and/or sensitivity analysis. The uncertainty in a variable represents the lack of knowledge about the variable, and is often represented with probability distributions. Its objective is to assess the variability in the model output, that is, how uncertain the output is given the uncertainties related to the inputs and structure of the model. By contrast, the sensitivity of a variable in a model is defined as the rate of change in the model output with respect to changes in the variable. A model may be insensitive to an uncertain variable. Conversely, a parameter to which a model is very sensitive may not be uncertain.

Details of the uncertainty and sensitivity analysis are included in Chapter 4.

2.6 Step 6: Document the Analysis

The amount of information required and generated by a fire modeling analysis can vary widely. Simple algebraic models may not require a large number of inputs, and the complete analysis, including output results, can be documented on a single piece of paper. On the other hand, some fire modeling exercises may require use of multiple computer models, where outputs from one are inputs to others. These cases, for the most part, will require a significant number of input parameters and will produce outputs requiring documentation. Regardless of the amount of information required or generated by the analysis, proper documentation is vital to identifying the important findings of the exercise and providing clear, focused conclusions.

Documentation of the fire scenario selection and description process should include enough information so that the final report is useful in current and future applications. This is particularly relevant in the commercial nuclear industry, where compartment and equipment layouts or processes do not change much over time. It is likely that fire scenarios analyzed for one application may be useful for other applications as well; the key, however, is to develop and maintain good documentation of the selected fire scenarios, including all the technical elements discussed in this section. The *SFPE Engineering Guide to Substantiating a Fire Model for a Given Application* (SFPE, 2010) provides general guidance on information to be included in fire modeling analyses.

It is likely that the information necessary for documenting the fire scenario selection will be gathered from a combination of observations made during engineering walkdowns and a review of existing plant documents and/or drawings. The documentation process then involves compiling the information from different sources into a well-organized package that can be used in future applications and for NRC regional inspections. The documentation package may consist of:

- Marked-up plant drawings. Plant layout, detection, suppression, cable tray, HVAC, and conduit drawings are often marked to highlight the location of the compartment, the ignition sources, the targets, the ventilation flow paths, and the fire protection features. The drawings also serve as sources of fire model input values, such as compartment dimensions, ventilation flow rates, and relative locations of fire protection systems or targets.

- Design basis documents (DBDs). Design basis documents (DBDs) provide in-depth assessments of plant features in various operation modes, such as the HVAC system.

- Sketches. Sketches are perhaps one of the most useful ways of documenting a fire scenario. A sketch typically consists of a drawing illustrating the ignition source, intervening combustibles, targets, and fire protection features. A first draft of the sketch is usually prepared during walkdowns. The analyst should take the opportunity to include details such as raceways and conduit identifications (IDs), and other information relevant to the fire modeling analysis. Pictures often supplement sketches.

- Write-ups and input tables. Write-ups and input tables are used to compile the information collected from drawings and walkdowns in an organized way. The write-up should include a brief scenario description and detailed documentation supporting quantitative inputs to the fire modeling analysis, as well as any relevant sketches or pictures associated with each scenario.

- Software versions, descriptions, and input files. The documentation package should include the version numbers of any software, brief descriptions of the software, and copies of the input files.

The examples presented in Appendices A through H of this guide illustrate techniques for the proper documentation of fire modeling calculations using the format described. In conclusion, a properly documented analysis should enable someone else to reproduce the results from the information contained within the documentation.

2.7 Summary

This chapter described a recommended process for conducting and documenting a fire modeling analysis. Chapter 3 provides guidance on selecting the appropriate fire modeling tool and input parameters for typical commercial NPP applications. Fire model uncertainty is addressed in Chapter 4 of this document. Specific fire modeling examples evaluated using the process described in this Chapter are provided in Appendices A through H.

3
GUIDANCE ON FIRE MODEL SELECTION AND IMPLEMENTATION

This chapter contains a catalogue of typical nuclear power plant (NPP) fire scenarios, including relevant physical phenomena, model selection criteria, and suggested modeling strategies. In particular, the chapter provides recommendations as to the appropriate use of empirical correlations, zone models, and CFD calculations.

3.1 Model Implementation of Fire Scenario Elements

This section provides a description of fire modeling elements typically present in commercial NPP scenarios. The following fire modeling elements are described:

- Heat Release Rate
- Plant Area Configuration
- Ventilation Parameters
- Targets
- Intervening Combustibles

3.1.1 Heat Release Rate

For most fire model applications, the heat release rate (HRR) is the most important parameter to specify. All enclosure fire models solve some form of the energy conservation equation (i.e., the energy from the fire increases the temperature and drives hot air out and cold air into the enclosure). The models essentially redistribute the energy from the fire throughout the enclosure. For most NPP applications, the HRR is specified by the analyst in the form of a time history. This time versus HRR curve typically has four stages: incipient, growth, steady burning at peak intensity, and decay.

During the incipient stage, the fire burns at a low intensity (i.e., smoldering insulation or a small trash can fire). The duration of this stage may vary from seconds to hours, and the energy release is relatively low. Because of the uncertainty in the intensity of the fire during this stage, and the exact time that the fire will transition to a significant fire, the incipient stage is often not considered in the analysis.

Depending on the combustible and its arrangement, the growth to a fully developed fire will vary from seconds to minutes. Unless experimental data is available, the HRR is usually specified to increase following a so-called t^2 growth profile (Karlsson and Quintiere, 2000). The basic idea behind this approximation is that the burning surface area of a growing fire increases as the square of the time from the beginning of the growth period.

The steady burning phase occurs when the fuel reaches its maximum burning rate. In most cases, the peak HRR is obtained from experimental data. Alternatively, it can be estimated by multiplying the heat of combustion by the maximum burning rate of the fuel, if known. The peak

HRR may be limited by the available air supply. Estimates of the maximum HRR inside a compartment with a given ventilation rate are available in fire protection engineering handbooks.

Following the steady burning phase, the HRR is usually specified to decrease to zero linearly. The duration of this phase depends on the amount of fuel available. In fact, the integral of the entire HRR time profile (in units of kJ) divided by the heat of combustion of the fuel (in units of kJ/kg) should equal to the mass of combustibles (in units of kg).

In addition to the HRR time history, the following parameters may be important depending on the model or the scenario.

- Fire elevation: The elevation of the base of the fire, measured from the floor. It is important in scenarios involving targets in the fire plume where the relative distance between the fire and the target strongly influences the exposing temperature. It is also important because the height of the fire relative to the hot gas layer (HGL) influences the air entrainment into the plume, the position of the HGL, and, potentially, the actual HRR (since air entrained from the HGL is oxygen-depleted).

- Fire location: In scenarios where the fire is located near a wall or corner, the plume is expected to entrain less air, resulting in higher plume temperatures (Karlsson and Quintiere, 2000, p. 72).

- Fuel mass: This parameter is an important factor in determining the burning duration

- Soot and product yields: The *yield* of a combustion product is the mass of the product generated per unit mass of fuel consumed. In particular, the soot yield is an important factor in radiative heat transfer (e.g., targets immersed in the HGL), visibility calculations, and smoke detector response estimates. The yields of toxic gases can also be important in habitability calculations

- Radiative fraction: The fraction of energy emitted in the form of thermal radiation. For most materials, the radiative fraction is approximately one third. That is, one third of the total HRR radiates in all directions and two thirds is convected upwards into the smoke plume

3.1.2 Plant Area Configuration

The plant area configuration refers to the geometrical layout and construction of the enclosure. Each of these elements is described in detail next.

Compartment Geometry

Compartment geometry refers to the physical layout of the volume in which the fire is postulated. The length, width, and height of the room are the typical inputs required by the model. The size of a compartment is an important factor in the volume, and is used to solve the fundamental conservation equations. Algebraic and zone models employ considerable simplifications of the geometry, while CFD models attempt to replicate as much of the geometry as possible.

Compartment Boundary Materials

Boundary (e.g., wall or ceiling) materials are characterized with thermophysical properties, which include the density, specific heat, and thermal conductivity of the material. In the majority of commercial nuclear power plant (NPP) applications, the wall material is concrete. Other materials may include steel, gypsum board, etc. Properties for these materials are often available in "drop down" menus in the fire models or in fire protection engineering handbooks. Table 3-1 provides typical properties for materials commonly found in NPPs. These properties were used in the examples described later in this Guide. It should be noted that these properties, as input parameters to the models, may also be available from sources other than the ones listed in Table 3-1. Regardless of the source of the information, users should always check that the material property values are appropriate for their specific application, as the resulting fire conditions may be sensitive to these parameters.

Table 3-1. Material properties.

Material	Thermal Conductivity (W/m/K)	Density (kg/m³)	Specific Heat (kJ/kg/K)	Source
Brick	0.8	2600	0.8	NUREG-1805, Table 2-3
Concrete	1.6	2400	0.75	NUREG-1805, Table 2-3
Copper	386	8954	0.38	SFPE Handbook, Table B.6
Gypsum	0.17	960	1.1	NUREG-1805, Table 2-3
Plywood	0.12	540	2.5	NUREG-1805, Table 2-3
PVC	0.192	1380	1.289	NUREG/CR-6850, Appendix R
Steel	54	7850	0.465	NUREG-1805, Table 2-3
XLP	0.235	1375	1.390	NUREG/CR-6850, Appendix R

3.1.3 Ventilation Effects

Ventilation effects include natural ventilation through vertical or horizontal openings, the effects of leakage paths, and/or the effects of mechanical ventilation. Each of these elements is described next.

Vertical Openings

Vertical openings are usually doors, but they can also be other wall openings, such as windows or passive ventilation ducts. In some cases, a compartment will have more vertical openings than the number that can be specified in a simplified model. For example, the McCaffrey, Quintiere, and Harkleroad (MQH) correlation for calculating HGL temperature uses only one opening when calculating the ventilation factor which, is defined as the product of the opening area and the square root of the opening height $A_o\sqrt{H_o}$ (Karlsson et al., 2000; Drysdale, 2011). Buchanan (2001) suggests the following method for calculating an effective height and area:

$$H_o = \frac{\sum_i A_{o,i}H_{o,i}}{A_o} \equiv \frac{\sum_i A_{o,i}H_{o,i}}{\sum_i A_{o,i}} \tag{3-1}$$

Here $A_{o,i}$ and $H_{o,i}$ are the individual door areas and heights. The effective width of multiple vertical openings can be estimated by the ratio A_o/H_o.

Regarding doors (and other operable openings), consideration should be given to the doors being opened (or closed) during a fire. For example, when the fire brigade arrives, they will open the doors to the fire area to gain access, which will affect the ventilation and possibly result in smoke spread.

Leakage Paths

The doors of most compartments in commercial NPPs are normally closed, but are not perfectly sealed. Consequently, the resulting pressure and the rate of pressure increase are often kept very small by gas leaks through openings in the walls and cracks around doors, or "leakage paths." Leakage paths must be specified in compartments with closed doors during the fire event unless the analysis considers a completely sealed enclosure where pressure rise is an important variable. By contrast, compartments with at least one open door or window can maintain pressure close to ambient during the fire event. Leakage paths therefore do not need to be specified, since the leakage opening area is negligible when compared with the opening areas of doors and windows.

Horizontal Openings

Horizontal openings consist of hatches or stairwells. For modeling purposes, the areas of horizontal openings can simply be added. Any zone model should provide similar answers with single or multiple horizontal openings as long as the total opening area is the same. Note that the Consolidated Fire Growth and Smoke Transport (CFAST) model only allows for a single connection between any pair of compartments included in a simulation. For a CFD model, no special provisions are necessary to describe a horizontal opening.

Mechanical Ventilation

Mechanical ventilation refers to any air injected into or extracted from a compartment by mechanical means. This has a number of practical applications, such as extracting smoke from the HGL (e.g., a smoke purge system). The ventilation rate and the vent position are the two most important mechanical ventilation parameters. For some applications, the velocity of the airflow may also be important. These mechanically induced flows have the potential to alter the fire-induced flows. Mechanical ventilation often consists of a supply and an exhaust system that are maintained to achieve a certain pressure level.

3.1.4 Targets

A target is an object of interest that can be affected by the fire-generated conditions and typically consists of cables in conduits, cables in raceways, or plant equipment. Targets are characterized by their location, damage criteria, and thermophysical properties.

A target's location simply refers to its location relative to the fire. The location is represented by three-dimensional coordinates within the volume of the room in which the fire conditions are simulated. Where the target faces in a particular direction, an orientation vector to indicate that direction needs to be entered.

The damage criteria refer primarily to a damage/response threshold. In general, the damage criteria for scenarios involving cable damage is expressed in terms of damage temperature or incident heat flux.

The models within the scope of this Guide require specification of the target's thermophysical properties, primarily the density, specific heat, and thermal conductivity, for the analysis. These parameters are used to estimate heat conducted into the targets. The predicted time for the gas temperature surrounding a target to reach a specific limit is usually less than the time it takes the target to reach the same limit because the heat conduction inside the target will delay the temperature rise at the surface during the heating process.

Information on target damage thresholds and target thermophysical properties can be found in documents such as NUREG-1805, NUREG/CR-6931, and NUREG/CR-6850 (EPRI 1011989).

3.1.5 Intervening Combustibles

In many cases, commercial NPP fire scenarios do not require burning targets to be modeled because it is sufficient to determine only when the target is damaged. This is clearly not the case with intervening combustibles, whose flammability characteristics need to be incorporated into the model so that the fire progression is considered. Therefore, the intervening combustibles should be described not only in terms of their proximity to the fire and the targets, but also in terms of their relevant thermophysical and flammability properties.

In many cases, intervening combustibles consist of cables in ladder back trays. Representing intervening combustibles, like cables, in fire models presents technical challenges that the analyst should consider, including (1) obtaining the necessary geometric and thermophysical properties representing the intervening combustible and (2) the ability of the computer tools to model the fire phenomena (e.g., fire propagation). Because of these challenges, simplified models for determining the contribution to the HRR due to flame spread and fire propagation through cable trays have been developed. Appendix R of NUREG/CR-6850 (EPRI 1011989) provides guidance on the calculation of fire spread and HRRs for cable trays. Additionally, research is underway to develop improved methods for predicting the HRR and flame spread of electrical cables. A simple model, Flame Spread over Horizontal Cable Trays (FLASH-CAT) which predicts flame spread over cables, has been developed as part of the Cable Heat Release, Ignition, and Spread in Tray Installations during Fire (CHRISTIFIRE) project (NUREG/CR-7010, vol. 1, 2012), sponsored by the U.S. Nuclear Regulatory Commission (NRC) and conducted by the National Institute of Standards and Technology (NIST).

3.2 Guidance on Model Selection and Analysis

This section provides guidance on model selection and analysis of specific fire scenarios. Each subsection is devoted to a specific fire scenario, as listed in Table 3-2. In addition, Figure 3-1 provides a pictorial representation of each of these scenarios. The circled numbers are intended to direct the reader to the section in which the scenario is described. Please note: these pictorial representations are not drawn to scale and are used for illustrative purposes only.

Table 3-2. Listing of generic scenarios described in this chapter.

Number	Chapter Section	Scenario Description
1	3.2.1	Scenarios consisting of determining time to damage of cables above the ignition source located inside the flames or the fire plume.
2	3.2.2	Scenarios consisting of determining time to damage of cables located inside or outside the HGL. This scenario also includes a secondary fuel source (i.e., propagation to cable trays).
3	3.2.3	Scenarios consisting of determining time to damage of cables located in a room adjacent to the room of fire origin.
4	3.2.4	Scenarios consisting of determining time to damage of cables located inside or outside the HGL in rooms with complex geometries.
5	3.2.5	Scenarios consisting of determining time to loss of habitability of the main control room.
6	3.2.6	Scenarios consisting of determining time to smoke or heat detector activation.
7	3.2.7	Scenarios consisting of determining temperature of structural elements.

GUIDANCE ON FIRE MODEL SELECTION AND IMPLEMENTATION

Each of the sections listed above is organized as follows:

- A sketch capturing most of the technical elements relevant to the analysis. A legend summarizing the different elements presented in the sketches is provided in Figure 3-2.

- A scenario objective stating the purpose of the modeling exercise in engineering terms.

- A description of the relevant technical fire scenario elements, such as mechanical ventilation, room geometry, etc. Recall that fire scenario elements refer to the different characteristics of the fire scenario that are relevant to the analysis, and should be properly represented in the model.

- A modeling strategy section summarizing the recommended steps for performing the calculation.

- A section listing fire model recommendations for the analysis.

- A section referencing relevant detailed fire modeling examples documented in the Appendix section of this guide.

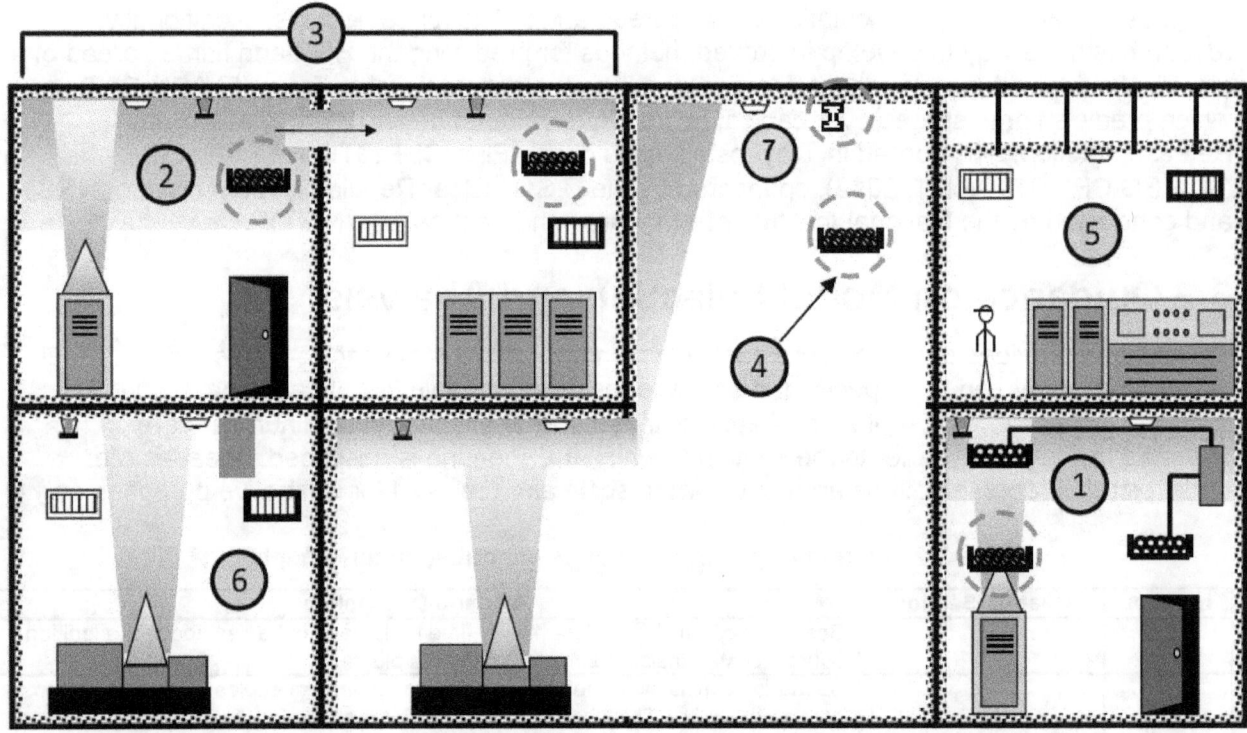

Figure 3-1. Pictorial representation of the fire scenario and corresponding technical elements described in this section.

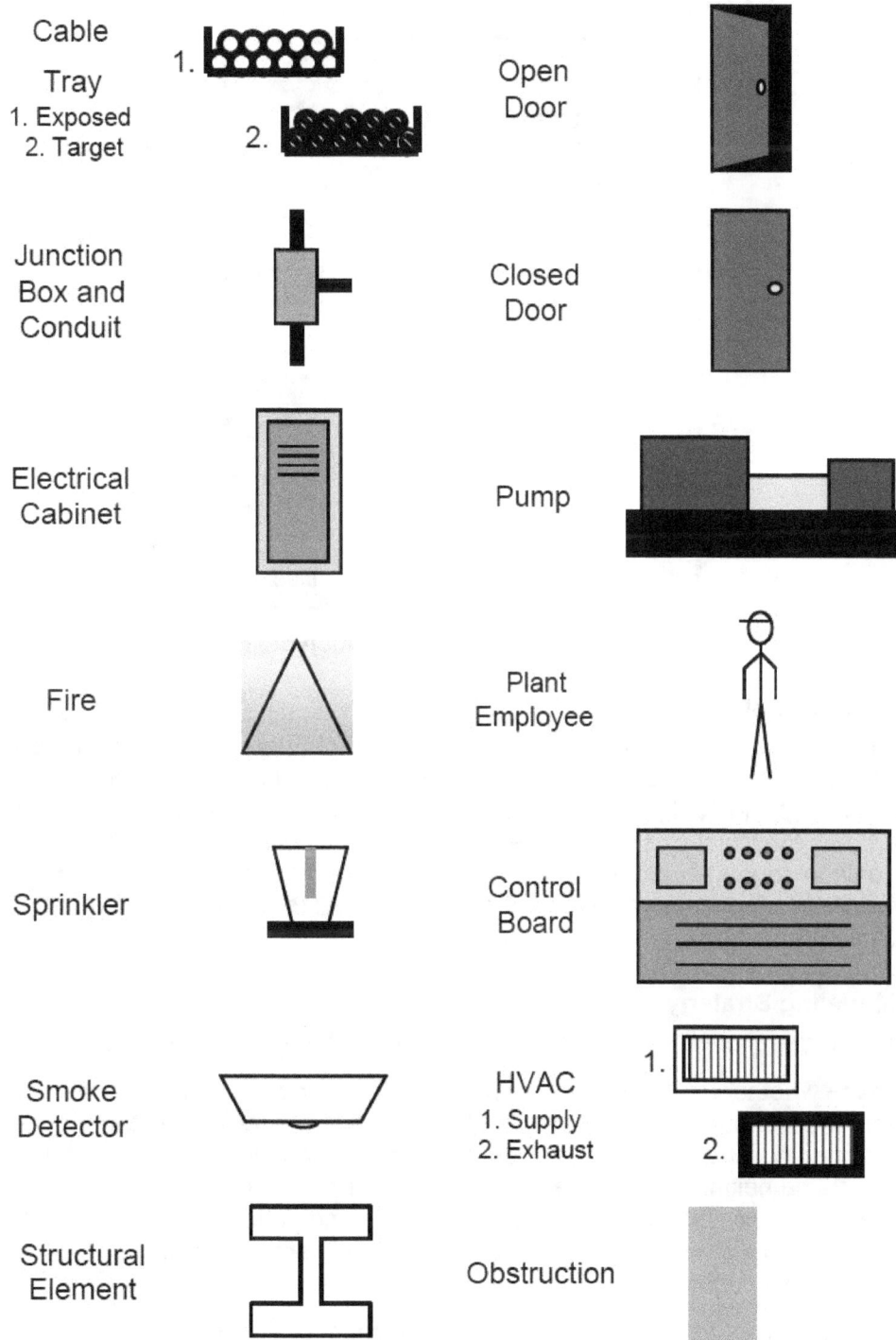

Figure 3-2. Legend for fire modeling sketches presented in this chapter.

3.2.1 Targets in the Flames or Plume

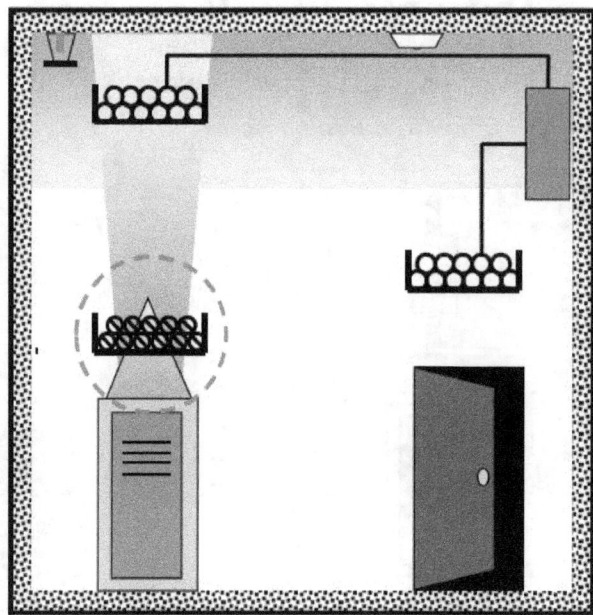

Figure 3-3. Pictorial representation of scenario 1.

3.2.1.1 General Objective

The objective of this scenario is to calculate the time to damage for a target immediately above a fire, as indicated by the dashed circle in Figure 3-3. For this scenario, the target is an electrical raceway and the fire source is an electrical cabinet. This scenario is often encountered as the initial part of a progression of target damage (i.e., the first item ignited, aside from the ignition source). Consequently, the characterization of the first item ignited after the ignition source is important in subsequent estimations of fire propagation through other intermediate combustibles.

3.2.1.2 Modeling Strategy

The recommended modeling strategy is summarized in the following steps:

1. Determine whether the target cable, which is directly above the fire, is within the flame zone or within the fire plume. The target should be considered inside the flame zone if it is located directly above the base of the fire and its distance from the base of the fire is less than the flame height. If the target is above the fire but is not within the flame zone, then it is considered to be within the fire plume. It should be noted that unobstructed fire plumes will increase in diameter as a function of height. Consequently, a target does not need to be directly above the ignition source to be immersed in the fire plume.

2. Calculate the time to damage by finding the minimum of either:

    a. The time at which the flame reaches the target. This is achieved by calculating the flame height as a function of time using the HRR profile (HRR vs. time) and fire diameter as input.

    b. The time it takes the fire plume temperature to exceed the target damage temperature. This is achieved by calculating the plume temperature at the specified height as a function of time, using the HRR profile as an input. This approach can be considered conservative, as it predicts cable damage occurs when the gas temperature surrounding the target reaches the damage temperature (i.e., heating of the cable is ignored). As an alternative, a potentially more accurate time to damage can be obtained by calculating the surface temperature of the cable as a function of time, given a heat flux profile generated by the flame or plume.

If non-target raceways are located between the ignition source and the target, the contributions of intervening combustibles need to be considered in the analysis. For example, consider a cabinet fire that ignites the first of a stack of trays overhead. The fire involving the combination of the cabinet and first tray may then ignite the second tray in the stack, and the fire may progress to damaging the target raceway. Considerations of the intervening combustibles in the analysis include the HRR contribution and the corresponding effects on the target heating time. Section 3.2.2 (Scenario 2) provides guidance on treatment of intervening combustibles.

In addition to the guidance provided above, the analyst should determine whether HGL effects are relevant to the scenario. The portion of the fire plume immersed in the HGL entrains air at higher temperatures (i.e., the HGL temperature) and is expected to have increased temperatures when compared with portions of the fire plume outside the HGL. In scenarios consisting of targets located relatively close to the ignition source (which is the case for the scenario discussed in this section), the HGL effects on the plume temperature are generally not considered, as the time to target damage is expected to be relatively short. For scenarios involving targets in the fire plume, located relatively far from the ignition source, the HGL effects on target heating should be considered. In the latter case, the room's geometry and ventilation (both natural and mechanical) conditions should be captured by the analysis.

3.2.1.3 Recommended Models
Algebraic Models

Both the FDT[s] and FIVE-Rev1 have models that can be useful for this scenario, provided that the configuration is within the correlation basis and that there are no significant HGL effects. Heskestad's flame height correlation is an alternative for determining flame height. Similarly, Heskestad's fire plume temperature correlation is an alternative for determining plume temperature and diameter (Heskestad, 2002).

The correlations listed above are particularly applicable for scenarios consisting of targets relatively close to the ignition source, where HGL effects are not considered in the analysis.

As noted above, the time to damage is the minimum of either the time at which the flame reaches the target, or the time it takes the fire plume temperature to exceed the target damage temperature. This is simply the time at which the HRR reaches the value required for either of the failure criteria. In both cases, the correlations are solved for a specific HRR at a specific time up to when the fire conditions suggest target damage (e.g., flames reach the location of the raceway or the plume temperature exceeds the damage temperature of the cables).

Another option, for scenarios where the flames do not reach the cables, would be to use the Thermally-Induced Electrical Failure (THIEF) model (NUREG/CR-6931, volume 3) to determine the surface temperature at the target. The THIEF model is included in NUREG-1805, Supplement 1 (2012).

Zone Models

Zone models can be used for this scenario. To do so, set up the necessary input file, which should include a "target" in the location of the electrical cable of interest, along with the corresponding thermophysical properties, so that the surface temperature of the cable can be tracked. Zone models have the ability to include HGL effects in their calculation of plume temperature, and are thus particularly appropriate for scenarios where the HGL temperature interacts with the fire plume at the location of the target.

Again, the time to damage is the minimum of either the time at which the flame reaches the target or the time it takes the fire plume temperature to exceed the target damage temperature. The zone models routinely calculate and report these values.

CFD Model

Although a CFD model could be used to analyze this scenario, the level of detail and resolution offered by a CFD calculation is generally not necessary. On the other hand, the CFD model would be particularly applicable if the scenario involved obstructions between the fire and the target inside the fire plume or if HGL effects are significant. The effects of these obstructions on the exposure conditions are not captured by algebraic models or zone models.

3.2.1.4 Detailed Examples

Readers are referred to Appendix B, which describes the analysis of an electrical cabinet fire in the switchgear room, and Appendix E, which describes the analysis of a transient fire in a cable spreading room.

3.2.2 Scenario 2: Targets Inside or Outside the Hot Gas Layer

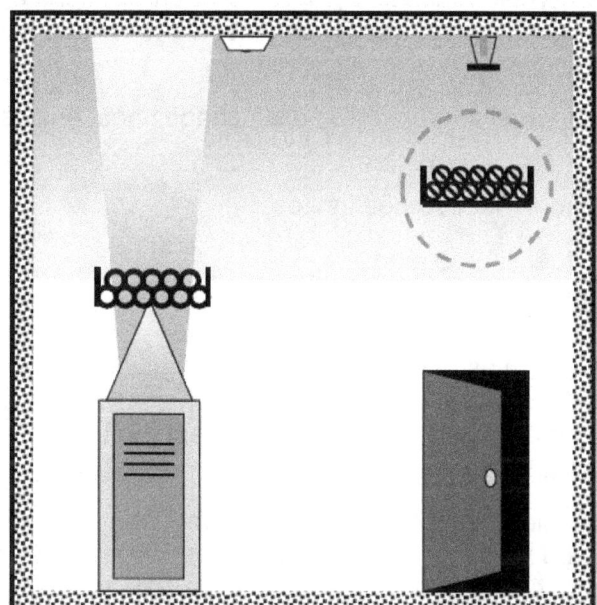

Figure 3-4. Pictorial representation of scenario 2.

3.2.2.1 General Objective

The objective of this scenario is to calculate the time to damage for a target inside or outside the HGL produced by a fire. The time to ignition of a secondary fuel source and the resulting contribution to the total HRR can also be determined. For the case shown in Figure 3-4, the target is a cable in an electrical raceway and the fire source is an electrical cabinet.

3.2.2.2 Modeling Strategy

Two levels of analysis can be employed: (1) algebraic models for the average room temperature as an indicator of the gas temperature surrounding the target, or (2) detailed heat transfer analysis for determining the target temperature.

The first strategy consists of determining the overall room temperature using an algebraic model (e.g., the MQH room temperature model) (McCaffrey et al., 1981). Such a calculation will indicate whether the target may be subjected to damaging temperatures and the time at which such temperatures may be observed. It should be noted that the room needs to be represented as a rectangular parallelepiped and the area of all the surfaces in the room must be conserved. In addition, if the target cable tray is relatively close, the target may be damaged by radiant heating. This can be assessed with simple point source estimates that only require the HRR of the fire, the separation distance between the fire and the target, and the damage criteria (i.e., critical heat flux for damage). The point source model may lead to over-predictions when used to predict heat flux to targets in very close proximity to the fire. In many applications, the over-predictions will clearly suggest target damage or ignition.

The second strategy is best addressed with a model capable of including detailed heat transfer analysis for determining the target's temperature. A raceway outside the fire plume may be exposed to HGL conditions if the smoke accumulating in the upper part of the room descends to the location of the raceway. Consequently, targets outside the fire plume are initially exposed to "lower layer" conditions. As the smoke continues to accumulate, the target is immersed in HGL conditions. As heat transfer conditions will be different for each case, a model with the ability to track the relevant/applicable heat transfer interaction and calculations as a function of time, such as a zone model or a CFD model, should be selected to handle this scenario at the desired level of resolution.

With regard to the secondary fuel source, three distinct additional analyses must be made to determine:

- time at which the secondary fuel source ignites,

- HRR of the secondary fuel source, and

- combined HRR of the primary and secondary fires.

The more detailed models, such as FDS, can handle the ignition and contribution of multiple fires, provided that the ignition criteria and source HRR characteristics are provided as input. Other models, especially the algebraic models, only accept the total HRR as a function of time, which is found by summing up the individual HRRs.

In the present example, consider a cable tray directly above the fire. The time to ignition of the cable tray can be determined via algebraic models that estimate the flame height and plume temperature as a function of time for the initial cabinet fire (see scenario previously discussed). Once the flames from the cabinet reach the cable tray, the cable ignites. The same is true when the plume temperature at the elevation of the cable tray reaches the ignition temperature of the cables. Both calculations should be completed, and the shorter time used as the ignition time.

The HRR from the cable tray can be added to the HRR of the cabinet to determine a combined HRR as a function of time. The resulting HRR profile can take into consideration both the fuel consumption and propagation to additional intervening combustibles as a function of time. In cases where fuel consumption is not considered, the resulting HRR profile is expected to overestimate the fire intensity. This total rate can then be used in the various models as an approximation of the HRR profile.

Appendix R of NUREG/CR-6850 (EPRI 1011989) addresses cable fires, including methods for calculating the HRR for a variety of cable configurations.

It should be noted that the simple summation of the two HRRs is a simplification of a complex phenomenon and only provides an approximation of the conditions created by the two separate fires.

3.2.2.3 Recommended Modeling Tools
Algebraic Models

Select the appropriate HGL (or room temperature) model and then collect the required inputs, including room size, opening sizes, boundary material properties, forced ventilation, and the HRR profiles for the initial and secondary fuel packages. For screening purposes, the use of algebraic models is recommended as long as the contributions of the first item ignited and intervening combustibles are considered. As was mentioned earlier, this approach will provide an approximation of the room temperature in which the target may be immersed. The methods used by algebraic models to address the secondary fire source are discussed above. Target damage due to radiant heating can be estimated using algebraic models; all that is required is the HRR of the fire (as a function of time), the separation distance between the fire and the target, and damage criteria (i.e., critical flux for damage).

Zone Models

Zone models provide a good alternative for modeling this scenario, as they provide the incident heat flux profile, the surface temperature, and the internal temperature of the target in one simulation. Set up the necessary input file with the required inputs, including room size, opening sizes, boundary material properties, HRR, fire diameter, and a target and fire location so that the cable's surface temperature can be predicted.

Zone models also have the benefit of being able to handle secondary fire sources as separate entities. Secondary fires can be ignited at a prescribed time, temperature, or heat flux. However, zone models have limited capabilities for handling obstructions.

Target damage due to radiant heating from the fire is easily handled by zone models, as long as there are no obstructions between the fire and the target that block radiant heat transfer. Zone models can also account for radiant heating of targets by the HGL.

CFD Model

The use of CFD models for this scenario is recommended for complex geometries capable of affecting the location of the HGL and the incident heat flux to the targets, or when greater accuracy of the ignition and contribution of secondary fires is warranted. For instance, obstructions between the ignition source and the target affect the heat balance at the surface of the target. The CFD model will require inputs similar to the ones collected for the zone models; however, the compartment geometry will need to be specified in greater detail.

Due to their detailed calculations, CFD models are best able to model secondary fire sources, including their ignition and subsequent contribution to the HRR within the enclosure.

Like zone models, CFD models can handle targets damaged by radiant heating from the fire and the HGL. CFD models can also include the effects of obstructions between the fire and the target.

3.2.2.4 Detailed Examples

Appendix C describes the analysis of a relatively large lubricating oil fire affecting a raceway in a pump room, and Appendix E describes the analysis of a transient fire in a cable spreading room.

3.2.3 Scenario 3: Targets Located in Adjacent Rooms

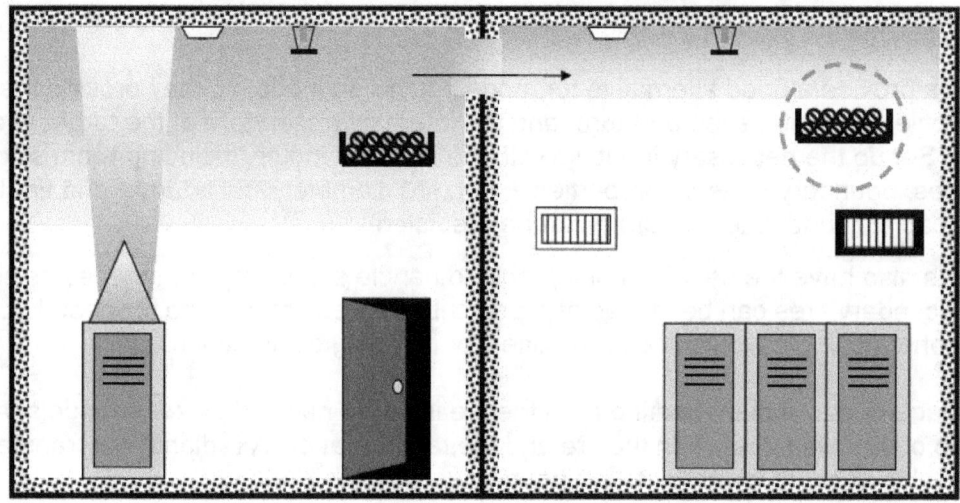

Figure 3-5. Pictorial representation of scenario 3.

3.2.3.1 General Objective

The objective of this scenario is to calculate the time to damage for a target in the HGL in a room adjacent to the room of fire origin. An opening connecting the room of origin to the adjacent room allows combustion products to enter the adjacent room. For the case shown in Figure 3-5, the target is a cable in an electrical raceway, the fire source is an electrical cabinet, and there is an unprotected opening in the wall.

3.2.3.2 Modeling Strategy

The recommended strategy for determining the temperature of targets located in a room adjacent to the room of fire origin consists of four basic steps:

1. Determine the following characteristics for the HGL in the room of fire origin and the adjacent compartment:

   a. Temperature as a function of time

   b. Depth as a function of time

2. Determine the incident heat flux surrounding the target cable.

3. Determine the surface and internal temperature of the target cable.

4. Compare the surface or internal temperature of the target with its damage temperature.

One way to approach this (using algebraic models or zone models) is to first model the HGL temperature in the room of origin and the resulting effect on remote targets in the room. If this approach indicates that target damage/ignition is unlikely in the room of origin, there would probably be little benefit in evaluating similar targets in adjacent spaces. However, if target damage is possible in the room of origin, the next step would be to model the room of origin and the adjacent room and determine whether the resulting HGL is capable of causing damage/ignition of the target(s).

3.2.3.3 Recommended Modeling Tools
Algebraic Models

Generally, algebraic models are not suitable for this calculation, as a model capable of tracking fire conditions in adjacent rooms is necessary. Zone and CFD models can provide this capability. As a screening tool, algebraic models could be used to model the HGL temperatures in the room of fire origin. If the estimated fire conditions in the room of origin are determined not to generate damage or ignition of targets, it can be concluded that targets in adjacent rooms are also not expected to be damaged.

Zone Models

The zone model is an appropriate tool for addressing this scenario. Zone models are efficient tools for scenarios involving relatively simple geometries (i.e., geometries and openings that can be easily represented in rectangular parallelepipeds without compromising the technical elements in the analysis). Consequently, the room geometry should be represented as accurately as possible. One of the primary outputs of zone models is the height and temperature of the HGL versus time in each of the rooms specified in the computational domain. Zone models are also capable of determining target temperature (not just the temperature of the gases surrounding the target), given the boundary conditions generated by the fire and the thermophysical properties of the target.

CFD Model

A CFD model would be particularly appropriate for addressing targets located in adjacent rooms in scenarios with complex geometries (i.e., geometries that can't be easily represented as rectangular parallelepipeds). CFD models can describe the geometry of the compartment in detail, including the opening(s) providing smoke migration paths to the adjacent room.

3.2.3.4 Detailed Examples
Readers are referred to Appendix G, which describes the analysis of targets in rooms remote from the fire room.

3.2.4 Scenario 4: Targets in Rooms with Complex Geometries

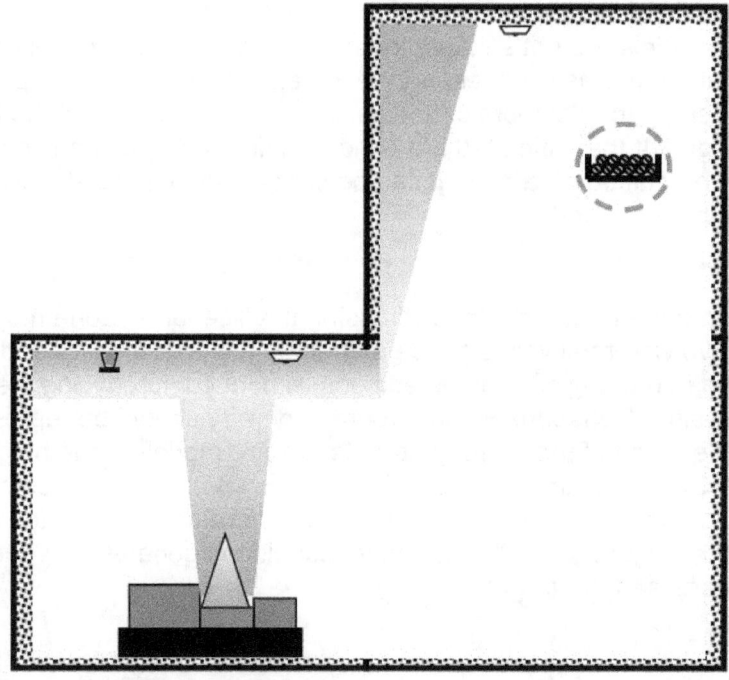

Figure 3-6. Pictorial representation of scenario 4.

3.2.4.1 General Objective

The objective of this scenario is to calculate the time to damage for a target in the HGL in a room with a complex geometry. For the case shown in Figure 3-6, the target is a cable in an electrical raceway and the fire source is an electrical cabinet.

3.2.4.2 Modeling Strategy

The first strategy involves using an algebraic or single compartment zone model to estimate the HGL temperature. This approach requires that the complex geometry be reduced to a single equivalent volume. In the case of two different ceiling heights, an empirical correlation or a single compartment zone model may underestimate the temperature of the smaller volume and overestimate the temperature of the larger. The fire's energy is conserved, but it is not expected to be uniformly distributed. For this reason, it is suggested that the model only be applied to the portion of the compartment where the fire is postulated. This reduction in room volume will result in a higher predicted HGL temperature. This strategy should not be applied if the purpose of the calculation is to predict the activation of a sprinkler or smoke detector because an overestimate of the HGL temperature will lead to an underestimate of the activation time.

The second strategy is to use a model capable of describing the complex geometry. A zone model can model the entire compartment as a collection of connected volumes. A CFD model can "block off" portions of the numerical grid to account for geometric obstructions. In complex geometries, HGL development can be significantly impacted by mixing associated with spilling and ventilation, and these can only be modeled by zone and CFD models.

3.2.4.3 Recommended Modeling Tools

Algebraic Models

Detailed analyses of complex geometries typically cannot be easily accomplished with algebraic models. However, for screening purposes, it is possible to use algebraic models. As mentioned earlier, this approach can provide an approximation of the HGL temperature in which the target may be immersed. To utilize this approach, first select the appropriate HGL (or room temperature) model and then collect the required inputs, including room size, opening sizes, boundary material properties, and HRR. Next, the complex geometry must be reduced to a single equivalent volume while maintaining total surface area (due to the importance of energy losses through the bounding surfaces) and ceiling height. It should be noted that the more complex the space, the less ideal the equivalent volume/area approximation becomes. Based on the estimates derived using the algebraic models, more detailed modeling may be indicated.

Zone Models

Zone models should also be used with caution when modeling this scenario. If the entire space is modeled, the interface between lower and upper compartments is treated as a big door. The entrainment correlations used by the zone model to handle vertical vents were not designed for such large, open "doors."

CFD Model

CFD models may be required when detailed analyses of complex geometries capable of affecting fire development and the location of the HGL and the incident heat flux to the target are desired. CFD models are expected to better estimate the overall compartment temperatures, both upper and lower, because the basic methodology can handle non-uniform ceilings.

3.2.4.4 Detailed Examples

Readers are referred to Appendix D, which consists of a switchgear fire in a room with a complex geometry, and Appendix H, which consists of a fire inside the containment annulus.

3.2.5 Scenario 5: Main Control Room Abandonment

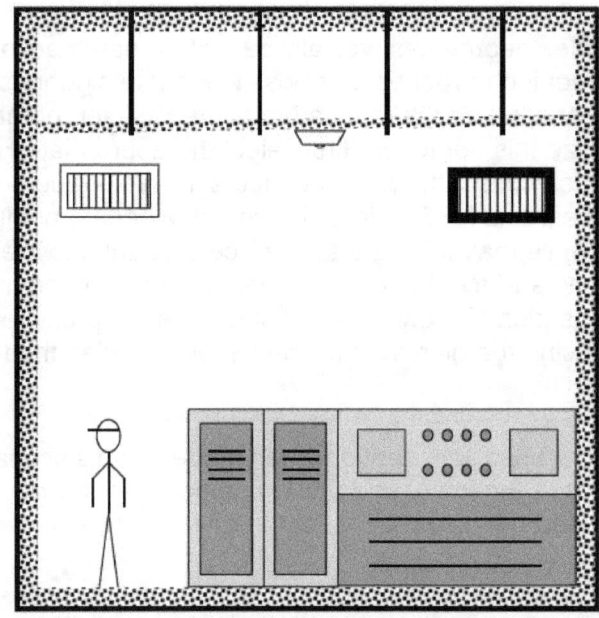

Figure 3-7. Pictorial representation of scenario 5.

3.2.5.1 General Objective

The objective of this scenario is to determine when control room operators will need to abandon the control room due to fire-generated conditions inside the room. This scenario consists of a fire, such as an electrical cabinet fire within the main control. A schematic diagram of this scenario is shown in Figure 3-7. Notice the presence of a suspended ceiling in the room.

3.2.5.2 Modeling Strategy

Main control room (MCR) abandonment is specified as being solely dependent on habitability conditions. As mentioned in the previous sections, control room operators are considered "targets" in this scenario, so it is necessary to establish the fire conditions that would force operators out of the control room. This can be considered the "abandonment criteria"; for example, visibility, temperature, heat flux, and toxicity are often the habitability indicators in these scenarios. Tracking these conditions can provide the time at which the operator may need to abandon the control room. Once the criteria have been established (see Chapter 11 of NUREG/CR-6850 (EPRI 1011989) for details on habitability conditions), the fire-generated conditions in the room can be calculated so that the abandonment time can be determined.

For MCR analyses, two ventilation conditions should be taken into consideration: (1) the ventilation system is turned off, causing hot gases and smoke to accumulate inside the control room and (2) the ventilation system is operating in smoke-purge mode.

3.2.5.3 Recommended Modeling Tools

Algebraic models

Algebraic models can be used to address individual abandonment criteria, but not all simultaneously. Algebraic models are typically based on empirical correlations, not the basic conservation laws of mass, momentum, and energy. For this reason, it is difficult to perform a sensitivity analysis for a collection of correlations because they may not capture the interdependent relationships of the predicted quantities.

Zone Models

Unlike algebraic models, zone models are capable of simultaneously tracking a number of relevant output variables (e.g., habitability conditions) in this scenario. They are also capable of modeling the impact of the various ventilation configurations required for modeling MCR abandonment. Zone models are a good tool for modeling fires in the MCR as long as compartment or zonal average conditions within a given enclosure provide the necessary insights to support the conclusions. Localized radiation conditions can also be estimated using zone models. Questions, such as average temperature, visibility conditions in the control room, and smoke management, are well handled by zone models. In contrast, questions associated with temperature, heat flux levels and visibility at specific locations within the control room (e.g., near a panel) are best handled by CFD models.

CFD Model

CFD models are also a good alternative to address this scenario, particularly if complex geometries are involved or localized fire conditions are needed. CFD models have the added advantage of handling rooms with complex geometries, intervening combustibles and obstructions, and varying ventilation conditions.

3.2.5.4 Detailed Examples

Readers are referred to Appendix A, which describes the analysis of a fire in an MCR.

GUIDANCE ON FIRE MODEL SELECTION AND IMPLEMENTATION

3.2.6 Scenario 6: Smoke Detection and Sprinkler Activation

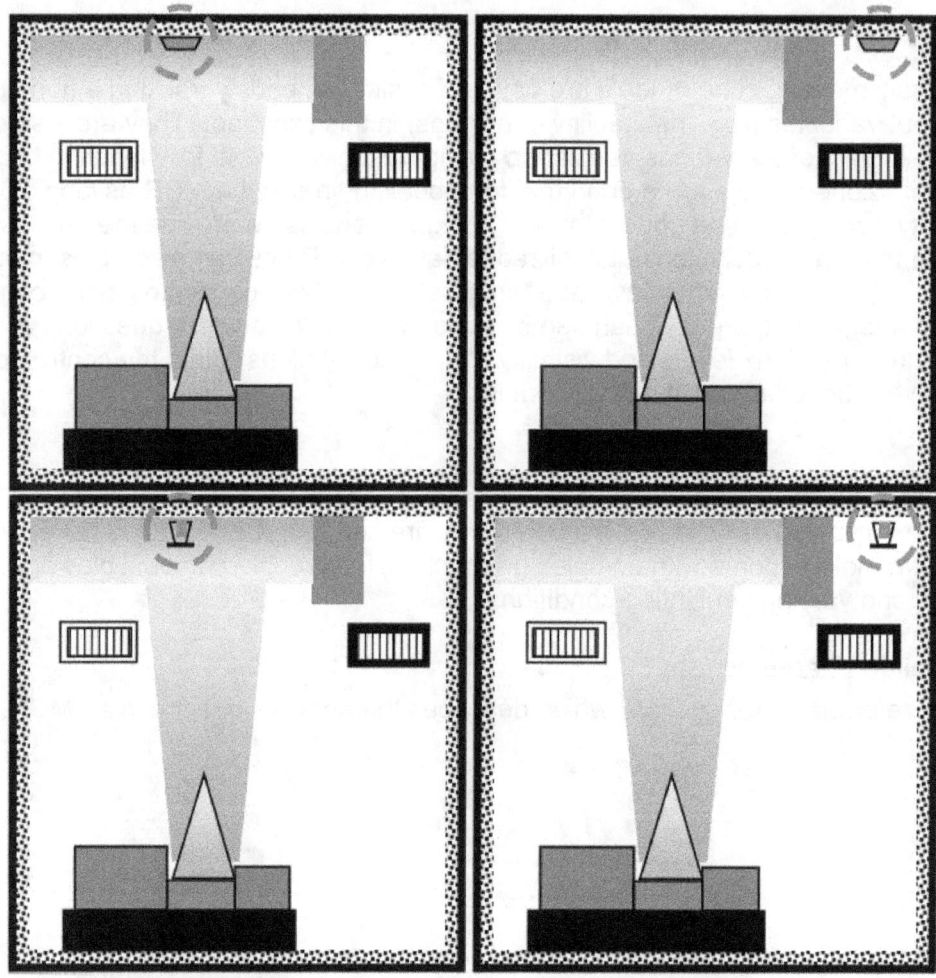

Figure 3-8. Pictorial representation of scenario 6.

3.2.6.1 General Objective

The objective of this scenario is to calculate the response time of a smoke or heat detector. The flow of heat and smoke from the fire to the detector may be obstructed by ceiling beams, ventilation ducts, etc. Failure of a detector to actuate in response to a fire can delay the response of either the fire brigade or an automatic suppression system. Typical scenarios are shown in Figure 3-8.

3.2.6.2 Modeling Strategy

For scenarios involving unobstructed smoke detector devices:

1. Determine the location of the detection device relative to the fire.

2. Select the detector response (activation) criteria. Chapter 11 of NUREG/CR-1805 contains guidance on estimating smoke detector response times.

3. Calculate the detection time using the appropriate model.

For scenarios involving obstructed smoke detector devices:

1. Determine the following characteristics of the HGL using all the necessary inputs for an HGL calculation, as described earlier in this chapter.

    a. Temperature as a function of time.

    b. Depth as a function of time. The smoke detector is expected to activate shortly after the HGL reaches the bottom of the obstruction and spills into the location of the device.

2. Select the detector response (activation) criteria. Chapter 11 of NUREG/CR-1805 contains guidance on estimating smoke detector response times.

3. Calculate the response time of the given smoke detector once the combustion products reach the detector.

For scenarios involving thermal devices (e.g., sprinklers, fusible links, or heat detectors), the process is similar. The only difference is that the thermal device needs to be characterized with relevant parameters, typically an activation temperature and the response time index (RTI). In addition, the selected model should account for the heating process of thermally thin elements (i.e., the heat detector device).

3.2.6.3 Recommended Modeling Tools
Algebraic models

Algebraic models can be used to determine time to heat or smoke detection when the fire-induced flows are not obstructed before reaching the detection device. By contrast, algebraic models are typically not suitable when fire-induced flows, such as fire plumes or ceiling jets, will be obstructed before reaching the detection device. In some cases, algebraic models that estimate the HGL temperature as a function of time may be used for rough estimates of activation times.

Zone Models

Zone models can address the different scenario conditions presented above; for instance, CFAST and MAGIC are capable of determining time to smoke or heat detection when no obstructions are present, and can simultaneously calculate smoke accumulation so that the time for smoke detection activation can be estimated. This would provide an approximation, as zone models do not directly account for complex geometries, including obstructions. These models are not recommended for determining time to heat detection in obstructed geometries, since the velocity of the gases impacting the heat detector is not available in zone model calculations. As mentioned above, in some cases the HGL temperature alone may be used as a rough indicator of smoke and heat activation times.

CFD Model
CFD models are good tools for estimating time to fire detection in complex geometries, including obstructions, as they can describe the compartment's complex geometries and mechanical ventilation conditions in detail.

3.2.6.4 Detailed Examples
Readers are referred to Appendices B and E, which discuss the calculation of the time to smoke and heat activation.

3.2.7 Scenario 7: Fire Impacting Structural Elements

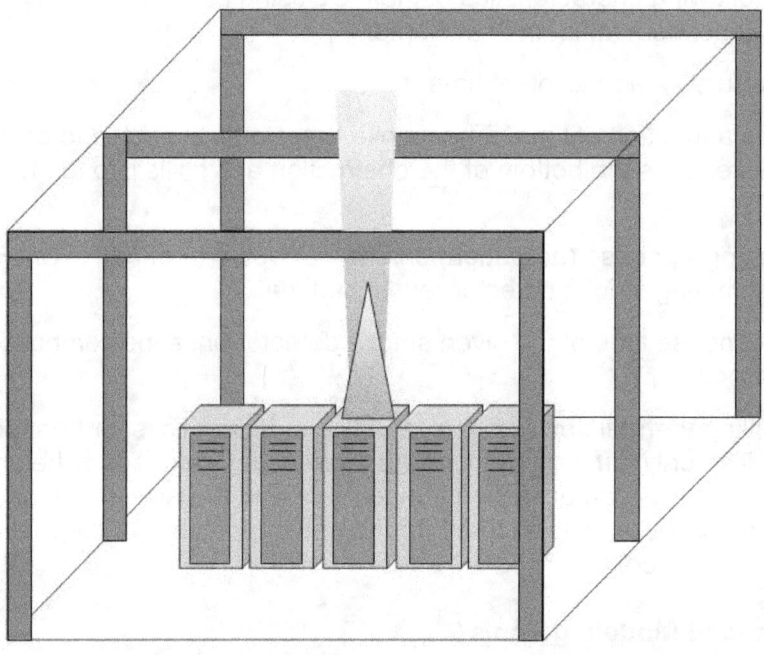

Figure 3-9. Pictorial representation of scenario 7.

3.2.7.1 General Objective

The objective of this scenario is to characterize the temperature of structural elements exposed to a nearby fire source. For the case shown in Figure 3-9, the exposure fire is an electrical cabinet.

3.2.7.2 Modeling Strategy

The fire modeling tools within the scope of this Guide should indicate whether the exposed structural element will reach damaging temperatures. However, this information is often not enough to determine whether the structural integrity of the compartment will be compromised by the exposing fire conditions. A more detailed structural analysis (i.e., one that involves complex temperature-dependent load-bearing calculations) may be needed if such a determination is necessary.

Considering the limitations listed above, the following general guidance is provided:

1. Determine whether the structural element is directly above the fire, within the ceiling jet, exposed to radiant heating, or within the HGL. The results of this determination will suggest which model or combination of models should be used.

2. Calculate the temperature of the structural element based on the fire conditions affecting it. This will require an initial estimate of the fire-generating conditions surrounding the structural element, and, subsequently, the temperature of the element itself.

3.2.7.3 Recommended Modeling Tools
Algebraic models

Provided that the fire conditions affecting the structural element are appropriately identified (e.g., a fire plume, a ceiling jet, flame radiation, or an HGL exposure without significant contributions from any of the other three exposure mechanisms), algebraic models may be capable of determining whether the structural element will be exposed to damaging conditions. For example, plume temperature correlations can be used to determine the gas temperature surrounding an element inside the fire plume. MQH calculations can indicate whether compartment temperatures are near the critical temperature of structural elements; however, these may provide overly conservative estimates, as the algebraic models do not account for the heating of those structural elements that typically have large masses. Point source radiation calculations can be used to estimate the heat flux to structural elements that are not directly in the plume but close enough to the fire to become significantly heated.

Zone Models

Zone models are an appropriate tool to address this scenario, as the input file can be developed to capture the relative location of the fire and the structural element(s) as long as compartment or zonal average conditions within a given enclosure are appropriate to support the analysis. Localized conditions, such as flame radiation or radiation from surfaces, can also be handled by zone models. Structural elements can be represented as a target, and the incident fire conditions can be tracked during the fire. Zone models are also capable of performing conduction heat transfer calculations for the structural element, resulting in a prediction of the temperature of the element itself.

CFD Model

CFD models are good tools for estimating temperatures in structural elements of complex geometries, including obstructions, as they can handle the compartment's complex geometries, fire development, and mechanical ventilation conditions in detail, as well as the localized heating of the structural elements.

3.2.7.4 Detailed Examples
Readers are referred to Appendix F, which describes the analysis of a lubricating oil fire's effect on structural elements.

4
UNCERTAINTY AND SENSITIVITY

The fire models discussed in this report are classified as *deterministic* to distinguish them from *stochastic* models. In essence, this means that each model takes as input a set of values, known as *input parameters*, that describe a specific fire scenario, and the model's algorithms then calculate the fire conditions within the compartment. The output of the models usually takes the form of time histories of the various predicted quantities of interest, such as temperature, heat flux or smoke concentration. In a sense, the model calculation is a virtual experiment because the design of a model simulation often involves the same thought process as the design of a physical experiment. The results of the calculation are likewise expressed in terms similar to those of an experiment, including an estimate of the uncertainty. The sources of uncertainty in a model prediction are different than those in an experimental measurement. According to NUREG-1855, Volume 1, *Guidance on the Treatment of Uncertainties Associated with PRAs in Risk-Informed Decision Making* (2009), there are three types of uncertainty associated with a model prediction:

Parameter Uncertainty: Input parameters are often chosen from statistical distributions or estimated from generic reference data. In either case, the uncertainty of these input parameters is propagated through the calculation, and the resulting uncertainty in the model prediction is known as the *parameter uncertainty*. The process of determining the extent to which the individual input parameters affect the results of the calculation is known as a *sensitivity analysis*.

Model Uncertainty: Idealizations of physical phenomena lead to simplifying approximations in the formulation of the model equations. In addition, the numerical solution of equations that have no analytical solution can lead to inexact results. Model uncertainty is estimated via the processes of *verification* and *validation* (V&V). The first seeks to quantify the error associated with the mathematical solution of the governing equations, typically through numerical analysis, while the second seeks to quantify the error associated with the simplifying physical approximations, typically through comparison of model predictions and full-scale experiments.

Completeness Uncertainty: This refers to the fact that a model may not be a complete description of the phenomena it is designed to predict. Some consider this a form of model uncertainty because most fire models neglect certain physical phenomena that are not considered important for a given application. For example, a model of sprinkler activation might neglect water condensation.

The purpose of this chapter is to provide relatively simple methods to assess model and parameter uncertainty. Completeness uncertainty is addressed, indirectly, by the same process used to address the model uncertainty. Model uncertainty is based primarily on comparisons of model predictions with experimental measurements as documented in NUREG-1824 (EPRI 1011999) and other model validation studies. Parameter uncertainty is addressed using simple techniques to propagate input parameter uncertainty and to conduct sensitivity analyses.

4.1 Validation of Fire Models

The use of fire models requires a good understanding of their limitations and predictive capabilities. For example, NFPA 805 (NFPA, 2001) states that fire models shall only be applied within the limitations of the given model and shall be verified and validated. The NRC Office of Nuclear Regulatory Research (RES) and the Electric Power Research Institute (EPRI) conducted a collaborative project for the V&V of the five selected fire models described in Chapter 2. The results of this project were documented in NUREG-1824 (EPRI 1011999), *Verification and Validation of Selected Fire Models for Nuclear Power Plant Applications.*

Twenty-six full-scale fire experiments from six different test series were used to evaluate the models' ability to estimate thirteen quantities of interest for fire scenarios that were judged to be typical of those that might occur in a nuclear power plant (NPP). The results of the study are summarized in Table 4-1. An explanation of this table is to follow.

Table 4-1. Results of the V&V study, NUREG-1824 (EPRI 1011999).

Output Quantity	FDT^s		FIVE-Rev1		CFAST		MAGIC		FDS		Exp
	δ	$\tilde{\sigma}_M$	δ	$\tilde{\sigma}_M$	δ	$\tilde{\sigma}_M$	δ	$\tilde{\sigma}_M$	δ	$\tilde{\sigma}_M$	$\tilde{\sigma}_E$
HGL Temperature Rise*	1.44	0.25	1.56	0.32	1.06	0.12	1.01	0.07	1.03	0.07	0.07
HGL Depth*	N/A		N/A		1.04	0.14	1.12	0.21	0.99	0.07	0.07
Ceiling Jet Temp. Rise	N/A		1.84	0.29	1.15	0.24	1.01	0.08	1.04	0.08	0.08
Plume Temperature Rise	0.73	0.24	0.94	0.49	1.25	0.28	1.01	0.07	1.15	0.11	0.07
Flame Height**	I.D.	I.D.	I.D.	I.D.	I.D.	I.D.	I.D.	I.D.	I.D.	I.D.	I.D.
Oxygen Concentration	N/A		N/A		0.91	0.15	0.90	0.18	1.08	0.14	0.05
Smoke Concentration	N/A		N/A		2.65	0.63	2.06	0.53	2.70	0.55	0.17
Room Pressure Rise	N/A		N/A		1.13	0.37	0.94	0.39	0.95	0.51	0.20
Target Temperature Rise	N/A		N/A		1.00	0.27	1.19	0.27	1.02	0.13	0.07
Radiant Heat Flux	2.02	0.59	1.42	0.55	1.32	0.54	1.07	0.36	1.10	0.17	0.10
Total Heat Flux	N/A		N/A		0.81	0.47	1.18	0.35	0.85	0.22	0.10
Wall Temperature Rise	N/A		N/A		1.25	0.48	1.38	0.45	1.13	0.20	0.07
Wall Heat Flux	N/A		N/A		1.05	0.43	1.09	0.34	1.04	0.21	0.10

I.D. indicates insufficient data for the statistical analysis.
N/A indicates that the model does not have an algorithm to compute the given Output Quantity.
Underlined values indicate that the data failed a normality test because of the relatively small sample size.
* The algorithm used to compute the layer temperature and depth for the model FDS is descr bed in NUREG-1824.
** All of the models except FDS use the Heskestad Flame Height Correlation (Heskestad, *SFPE Handbook*). These models were shown to be in qualitative agreement with the experimental observations, but there was not enough data to further quantify this assessment.

Models: Five fire models were selected for the study, based on the fact that they are commonly used in fire analyses of NPPs in the U.S. Two of the models consist of simplified engineering correlations (Fire Dynamics Tools (FDT^s) and Fire-Induced Vulnerability Evaluation (FIVE-Rev1)), two are "zone" models (Consolidated Fire Growth and Smoke Transport (CFAST) model (CFAST) and MAGIC), and one is a computational fluid dynamics (CFD) model (Fire Dynamics Simulator (FDS)).

Experiments: Six series of experiments (26 individual fire experiments in all) were selected for the NRC/EPRI fire model validation study (NUREG-1824/EPRI 1011999). Each series represented a typical fire scenario (for example, a fire in a switchgear room or turbine hall); however, the test parameters could not encompass every possible NPP fire scenario. Table 2-5 lists various normalized parameters that can be used to characterize fire scenarios and the ranges of the validation experiments. These parameters express, for instance, the size of the fire relative to the size of the room, or the relative distance from the fire to critical equipment. This information is important because typical fire models are not designed for fires that are very small or very large in relation to the volume of the compartment or very large in relation to the ceiling height.

For a given set of experiments and NPP fire scenarios, the user can calculate the relevant normalized parameters. These parameters will either be inside, outside, or on the margin of the validation parameter space. Consider each case in turn:

1. If the parameters fall within the ranges that were evaluated in the validation study, then Table 4-1 can be referenced directly.

2. If only some of the parameters fall within the range of the study, additional justification is necessary (see Section 2.3.7 for guidance). This is a common occurrence because realistic fire scenarios involve a variety of fire phenomena, some of which are easier to estimate than others. A case in point is the burning of electrical cabinets and cables. NUREG-1824 (EPRI 1011999) does not address these fires directly, even though some of the experiments used in the study were intended as mock-ups of control or switchgear room fires. For scenarios involving these kinds of fires, the heat release rates (HRR) are often taken from experiments rather than being predicted by a model. It has been shown, in NUREG-1824 (EPRI 1011999) and other validation studies, that the models can estimate the transport of smoke and heat with varying degrees of accuracy, but they have not been shown (at least not in NUREG-1824 (EPRI 1011999)) to estimate the details of the fire's ignition and growth. While this does not eliminate the models from the analysis, it still restricts their applicability to only some of the phenomena.

3. If the parameters fall outside the range of the study, a validation determination cannot be made based on the results from the study. The modeler needs to provide independent justification for using the particular model. For example, none of the experiments considered in NUREG-1824 (EPRI 1011999) were under-ventilated. However, several of the models have been independently compared to under-ventilated test data, and the results have been documented either in the literature or in the model documentation. As another example, suppose that the selected model uses a plume, ceiling jet, or flame height correlation outside the parameter space of NUREG-1824 (EPRI 1011999) but still within the parameter space for which the correlation was originally developed. In such cases, appropriate references are needed to demonstrate that the correlation is still appropriate, even if not explicitly validated in NUREG-1824 (EPRI 1011999). It is expected that the V&V effort, as documented in NUREG-1824 (EPRI 1011999), will be updated over time to include comparisons with additional test data and new versions of the models. These updates will expand the validation range shown in Table 2-5 and ensure the availability of V&V information for the latest versions of the models documented in this report.

UNCERTAINTY AND SENSITIVITY

Predicted Quantities: The experimental data for the validation study consisted of measurements of one or more of the 13 physical quantities listed in the table. The FDT[s] and FIVE-Rev1 do not possess algorithms to estimate every quantity; in the cases in which there was no estimate, the table cell is labeled N/A.

Statistics: For each model and output quantity, a summary plot of the results is presented in NUREG-1824 (EPRI 1011999). For example, Figure 4-1 compares the measured and predicted target temperatures for the model FDS. If a particular prediction and measurement are the same, the resulting point falls on the solid diagonal line. The longer-dashed off-diagonal lines indicate the experimental uncertainty. Roughly speaking, points within the longer dashed lines are said to be "within experimental uncertainty," and in such cases it is not possible to further quantify the accuracy of the prediction. Points falling outside the experimental uncertainty bounds cannot be said to be free of model uncertainty. At the time of the publication of NUREG-1824 (EPRI 1011999), the writing team decided to use the colors green and yellow to indicate the degree to which the model predictions are inside or outside of the experimental uncertainty bounds. However, since the writing of NUREG-1824 (EPRI 1011999), it was decided to replace the color system with a more quantifiable metric of model accuracy.

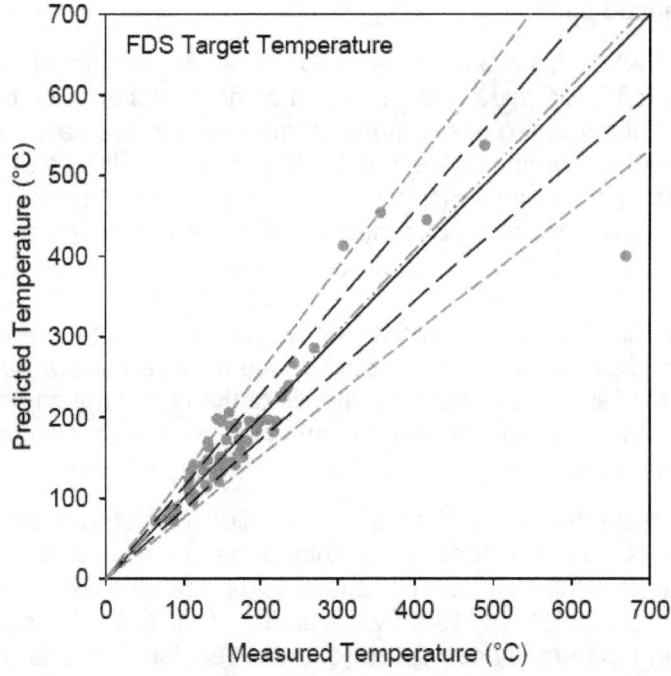

Figure 4-1. Sample set of results from **NUREG-1824 (EPRI 1011999)**.

Consider again Figure 4-1. To make better use of results such as these, two statistical parameters[8] have been calculated for each model and each predicted quantity. The first parameter, δ, is the *bias factor*. It indicates the extent to which the model, on average, under- or over-predicts the measurements of a given quantity. For example, the bias factor for the data shown in Figure 4-1 is 1.02. This means that the model has been shown to slightly

---

[8] The statistical parameters listed in Table 4-1 are based on the versions of the fire models used in the V&V study, circa 2006. As the models are improved and new validation data introduced, these values may change.

overestimate target temperatures by an average 2%, and this is represented by the red dash-dot line just above the diagonal. The bias factor for each model and each output quantity is listed in Table 4-1. Results of the V&V study, NUREG-1824 (EPRI 1011999)..

The second statistic[9] in Table 4-1 is the relative standard deviation of the model, $\tilde{\sigma}_M$, and the experiments, $\tilde{\sigma}_E$. These indicate the uncertainty or degree of "scatter" of the model and the experiments, respectively. Referring again to Figure 4-1, there are two sets of off-diagonal lines. The first set, shown as long-dashed black lines, indicate the experimental uncertainty. The slopes of these lines are $1 \pm 2\tilde{\sigma}_E$ (it is customary to express uncertainties in the form of "2-sigma" or 95% confidence intervals). The second set of off-diagonal lines, shown as short-dashed red lines, indicates the model uncertainty. The slopes of these lines are $\delta \pm 2\tilde{\sigma}_M$. If the model is as accurate as the measurements against which it is compared, the two sets of off-diagonal lines would merge. The extent to which the data scatters outside of the experimental bounds is an indication of the degree of model uncertainty.

The derivation of the statistical parameters in Table 4-1 is provided in the next section, while their use is described here. Suppose that a model prediction is denoted as M. The "true" value of the predicted quantity is specified as a normally distributed random variable with a mean, $\mu = M/\delta$, and a standard deviation, $\sigma = \tilde{\sigma}_M M/\delta$. Using these values, the probability of exceeding a critical value, x_c, is:

$$P(x > x_c) = \frac{1}{2}\operatorname{erfc}\left(\frac{x_c - \mu}{\sigma\sqrt{2}}\right) \tag{4-1}$$

Note that the *complementary error function* is defined as follows:

$$\operatorname{erfc}(x) = \frac{2}{\sqrt{\pi}}\int_x^\infty e^{-t^2}\,dt \tag{4-2}$$

It is a standard function in mathematical or spreadsheet programs like Microsoft Excel[10].

To summarize, the procedure for determining the probability that a quantity predicted by a model could exceed a critical value is as follows:

1. Express the model prediction as a rise above its ambient value. Call this number M. Note that the ambient value of most output quantities is zero. Temperature, oxygen concentration, and smoke layer height are exceptions. For these quantities, express the predicted value as a temperature *rise*, oxygen *decrease* below ambient, and layer *depth*.

2. Using the values of δ and $\tilde{\sigma}_M$ from Table 4-1, compute the mean, $\mu = M/\delta$, and standard deviation, $\sigma = \tilde{\sigma}_M M/\delta$, of the normal distribution for the quantity of interest.

3. Use the equation to compute the probability that the predicted quantity could exceed a critical value, x_c. Remember to also express this critical value as a rise above ambient in the same way as the predicted value, M.

A few examples of this procedure are included in Section 4.3.

---

[9] For some models/quantities, there was an insufficient amount of data to calculate the relative standard deviation of the distribution, in which case I.D. is shown in the Table.
[10] Excel 2007 does not evaluate $\operatorname{erfc}(x)$ for negative values of x, even though the function is defined for all real x. In such cases, use the identity $\operatorname{erfc}(-x) = 2 - \operatorname{erfc}(x)$.

4.2 Derivation of the Model Uncertainty Statistics

This section describes the derivation of the statistics listed in Table 4-1. These values summarize the results of the NRC/EPRI fire model validation study documented in NUREG-1824 (EPRI 1011999). This section is included for information only; there is no need for a model user to perform this type of calculation. McGrattan and Toman (2011) provide additional details on the development of these uncertainty calculations.

For each of the fire models and each of the output quantities that were evaluated in the study, a plot similar to that shown in Figure 4-1. Sample set of results from NUREG-1824 (EPRI 1011999). was produced. For each measurement point, a single experimental measurement was plotted against a single model prediction. The plot shows all the comparison points. The calculation of the statistics uses this set of measured and predicted values, along with an estimate of the experimental uncertainty. The purpose of the calculation is to "subtract off," in a statistical sense, the experimental uncertainty so that the model uncertainty can be estimated. This calculation is subject to the following constraints:

1. The experimental measurements are unbiased, and their uncertainty is normally distributed with a constant relative standard deviation, $\tilde{\sigma}_E$ (that is, the standard deviation as a fraction of the measured value). Table 4-2 provides estimates of relative experimental uncertainties for the quantities of interest in terms of 2-sigma (95 percent) confidence intervals.

2. The model error is normally distributed around the predicted value divided by a bias factor, δ. The relative standard deviation of the distribution is denoted as $\tilde{\sigma}_M$.

The computation of the estimated bias and scatter associated with model error proceeds as follows. Given a set of n experimental measurements, E_i, and a corresponding set of model predictions, M_i, compute the following:

$$\overline{\ln(M/E)} = \frac{1}{n}\sum_{i=1}^{n} \ln(M_i/E_i) \tag{4-3}$$

The standard deviation of the model error, $\tilde{\sigma}_M$, can be computed from the following equation:

$$\sqrt{\tilde{\sigma}_M^2 + \tilde{\sigma}_E^2} \cong \sqrt{\frac{1}{n-1}\sum_{i=1}^{n}\left[\ln(M_i/E_i) - \overline{\ln(M/E)}\right]^2} \tag{4-4}$$

The bias factor is:

$$\delta = \exp\left(\overline{\ln(M/E)} + \frac{\tilde{\sigma}_M^2 - \tilde{\sigma}_E^2}{2}\right) \tag{4-5}$$

For a given model prediction, M, the "true" value of the quantity of interest, is a normally distributed random variable with a mean of M/δ and a standard deviation of $\tilde{\sigma}_M(M/\delta)$.

Table 4-2. Experimental uncertainty of the experiments performed as part of the validation study in NUREG-1824 (EPRI 1011999).

Quantity	$2\widetilde{\sigma}_E$
HGL Temperature Rise	0.14
HGL Depth	0.13
Ceiling Jet Temperature Rise	0.16
Plume Temperature Rise	0.14
Gas Concentration	0.09
Smoke Concentration	0.33
Pressure (no forced ventilation)	0.40
Pressure (with forced ventilation)	0.80
Heat Flux	0.20
Surface or Target Temperature	0.14

There are a few issues to consider when using this procedure:

1. All values need to be positive, and each value needs to be expressed as an increase over its ambient value. For example, the oxygen concentration should be expressed as a positive number (i.e., the decrease in concentration below its ambient value). Thus, for an ambient oxygen concentration of 21 % and a predicted value of 14 %, the predicted decrease in oxygen concentration is 7 %.

2. If the measurement uncertainty is overestimated, the model error will be underestimated. If the model error is less than the experimental uncertainty, the latter should be reevaluated. The model cannot be shown to have less error than the uncertainty of the experiment with which it is compared.

3. The procedure requires that the quantity $\ln(M/E)$ is normally distributed. This is not necessarily true, especially in cases where there are an insufficient number of points in the sample. Figure 4-2 provides two examples in which the normality of the validation data is tested[11]. In cases where the data is not normally distributed, only the bias is reported.

4. The validation data used to develop the accuracy statistics may be fairly sparse over a particular range. For example, in Figure 4-1, for temperatures below 300 °C (570 °F), there are a sufficient number of points to derive the statistics. For temperatures above 300 °C (570 °F), however, there are less data points and these few points do not necessarily follow the trend. In such cases, it is left to the discretion of the analyst to determine if it is appropriate to calculate the model uncertainty using the derived statistics if the temperature is clearly outside of the range. Depending on the application, it might be sufficient to indicate the model tends to over-predict these higher temperatures, as is this case, or the reviewer might request additional evidence that the model can be applied for this range of temperatures.

---

[11] The Kolmogorov-Smirnov test for normality has been applied using the software package SigmaPlot®10, Systat Software, Inc. The default P value of 0.05 was used.

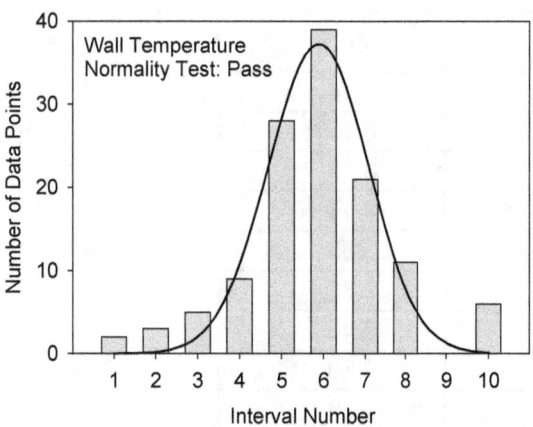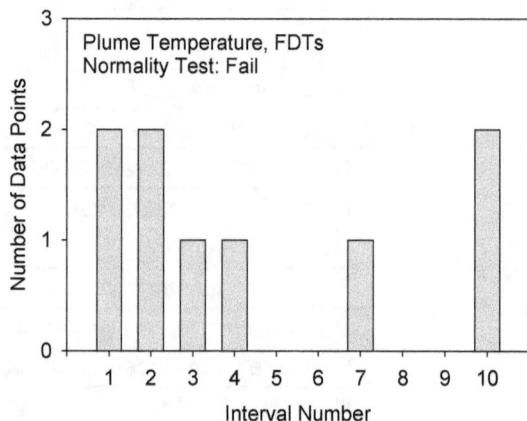

Figure 4-2. Two examples demonstrating how the validation data is tested for normality.

4.3 Calculation of Model Uncertainty

This section contains a few exercises to explain the procedure for calculating model uncertainty. These examples consider model uncertainty only; that is, the input parameters are not subject to uncertainty.

4.3.1 Example 1: Target Temperature

Suppose that cables within a compartment fail if their surface temperature reaches 330 °C (625 °F). The model FDS predicts that the maximum cable temperature due to a fire in an electrical cabinet is 300 °C (570 °F). What is the probability that the cables could fail? For this exercise, the model input parameters are not subject to uncertainty, only the model itself.

Step 1: Subtract the ambient value of the cable temperature, 20 °C (68 °F) to determine the predicted temperature _rise_. Refer to this value as the _model prediction_:

$$M = 300\ °C - 20\ °C = 280\ °C \tag{4-6}$$

Step 2: Refer to Table 4-1, which indicates that, on average, FDS over-predicts target temperatures with a bias factor, δ, of 1.02. Calculate the _adjusted model prediction_:

$$\mu = \frac{M}{\delta} = \frac{280\ °C}{1.02} \cong 275\ °C \tag{4-7}$$

Referring again to Table 4-1, calculate the standard deviation of the distribution:

$$\sigma = \tilde{\sigma}_M \left(\frac{M}{\delta}\right) = 0.13 \left(\frac{280\ °C}{1.02}\right) \cong 36\ °C \tag{4-8}$$

Step 3: Calculate the probability that the actual cable temperature would exceed 330 °C[12]:

---

[12] In the result of Equation 4-9, the precise value for μ and σ were used rather than the rounded values shown.

$$P(T > 330\,°\text{C}) = \frac{1}{2}\,\text{erfc}\left(\frac{T - T_0 - \mu}{\sigma\sqrt{2}}\right) = \frac{1}{2}\,\text{erfc}\left(\frac{330\,°\text{C} - 20\,°\text{C} - 275\,°\text{C}}{36\,°\text{C}\,\sqrt{2}}\right) \cong 0.16 \qquad (4\text{-}9)$$

The process is shown graphically in Figure 4-3. The area under the "bell curve" for temperatures higher than 330 °C (625 °F) represents the probability that the actual cable temperature would exceed that value. Note that this estimate is based only on the model uncertainty.

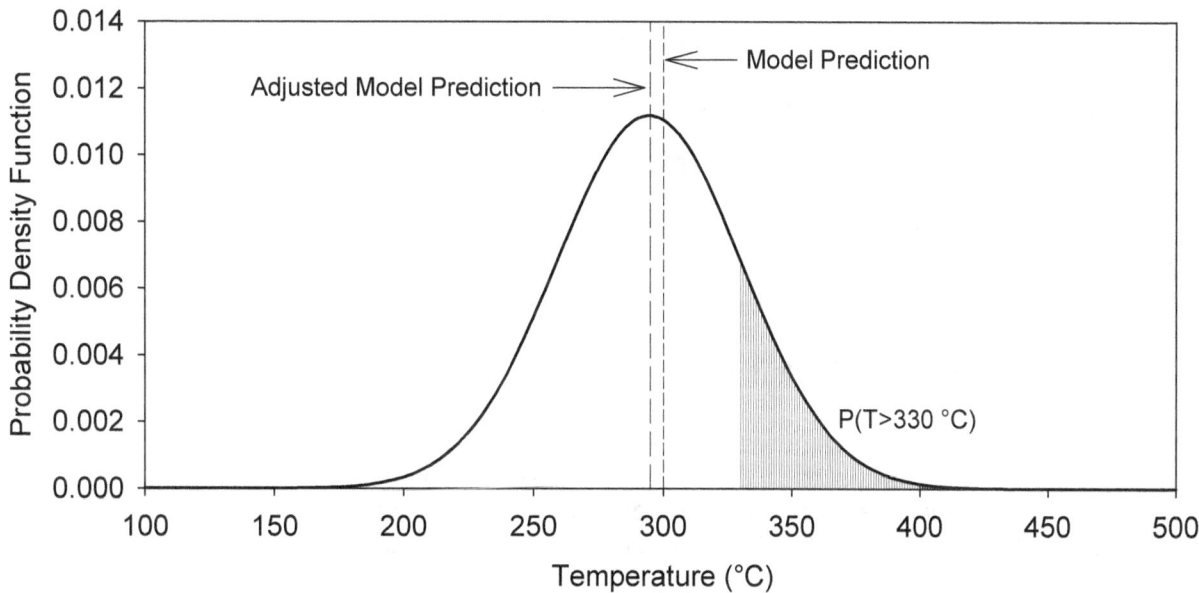

Figure 4-3. Normal distribution of the "true" value of the cable temperature in a hypothetical fire.

4.3.2 Example 2: Critical Heat Flux

As part of a screening analysis, the model MAGIC is used to estimate the radiant heat flux from a fire to a nearby group of thermoplastic (TP) cables. Using, for example purposes, data from NUREG/CR-6850 (EPRI 1011989), Appendix H, one of the damage criteria for TP cables is a radiant heat flux to the target cable that exceeds 6 kW/m². The model, by coincidence, predicts a heat flux of 6 kW/m². What is the probability that the actual heat flux from a fire will be 6 kW/m² or greater? For this exercise, the model input parameters are not subject to uncertainty, only the model itself.

Step 1: Unlike in the previous example, there is no need to subtract an ambient value of the heat flux (it is zero). Thus, the *model prediction* is:

$$M = 6\,\text{kW/m}^2 \qquad (4\text{-}10)$$

Step 2: Refer to Table 4-1, which indicates that, on average, MAGIC over-predicts radiant heat flux with a bias factor, δ, of 1.07. Calculate the *adjusted model prediction*:

$$\mu = \frac{M}{\delta} = \frac{6}{1.07} \cong 5.6\,\text{kW/m}^2 \qquad (4\text{-}11)$$

Referring again to Table 4-1, calculate the standard deviation of the distribution:

$$\sigma = \tilde{\sigma}_M \left(\frac{M}{\delta}\right) = 0.36 \left(\frac{6}{1.07}\right) \cong 2.0 \text{ kW/m}^2 \qquad (4\text{-}12)$$

Step 3: Calculate the probability that the actual heat flux, $\dot{q}''$, will exceed the critical value of the heat flux, $\dot{q}_c''$:

$$P(\dot{q}'' > 6 \text{ kW/m}^2) = \frac{1}{2}\text{erfc}\left(\frac{\dot{q}_c'' - \mu}{\sigma\sqrt{2}}\right) = \frac{1}{2}\text{erfc}\left(\frac{6 \text{ kW/m}^2 - 5.6 \text{ kW/m}^2}{2.0 \text{ kW/m}^2 \sqrt{2}}\right) \cong 0.42 \qquad (4\text{-}13)$$

Even though the model predicts a peak radiant heat flux equal to the critical value, there is only a 42% chance that the actual heat flux would exceed this value. This is mainly due to the fact that MAGIC has been shown to overestimate the heat flux by about 7%.

It is important to note that this calculation of model uncertainty does not take into account the input parameters, such as the HRR of the fire. It is only an assessment of how well the model MAGIC can estimate the radiant heat flux to a target. Another key point is that the damage threshold, temperature in the first example and heat flux in the second, is known exactly, i.e., without uncertainty.

4.4 Parameter Uncertainty

The previous sections describe how to express the uncertainty of a model prediction resulting from the inherent limitations of the model itself. However, in most cases, the larger source of uncertainty is the input parameters of the model, not the model itself. This section suggests ways to assess the impact of this kind of uncertainty on the final prediction.

Parameter uncertainty is addressed in this chapter in two ways: (1) parameter uncertainty propagation, and (2) sensitivity analysis. It is recognized that one or both approaches may be necessary for a given application. Depending on the complexity of the model, the number of input parameters will range from one to several dozen. The choice of which to analyze is based on (1) the importance of the parameter to the outcome, and (2) its degree of uncertainty. In almost all cases, the heat release rate of the fire dominates all other parameters in terms of these two criteria.

The propagation of parameter uncertainty refers to the process of propagating the probability distributions of one or more input parameters through the model in order to determine the distributions of the model output quantities. Guidance on uncertainty propagation is provided in Section 4.4.1.

Sensitivity analysis refers to the process of determining the "rate of change" of the model output with respect to variations in the magnitude in one or more of the model inputs, usually one parameter at a time. Guidance on sensitivity analysis is provided in section 4.4.2.

4.4.1 Parameter Uncertainty Propagation

Typically, fire models are run using a discrete set of input parameters that describe a single, specific fire scenario. However, for some Fire PRA applications, it may be necessary to consider the range of consequences due to the variability that can result from that specific fire scenario within a particular compartment. If the key input parameters can be expressed in the form statistical distributions, then the model output quantities may also be expressed as distributions. In this way, it is possible to determine the probability of exceeding a critical temperature, heat flux, or some other critical value.

Notarianni and Parry (*SFPE Handbook*, 4th edition) discuss a number of techniques for propagating parameter uncertainty. The most common are Monte Carlo methods in which the fire model is run repeatedly with randomly chosen input parameters, based on specified probability distributions. For simple algebraic models, this technique is relatively simple and there are various software packages available to help. For zone models, the technique becomes more complicated because it requires more time to set up and run the model, but it is still practical if the number of parameters is reduced and the ranges of the parameters are appropriately discretized into "bins". For CFD models, the technique is not practical except in special cases where the number of model runs can be reduced to a relatively small number. The increased accuracy afforded by the CFD model is often unwarranted given that the uncertainty in the input parameters is typically greater than that of the models themselves.

Because the HRR is the most important input parameter in most fire model analyses, and because NUREG/CR-6850 (EPRI 1011989) provides distributions of the HRR for a variety of combustibles within an NPP, parameter uncertainty propagation for fire modeling may involve only the HRR distribution applied within an algebraic model. In fact, Appendix E of NUREG/CR-6850 (EPRI 1011989) provides data which can be used to illustrate a simple technique to propagate the HRR distribution.

Suppose, for example, that as part of a Fire PRA the problem is to determine the probability of flames extending above an electrical cabinet to a particular height, threatening a cable tray. To answer this question, the flame height needs to be represented as a probability distribution. Figure 4-4 displays the distribution[13] of peak heat release rates, $\dot{Q}$, for vertical cabinets with more than one bundle of unqualified cable (NUREG/CR-6850, Appendix E). The probability density function (pdf) is denoted $g(\dot{Q}; \alpha, \beta)$, where α and β (2.6 and 67.8) are parameters of this particular gamma distribution.

---

[13] NUREG/CR-6850 specifies gamma distributions for the various types of combust bles found within an NPP. Microsoft Excel® provides a built-in function (GAMMA.DIST) that calculates the probability density function given the parameters α and β.

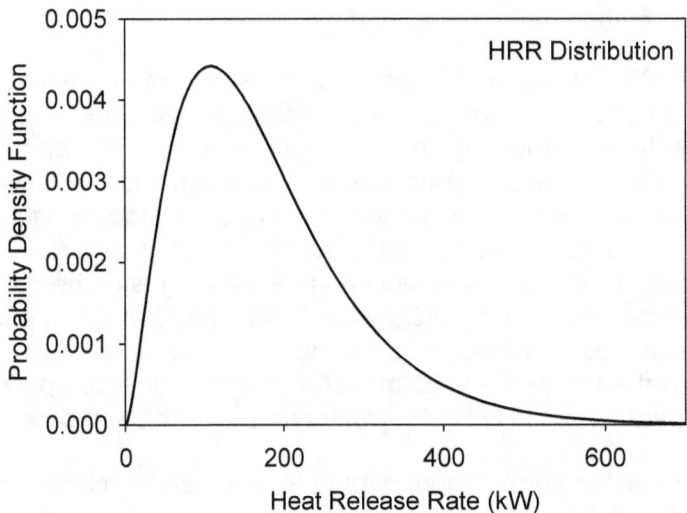

Figure 4-4. Distribution of HRR for an electrical cabinet fire.

For convenience, a spreadsheet can be used to take each value of $\dot{Q}$, from 1 kW to 600 kW, and compute a corresponding flame height, L_f (in meters), using Heskestad's correlation:

$$L_f = 0.235 \, \dot{Q}^{2/5} - 1.02 \, D \qquad (4\text{-}14)$$

The diameter of the fire, D, is fixed at 0.48 m (1.6 ft), based on the equivalent diameter of the vent. Whereas Appendix E of NUREG/CR-6850 recommends dividing the range of $\dot{Q}$ into 15 "bins", it is just as easy for this example to compute the flame height for 600 values of $\dot{Q}$ (each bin has a width of 1 kW). The pdf for the flame height, $f(L_f)$, is related to the pdf for $\dot{Q}$ by the following expression:

$$f(L_f) = \frac{g(\dot{Q}; \alpha, \beta)}{\left| \frac{dL_f}{d\dot{Q}} \right|} = g(\dot{Q}; \alpha, \beta) \, \frac{\dot{Q}^{3/5}}{0.094} \qquad (4\text{-}15)$$

This distribution is shown in Figure 4-5. Note that when the derivative in Equation (4-15) is not easily written in closed form, it is sufficient to calculate the bin width of the model output divided by that of the model input. In this example, the bin width of the model input parameter, $\dot{Q}$, is 1 kW and the bin width of the model output parameter is the difference in flame heights for two successive values of $\dot{Q}$.

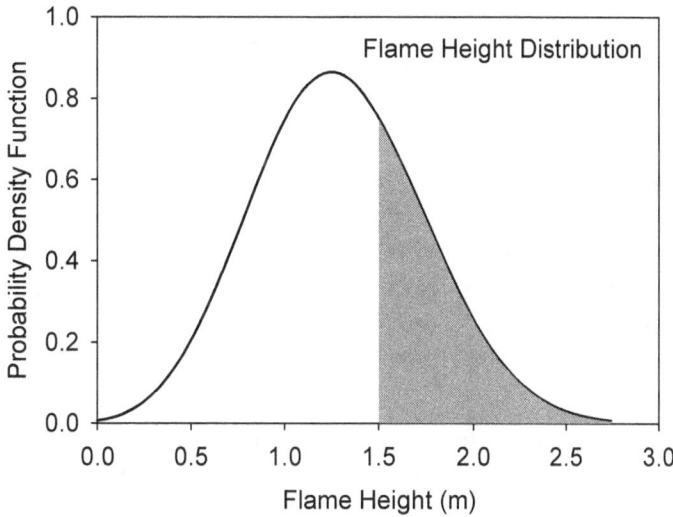

Figure 4-5. Distribution of flame heights for the entire range of cabinet fires.

Once the pdf for the flame height is calculated, it can be used to determine the probability of the flames reaching a certain height. In this case, the cable tray is 1.5 m (4.9 ft) above the top of the cabinet. The probability that the flames from a randomly chosen fire will reach the cables is given by the area beneath the curve in Figure 4-5 for flame heights greater than 1.5 m (4.9 ft). In this example, it is approximately 0.31.

4.4.2 Sensitivity Analysis

The algebraic calculations described in this report are perhaps the models more amenable to parameter uncertainty studies in practical applications. The more complex fire models discussed in this report can require dozens of physical and numerical input parameters for a given fire scenario. However, only a few of these parameters, when varied over their plausible range of values, will significantly impact the results. For example, the thermal conductivity of the compartment walls will not significantly affect a predicted cable surface temperature. Table 4-3 lists the input parameters whose impact on the given output quantity significantly outweighs all the other parameters. The HRR is almost always one of these.

In Volume 2 of NUREG-1824 (EPRI 1011999), Hamins quantifies the functional dependence of these key input parameters (see Table 4-3). These relationships are based either on the governing mathematical equations or on algebraic models. The basic mathematical form of the relationship is:

$$\text{Output Quantity} = \text{Constant} \times (\text{Input Parameter})^{\text{Power}} \qquad (4\text{-}16)$$

The exact value of the Constant is not important; rather, it is the Power that matters. The larger its absolute value, the more important the Input Parameter. According to the McCaffrey, Quintiere, and Harkleroad (MQH) correlation, for example, the hot gas layer (HGL) temperature rise in a compartment fire is proportional to the HRR raised to the two-thirds power:

$$T - T_0 = C\dot{Q}^{2/3} \qquad (4\text{-}17)$$

What is important here is the amount that the HGL temperature, ΔT, changes due to a shift in the HRR, $\Delta \dot{Q}$. It is the two-thirds power dependence, as found in Table 4-3, that matters. To

UNCERTAINTY AND SENSITIVITY

see why, take the first derivative of T with respect to $\dot{Q}$ and write the result in terms of differentials:

$$\frac{\Delta T}{T - T_0} \cong \frac{2}{3} \frac{\Delta \dot{Q}}{\dot{Q}} \qquad (4\text{-}18)$$

This is a simple formula with which one can readily estimate the relative change in the model output quantity, $\Delta T/(T - T_0)$, due to the relative change in the model input parameter, $\Delta \dot{Q}/\dot{Q}$. The uncertainty in a measured quantity is often expressed in relative terms[14]. Suppose that the uncertainty in the HRR of the fire, $\Delta \dot{Q}/\dot{Q}$, is 0.15, or 15%. The expression above indicates that a 15% increase in the HRR should lead to a 2/3 x 15 = 10% increase in the prediction of the HGL temperature. The result is equally valid for a reduction; if the HRR is reduced by 15%, the HGL temperature is reduced by 10%.

Table 4-3. Sensitivity of model outputs from Volume 2 of NUREG-1824 (EPRI 1011999).

Output Quantity	Important Input Parameters	Power Dependence
HGL Temperature	HRR Surface Area Wall Conductivity Ventilation Rate Door Height	2/3 -1/3 -1/3 -1/3 -1/6
HGL Depth	Door Height	1
Gas Concentration	HRR Production Rate	1/2 1
Smoke Concentration	HRR Soot Yield	1 1
Pressure	HRR Leakage Rate Ventilation Rate	2 2 2
Heat Flux	HRR	4/3
Surface/Target Temperature	HRR	2/3

This relationship is based on experimental data, and has nothing to do with any particular model; however, an effective way to check a fire model is to take a simple compartment fire simulation, vary the HRR, and ensure that the change in the HGL temperature agrees with the correlation. Consider the two curves shown in Figure 4-6. For Benchmark Exercise #3 of the International Collaborative Fire Model Project (ICFMP), Test 3 was simulated with FDS, using HRR values of 1000 kW and 1150 kW. An examination of the peak values confirms that the relative change in the HGL temperature (10%) is two-thirds the relative change in the HRR

[14] Note that a differential relationship is only approximate. This method of relating input parameters to output quantities is valid for relative differences that are less than approximately 30% in absolute value.

(15%), consistent with the empirical result of the MQH correlation. Even though FDS is a much more complicated model than the simple expression shown above, it still exhibits the same functional dependence on the HRR.

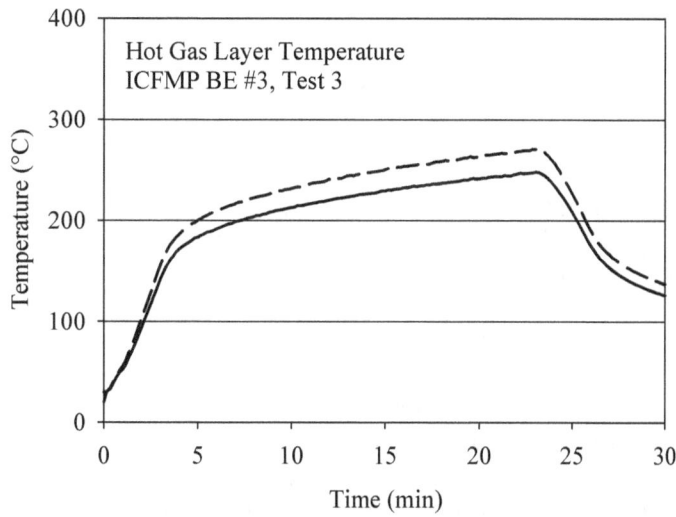

Figure 4-6. FDS predictions of HGL Temperature as a function of time due to a 1,000 kW fire (solid line) and a 1,150 kW fire (dashed).

This section illustrates the usefulness of sensitivity analysis. As an example, consider that NFPA 805 uses the term *Maximum Expected Fire Scenario* (MEFS) to describe a severe fire that could be "reasonably anticipated" to occur within a compartment and the term *Limiting Fire Scenario* (LFS) to describe a severe fire that exceeds one or more performance criteria. The analyst is often asked to determine the model inputs for both of these scenarios. For MEFS, input parameters can be chosen from distributions for a particular percentile value. Gallucci (2011) discusses the issues important in determining the appropriate choice of HRR distribution. The development of the LFS is essentially a sensitivity analysis performed to identify which combinations of input parameters or variables are critical to the analysis. The particular variables to be evaluated depend entirely on the problem being analyzed. At a minimum, the following parameters should be varied until failure conditions result: HRR, the fire growth rate or the flame spread rate, the flame radiative fraction or the radiative power, and the location of the fuel package relative to the target (if variable).

Suppose, for example, that as part of an analysis the problem is to determine the minimum HRR needed to cause damage in a particular compartment whose HGL temperature is not to exceed 500 °C (930 °F). The geometrical complexity of the compartment rules out the use of the algebraic and zone models, and that FDS has been selected for the simulation.

Step 1: Determine a reasonable, but conservative, estimate of what might be the maximum fire that could occur in the compartment. Using data from NUREG/CR-6850, for this example, suppose that a 98th percentile HRR for the electrical cabinet fire, 702 kW, has been chosen for this representative estimate. Choose a model and calculate the peak HGL temperature.

Step 2: The model chosen is FDS and it predicts 450 °C (840 °F) for the selected fire scenario. Adjust the prediction to account for the model bias, δ (see Table 4-1):

$$T_{adj} = T_0 + \frac{T - T_0}{\delta} = 20\ °C + \frac{450\ °C - 20\ °C}{1.03} \cong 437\ °C \tag{4-19}$$

Step 3: Calculate the change in HRR required to increase the HGL temperature to 500 °C (930 °F):

$$\Delta \dot{Q} \approx \frac{3}{2} \dot{Q} \frac{\Delta T}{T_{adj} - T_0} = \frac{3}{2} \times 702\ kW \times \frac{500\ °C - 437\ °C}{437\ °C - 20\ °C} \cong 159\ kW \tag{4-20}$$

This calculation suggests that adding an additional 159 kW to the original 702 kW will produce an HGL temperature in the vicinity of 500 °C (930 °F). This result can be double-checked by re-running the model with the modified input parameters.

Table 4-3 lists several other parameters besides the HRR that can affect the HGL temperature. Following the example just discussed, similar calculations can be performed in which these other parameters are varied to determine how else the minimum damage threshold might be reached. For example, suppose that the surface area, A, of the compartment is 400 m^2 (4300 ft^2). How much would the surface area have to increase or decrease to raise the HGL temperature to 500 °C (930 °F)? If the thermal conductivity of the walls, k, is 0.1 W/m/K, how much would it have to change? If the ventilation rate is 1 m^3/s (2100 cfm), how much would it have to change? If the door height, h, is 2 m (6.6 ft), how much would it have to change? Following the example for the HRR, the required changes in these parameters can be calculated as follows:

$$\Delta A \cong -3A \frac{\Delta T}{T_{adj} - T_0} = -3\ (400\ m^2)\ \frac{500 - 437}{417} \cong -181\ m^2 \tag{4-21}$$

$$\Delta k \cong -3k \frac{\Delta T}{T_{adj} - T_0} = -3\ (0.1\ W/m/K)\ \frac{500 - 437}{417} \cong -0.045\ W/m/K \tag{4-22}$$

$$\Delta \dot{V} \cong -3\dot{V} \frac{\Delta T}{T_{adj} - T_0} = -3\ (1\ m^3/s)\ \frac{500 - 437}{417} \cong -0.45\ m^3/s \tag{4-23}$$

$$\Delta h \cong -6h \frac{\Delta T}{T_{adj} - T_0} = -6\ (2\ m)\ \frac{500 - 437}{417} \cong -1.8\ m \tag{4-24}$$

For this example, to increase the HGL temperature by 63 °C (145 °F), one could increase the HRR by 159 kW, decrease the surface area of the compartment by 181 m^2 (1948 ft^2), decrease the thermal conductivity of the walls by 0.045 W/m/K, decrease the ventilation rate by 0.45 m^3/s (950 cfm), or decrease the door height by 1.8 m (5.9 ft). Of course, some of these options are not physically possible. Room dimensions and thermal properties are not subject to significant change, but the HRR and ventilation rates can vary significantly. Also note that if the relative change in the parameter values exceeds 30 %, it is recommended that further calculations be performed to confirm the estimated quantity changes.

4.5 Chapter Summary

This chapter describes the three forms of uncertainty related to fire modeling: *parameter*, *model*, and *completeness* uncertainty. Model and completeness uncertainty are closely related, and it would be impractical to evaluate them separately. The most practical way to quantify their combined effect is to compare model predictions with as many experimental measurements as possible in order to develop a robust statistical description of the model's accuracy. The five models considered in this repor underwent a validation study (NUREG-1824 (EPRI 1011999)) in which their predictions were compared with measurements from a variety of full-scale experiments. It is possible to take a given model's prediction of a given quantity and determine a distribution for the "true" value of this quantity. Rather than reporting the result of a calculation as a single value, it is preferable to report the probability that the true value of a predicted quantity will exceed a given critical value.

Treatment of parameter uncertainty is dependent on the application. Some analyses, probabilistic risk assessments (PRAs) for example, employ relatively simple techniques to propagate the distribution of the HRR through the model. More complex forms of PRAs can involve a broad statistical sampling of input parameters from specified distributions. From a fire modeling perspective, some modeling tools are more amenable to the evaluation of parameter uncertainty given the complexity and computational times of some of these tools.

Deterministic applications usually consider "worst case" or "bounding" analyses, in which extreme, yet plausible, input parameters are used. In mathematical terms, all of these applications involve selecting parameters from relatively narrow or broad regions of the parameter "space." It is often prohibitive to consider all possible combinations of input parameters, which is why a simple form of sensitivity analysis, outlined in this chapter, can be used to extend the range of outcomes. Algebraic models may initially be used to indicate the extent to which all of the output quantities of interest are sensitive to changes in the specified HRR, reducing the need to re-run model simulations for an extensive number of values.

Over time, the V&V effort, documented in NUREG-1824 (EPRI 1011999), will be expanded to include comparisons with additional test data and new versions of the models. These updates will expand the validation range shown in Table 2-5 and ensure the availability of V&V information for the latest versions of the models documented in this report. Model users should refer to the latest version of the V&V when conducting their fire modeling analyses. In general, the techniques documented in this chapter for analyzing uncertainty and sensitivity will remain applicable regardless of the V&V version being used for the analysis.

UNCERTAINTY AND SENSITIVITY

4-18

5
REFERENCES

10 CFR 50, "Voluntary Fire Protection Requirements for Light-Water Reactors," 10 CFR Part 50, Section 50.48(c), RIN 3150-AG48, *Federal Register*, Volume 69, Number 115, U.S. Nuclear Regulatory Commission, Washington, D.C., June 16, 2004.

AISC, *Steel Construction Manual*, 13th edition, American Institute of Steel Construction, New York, New York, 2006.

ASME/ANS RA-Sa-2009, *Standard for Level 1/Large Early Release Frequency Probabilistic Risk Assessment for Nuclear Power Plant Applications, Addendum A*, ASME/ANS RA-Sa-2009, American Society of Mechanical Engineers, 2009.

ASTM E119-10a, *Standard Test Methods for Fire Tests of Building Construction Materials*, American Society for Testing and Materials, West Conshohocken, PA, 2010.

ASTM E1355–05a (2005), *ASTM Standard Guide for Evaluating the Predictive Capability of Deterministic Fire Models*, American Society for Testing and Materials, West Conshohocken, PA, 2005.

Babrauskas, V*., Ignition Handbook*, Fire Science Publishers/Society of Fire Protection Engineers, Issaquah, WA, 2003.

Buchanan, A. H., *Structural Design for Fire Safety*, John Wiley and Sons, LTD, Chichester, England, 2001.

Drysdale, D., *An Introduction to Fire Dynamics*, John Wiley and Sons, 3rd Edition, Chichester, pp. 14, 283-284, 2011.

EPRI 1002981, *Fire Modeling Guide for Nuclear Power Plant Applications*, Electric Power Research Institute, Palo Alto, CA, August 2002.

Forney, G. P., *User's Guide for Smokeview Version 5 - A Tool for Visualizing Fire Dynamics Simulation Data, Volume II: Technical Reference Guide*, NIST Special Publication 1017-2, National Institute of Standards and Technology, Gaithersburg, MD, October 2010.

Gallucci, R.H.V., "Examination of the Efficacy of the NFPA-805 "Fire Modeling" Approach (Comparison between "Maximum Expected" and "Limiting" Fire Scenarios)," ANS PSA 2011 International Topical Meeting on Probabilistic Safety Assessment and Analysis, Wilmington, NC, March 13-17, 2011, American Nuclear Society, LaGrange Park, IL, 2011.

Gay, L., *User Guide of MAGIC Software V4.1.1*, EdF HI82/04/022/B, Electricité de France, France, April 2005.

Gay, L., C. Epiard, and B. Gautier, *MAGIC Software Version 4.1.1: Mathematical Model*, EdF HI82/04/024/B, Electricité de France, France, November 2005.

Gay, L., J. Frezabeu, and B. Gautier, *Qualification File of Fire Code MAGIC Software Version 4.1.1: Mathematical Model*, EdF HI82/04/024/B, Electricité de France, France, November 2005.

REFERENCES

Grove, B. S., and J.G. Quintiere, "Calculating Entrainment and Flame Height in Fire Plumes of Axisymmetric and Infinite Line Geometries," *Journal of Fire Protection Engineering*, Volume 12, August, 2002, pp. 117 - 137.

Gottuk, D. T., and Lattimer, B. Y., "Section 2, Chapter 2-5, Effect of Combustion Conditions on Species Production," *SFPE Handbook of Fire Protection Engineering*, 4[h] Edition (P. J. DiNenno, Editor-in-Chief), National Fire Protection Association and The Society of Fire Protection Engineers, Quincy, MA, 2008.

Heskestad, G., Section 2, Chapter 2-1, "Fire Plumes, Flame Height, and Air Entrainment," *SFPE Handbook of Fire Protection Engineering*, 4[h] Edition (P. J. DiNenno, Editor-in-Chief), National Fire Protection Association and The Society of Fire Protection Engineers, Quincy, MA, 2008.

Holman, J. P., *Heat Transfer*, 7th edition, McGraw-Hill, New York, 1990.

Janssens, M., Section 3, Chapter 3-2, "Calorimetry," *SFPE Handbook of Fire Protection Engineering*, 4th Edition (P. J. DiNenno, Editor-in-Chief), National Fire Protection Association and The Society of Fire Protection Engineers, Quincy, MA, 2008.

Jones, W., R. Peacock, G. Forney, and P. Reneke, *CFAST: An Engineering Tool for Estimating Fire Growth and Smoke Transport, Version 5 - Technical Reference Guide*, NIST Special Publication 1030, National Institute of Standards and Technology, Gaithersburg, MD, 2004.

Karlsson, B., and J. Quintiere, *Enclosure Fire Dynamics*, CRC Press, Boca Raton, Florida, 2000.

Köylu, U. O., and G.M. Faeth, "Carbon Monoxide and Soot Emissions from Liquid-Fueled Buoyant Turbulent Diffusion Flames," *Combustion and Flame*, 87:61-76, 1991.

McCaffrey, B. J., J. G. Quintiere, and M. F. Harkleroad, "Estimating Compartment Temperature and Likelihood of Flashover Using Fire Test Data Correlation," *Fire Technology*, Volume 17, No. 2, pp. 98-119, Quincy, MA, 1981.

McGrattan, K. et al., *Fire Dynamics Simulator (Version 5) Technical Reference, Volume 3: Validation*, NIST Special Publication 1018-5, National Institute of Standards and Technology, Gaithersburg, MD, 2010.

McGrattan, K., B. Klein, S. Hostikka, and J. Floyd, *Fire Dynamics Simulator (Version 5) User's Guide*, NIST Special Publication 1019-5, National Institute of Standards and Technology, Gaithersburg, MD, 2009.

McGrattan, K., and B. Toman, "Quantifying the predictive uncertainty of complex numerical models," *Metrologia*, 48:173-180, 2011.

Mulholland, G. W., and C. Croarkin, "Specific Extinction Coefficient of Flame-Generated Smoke," *Fire and Materials*, 24:227-230, 2000.

NFPA, *Fire Protection Handbook*, National Fire Protection Association, 20th Ed., A. E. Cote (Editor), 2008.

NFPA 70 (NEC 2008), *National Electric Code*, National Fire Protection Association, Quincy, MA, 2008.

NFPA 805, *Performance-Based Standard for Fire Protection for Light Water Reactor Electric Generating Plants*, National Fire Protection Association, Quincy, MA, 2001.

NEI 00-01 (2009), *Guidance for Post Fire Safe Shutdown Circuit Analysis*, Revision 2, Nuclear Energy Institute, Washington, D.C., May, 2009.

NEI 00-01 (2010), *Guidance for Post Fire Safe Shutdown Circuit Analysis*, Draft Revision 3, Nuclear Energy Institute, Washington, D.C., 2010.

NEI 04-02 (2009), *Guidance for Implementing a Risk-Informed Performance-Based Fire Protection Program Under 10 CFR 50.48(c)*, Rev. 1, Nuclear Energy Institute, Washington, D.C., September, 2009.

NIST NCSTAR 1-5F, *Federal Building and Fire Safety Investigation of the World Trade Center Disaster: Computer Simulation of the Fires in the World Trade Center Towers*, National Institute of Standards and Technology, Gaithersburg, MD, 2005.

NRC, Regulatory Guide 1.189, *Fire Protection for Nuclear Power Plants*, U.S. Nuclear Regulatory Commission, Bethesda, MD, April, 2009.

NUREG-1805, *Fire Dynamics Tools (FDT^s): Quantitative Fire Hazard Analysis Methods for the U.S. Nuclear Regulatory Commission Fire Protection Inspection Program*, U.S. Nuclear Regulatory Commission, Washington, D.C., December 2004.

NUREG-1824, *Verification and Validation of Selected Fire Models for Nuclear Power Plant Applications, Volume 1: Main Report*, U.S. Nuclear Regulatory Commission, Office of Nuclear Regulatory Research (RES), Washington, D.C., 2007, and EPRI 1011999, Electric Power Research Institute (EPRI), Palo Alto, CA.

NUREG-1824, *Verification and Validation of Selected Fire Models for Nuclear Power Plant Applications, Volume 3: Fire Dynamics Tools (FDT^s)*, U.S. Nuclear Regulatory Commission, Office of Nuclear Regulatory Research (RES), Washington, D.C., 2007, and EPRI 1011999, Electric Power Research Institute (EPRI), Palo Alto, CA.

NUREG-1824, *Verification and Validation of Selected Fire Models for Nuclear Power Plant Applications, Volume 4: Fire-Induced Vulnerability Evaluation (FIVE-Rev1)*, U.S. Nuclear Regulatory Commission, Office of Nuclear Regulatory Research (RES), Washington, D.C., 2007, and EPRI 1011999, Electric Power Research Institute (EPRI), Palo Alto, CA.

NUREG-1824, *Verification and Validation of Selected Fire Models for Nuclear Power Plant Applications, Volume 5: Consolidated Fire Growth and Smoke Transport Model (CFAST)*, U.S. Nuclear Regulatory Commission, Office of Nuclear Regulatory Research (RES), Washington, D.C., 2007, and EPRI 1011999, Electric Power Research Institute (EPRI), Palo Alto, CA.

NUREG-1824, *Verification and Validation of Selected Fire Models for Nuclear Power Plant Applications, Volume 6: MAGIC*, U.S. Nuclear Regulatory Commission, Office of Nuclear Regulatory Research (RES), Washington, D.C., 2007, and EPRI 1011999, Electric Power Research Institute (EPRI), Palo Alto, CA.

NUREG-1824, *Verification and Validation of Selected Fire Models for Nuclear Power Plant Applications, Volume 7: Fire Dynamics Simulator (FDS)*, U.S. Nuclear Regulatory Commission, Office of Nuclear Regulatory Research (RES), Washington, D.C., 2007, and EPRI 1011999, Electric Power Research Institute (EPRI), Palo Alto, CA.

NUREG-1855 Volume 1, *Guidance on the Treatment of Uncertainties Associated with PRAs in Risk-Informed Decision Making*, U.S. Nuclear Regulatory Commission, Office of Nuclear Regulatory Research (RES), Washington, D.C., 2009.

NUREG/CR-4680, *Heat and Mass Release Rate for Some Transient Fuel Source Fires: A Test Report*, U.S. Nuclear Regulatory Commission, Washington, D.C., December 2004.

REFERENCES

NUREG/CR-6738, *Risk Methods Insights Gained from Fire Incidents*, U.S. Nuclear Regulatory Commission, Office of Nuclear Regulatory Research (RES), Washington, D.C., 2001.

NUREG/CR-6850, *EPRI/NRC-RES Fire PRA Methodology for Nuclear Power Facilities: Volume 1: Summary and Overview*, U.S. Nuclear Regulatory Commission, Office of Nuclear Regulatory Research (RES), Washington, D.C., 2005, and EPRI 1011989, Electric Power Research Institute (EPRI), Palo Alto, CA.

NUREG/CR-6850, *EPRI/NRC-RES Fire PRA Methodology for Nuclear Power Facilities: Volume 2: Detailed Methodology*, U.S. Nuclear Regulatory Commission, Office of Nuclear Regulatory Research (RES), Washington, D.C., 2005, and EPRI 1011989, Electric Power Research Institute (EPRI), Palo Alto, CA.

NUREG/CR-6850 Supplement 1, *Fire Probabilistic Risk Assessment Methods Enhancements*, U.S. Nuclear Regulatory Commission, Office of Nuclear Regulatory Research (RES), Washington, D.C., 2010, and EPRI 1019259, Electric Power Research Institute (EPRI), Palo Alto, CA.

NUREG/CR-6931, *Cable Response to Live Fire (CAROLFIRE), Volume 1: Test Descriptions and Analysis of Circuit Response Data*, U.S. Nuclear Regulatory Commission, Washington, D.C., 2007.

NUREG/CR-6931, *Cable Response to Live Fire (CAROLFIRE), Volume 2: Cable Fire Response Data for Fire Model Improvement*, U.S. Nuclear Regulatory Commission, Washington, D.C., 2007.

NUREG/CR-6931, *Cable Response to Live Fire (CAROLFIRE), Volume 3: Thermally-Induced Electrical Failure (THIEF) Model*, U.S. Nuclear Regulatory Commission, Washington, D.C., 2007.

NUREG/CR-7010, *Cable Heat Release, Ignition, and Spread in Tray Installations during Fire (CHRISTIFIRE), Phase 1: Horizontal Trays*, National Institute of Standards and Technology, Gaithersburg, MD, 2012.

Peacock, R., K. McGrattan, B. Klein, W. Jones, and P. Reneke, *CFAST – Consolidated Model of Fire Growth and Smoke Transport (Version 6) – Software Development and Model Evaluation Guide*, Special Publication 1086, National Institute of Standards and Technology, Gaithersburg, MD, 2008.

Peacock, R., W. Jones, P. Reneke, and G. Forney, *CFAST: An Engineering Tool for Estimating Fire Growth and Smoke Transport, Version 6 – User's Guide*, Special Publication 1041, National Institute of Standards and Technology, Gaithersburg, MD, 2008.

Quintiere, J. G., *Principles of Fire Behavior*, Delmar Publishers, 1998.

Quintiere, J. G., *Fundamentals of Fire Phenomena*, John Wiley, 2006.

Salley, M.H., J. Dreisbach, K. Hill, R. Kassawara, B. Najafi, F. Joglar, A. Hamins, K. McGrattan, R. Peacock; and B. Gautier, "Verification and Validation--How to Determine the Accuracy of Fire Models," *Fire Protection Engineering*, vol. 34, pp. 34 - 44, Spring 2007.

Schiffiliti, R., and W. Pucci, *Fire Modeling, State of the Art*, Fire Detection Institute, 1996.

SFPE, *SFPE Engineering Guide to Assessing Flame Radiation to External Targets from Pool Fires*, SFPE Engineering Guide, Society of Fire Protection Engineers, Bethesda, MD, March 1999.

SFPE, *SFPE Engineering Guide to Fire Exposures to Structural Elements*, SFPE Engineering Guide, Society of Fire Protection Engineers, Bethesda, MD, November, 2005.

SFPE, *SFPE Engineering Guide to Piloted Ignition of Solid Materials Under Radiant Exposure*, SFPE Engineering Guide, Society of Fire Protection Engineers, Bethesda, MD, January, 2002.

SFPE, *SFPE Engineering Guide to Predicting Room of Origin Fire Hazards*, SFPE Engineering Guide, Society of Fire Protection Engineers, Bethesda, MD, November, 2007.

SFPE, *SFPE Engineering Guide to Substantiating a Fire Model for a Given Application*, SFPE Engineering Guide, Society of Fire Protection Engineers, Bethesda, MD, June, 2010.

SFPE, *SFPE Handbook of Fire Protection Engineering*, 4th Edition (P. J. DiNenno, Editor-in-Chief), National Fire Protection Association and The Society of Fire Protection Engineers, Quincy, MA, 2008.

SFPE, *SFPE Engineering Standard on Calculating Fire Exposures to Structures*, SFPE S.01.2001, Society of Fire Protection Engineers, Bethesda, MD, 2011.

Underwriters' Laboratories, *Single Station Fire Alarm Device*, UL 217, Underwriters' Laboratories, Northbrook, Illinois.

Wallis, G. B., "Draft Final NUREG-1824, Verification and Validation of Selected Fire Models for Nuclear Power Plant Applications," Memorandum from Advisory Committee on Reactor Safeguards to Luis Reyes, ML062980154, U.S. Nuclear Regulatory Commission, October, 2006.

Yuan, L., and G. Cox, "An Experimental Study of Some Line Fires," *Fire Safety Journal*, Volume 27, 1996, pp. 123 – 139.

REFERENCES

5-6

6
INTRODUCTION TO THE APPENDICES

Eight unique fire scenarios that typify different fire modeling applications likely to be encountered in commercial nuclear power plants (NPP), have been evaluated using the process and guidance of Chapters 2 – 4 of this document. The analysis of the eight example fire scenarios is summarized in Appendices A through H. Table 6-1 provides a summary of the individual fire scenarios as well as a cross-index between the eight appendix fire scenarios and the seven generic fire scenarios described in Chapter 3. In the scenarios, one or more aspects of Chapter 3 generic fire scenarios are incorporated. For example, Appendix B considers both localized exposure effects near the fire (Chapter 3.2.1) and hot gas layer (HGL) effects in a simple room (Chapter 3.2.2). Default values for many quantities (e.g., material properties, numerical parameters, physical parameters, empirical constants, etc.) are referenced throughout these application examples. Fire model users must understand and justify the use of default values for their applications. Fire model users are expected to assess the appropriateness of default values provided in the fire models and make changes or adjust values as necessary. Fire model user guides, handbooks, and other technical documentation are useful for this purpose.

Table 6-1. Appendix fire scenario descriptions and cross-index.

Appendix	Appendix Scenario Description	Chapter 3 Generic Fire Scenario
A	Determination of the time to main control room (MCR) abandonment given a low voltage panel fire.	3.2.5
B	Determination of the potential for a switchgear panel fire to damage cables above the fire and adjacent to the fire.	3.2.1, 3.2.2
C	Determination of the potential for a lubricant oil fire to damage cables protected with an electrical raceway fire barrier system (ERFBS).	3.2.2
D	Determination of the potential for an motor control center (MCC) panel fire to damage cables via the HGL in an irregularly shaped enclosure.	3.2.4
E	Determination of the potential for a transient trash fire to damage cables in a stack of horizontal cable trays located directly above the fire.	3.2.1
F	Determination of the potential for a lubricant oil fire to cause structural damage.	3.2.7
G	Determination of the potential for a large transient fuel package fire to damage cables in an adjacent space.	3.2.3
H	Determination of the potential for a cable tray fire to damage a nearby raceway with and without sprinklers.	3.2.1, 3.2.6

The structure of each appendix reflects the analysis sequence as well as the recommended means by which to document each type of fire modeling analysis. In particular, the appendices use the following format

- Fire modeling objective. This section clearly presents the issue that is being evaluated and the reason the analysis is being performed.

- Description of the fire scenario. This section provides a detailed description of the fire scenario being evaluated, including important parameters such as the room geometry, ventilation, fuel package data, material properties, and scenario specific information required for a successful analysis.

- Selection of the fire models. This section presents the basis for selecting the different types of fire models used to address the fire modeling objective. This includes considerations of the model capability as well as the available verification and validation (V&V) basis. Generally, two or more models are selected in order to highlight the strengths and weaknesses of the models for the given application. The selected models are not necessarily the only models that can be used, but the basis provided is broadly applicable to a given class of fire models.

- Estimation of the fire-generated conditions. This section presents model specific inputs and supporting calculations and identifies applicable model nuances for the scenario considered.

- Evaluation of results. The results of the fire models are presented in terms of the data necessary to support the fire modeling objective in this section. Model sensitivity and uncertainty are also described for each output parameter and model considered.

- Conclusions. A summary of the fire modeling analysis, including the outcome relative to the fire modeling goal is presented in this section.

The authors strove to create fire scenarios based on actual commercial NPP configurations in order to provide realistic fire scenario development and results. Be cautioned that the fire modeling results are not generic, and different applications will require careful consideration of the various input parameters, the fire model applicability, and the dominant exposure mechanism(s). However, the structure presented is generic, and when followed it is expected to lead to a fire modeling analysis that could be used to support safety related applications.

This report also contains a CD. On the CD you will find the complete NUREG-1934 final report in pdf format and the input files used for the fire model runs discussed in each appendix. In addition, the current versions of the FDT^s, CFAST and FDS that were used for each appendix are on the CD. The complete layout and contents of the CD are described in the README file included on the CD.

APPENDIX A
CABINET FIRE IN MAIN CONTROL ROOM

A.1 Modeling Objective

The purpose of the calculations described in this appendix is to approximate the length of time that the main control room (MCR) remains habitable after the start of a fire within a low-voltage control cabinet. These calculations follow the guidance provided in NUREG/CR-6850 (EPRI 1011989), Volume 2, Chapter 11, "Detailed Fire Modeling (Task 11)." MCR fire scenarios are treated differently than fires within other compartments, mainly because it is necessary to consider and evaluate forced abandonment in addition to equipment damage.

A.2 Description of the Fire Scenario

General Description: A fire ignites within a control cabinet containing XPE/neoprene cables. The door to the MCR is normally closed, and normal ventilation conditions are in place at the start of the fire. The typical ambient temperature is 20 °C (68 °F). Following guidance given in Chapter 11 of NUREG/CR-6850 (EPRI 1011989), two scenarios are considered, one in which the ventilation system is turned off and one in which the ventilation system is switched to smoke-purge mode at the start of the fire.

Geometry: Drawings of the MCR are shown in Figure A-1 and Figure A-2. The compartment has a variety of control cabinets in addition to typical office equipment, such as computer monitors on table tops. There is an open-grate suspended ceiling above the floor, a photograph of which is shown in Figure A-3. This suspended ceiling is specified to be of noncombustible construction for this scenario, but the combustibility of such ceilings should be verified for other MCR scenarios. One wall of the compartment is made of 0.9 m (3 ft)-thick concrete with no additional lining material. The other bounding walls are constructed of 1.6 cm (5/8 in) gypsum board supported by steel studs. The floor is a slab of concrete covered with low-pile carpet that is nominally 1.25 cm (0.5 in) thick. The ceiling is concrete with the same thickness as the floor (0.5 m (1.6 ft)), but with no lining material.

Materials: Thermal properties of various materials in the compartment are listed in Table 3-1. Carpet is not listed in the table, but, according to Table 6-5 of NUREG-1805, the thermal inertia ($k\rho c$) for "Carpet (Nylon/Wool Blend)" is 0.68 $(kW/m^2/K)^2$ s, its ignition temperature is 412 °C (774 °F), and its minimum heat flux for ignition is 18 kW/m^2. The steel housing of the electrical cabinets is nominally 1.5 mm (0.06 in) thick.

Ventilation: During normal operation, the mechanical ventilation system provides five air changes per hour (ACH). As shown in Figure A-1, ventilation is provided through six supply diffusers and two return vents of nominally the same size. The supply air to the compartment is equally distributed among the six supply vents, and the return air is drawn equally from the two returns. A 120 Pa overpressure (relative to the adjacent compartments) is maintained in the MCR during normal operation. Leakage from the compartment occurs via a 1.3 cm (0.5 in) high crack under the 0.9 m (3 ft) wide door. All other penetrations are sealed. Smoke-purge mode provides 25 air changes per hour and is activated manually by an MCR operator in accordance

CABINET FIRE IN MAIN CONTROL ROOM

with specified procedures, which is known for this control room to be 120 seconds after the start of a fire.

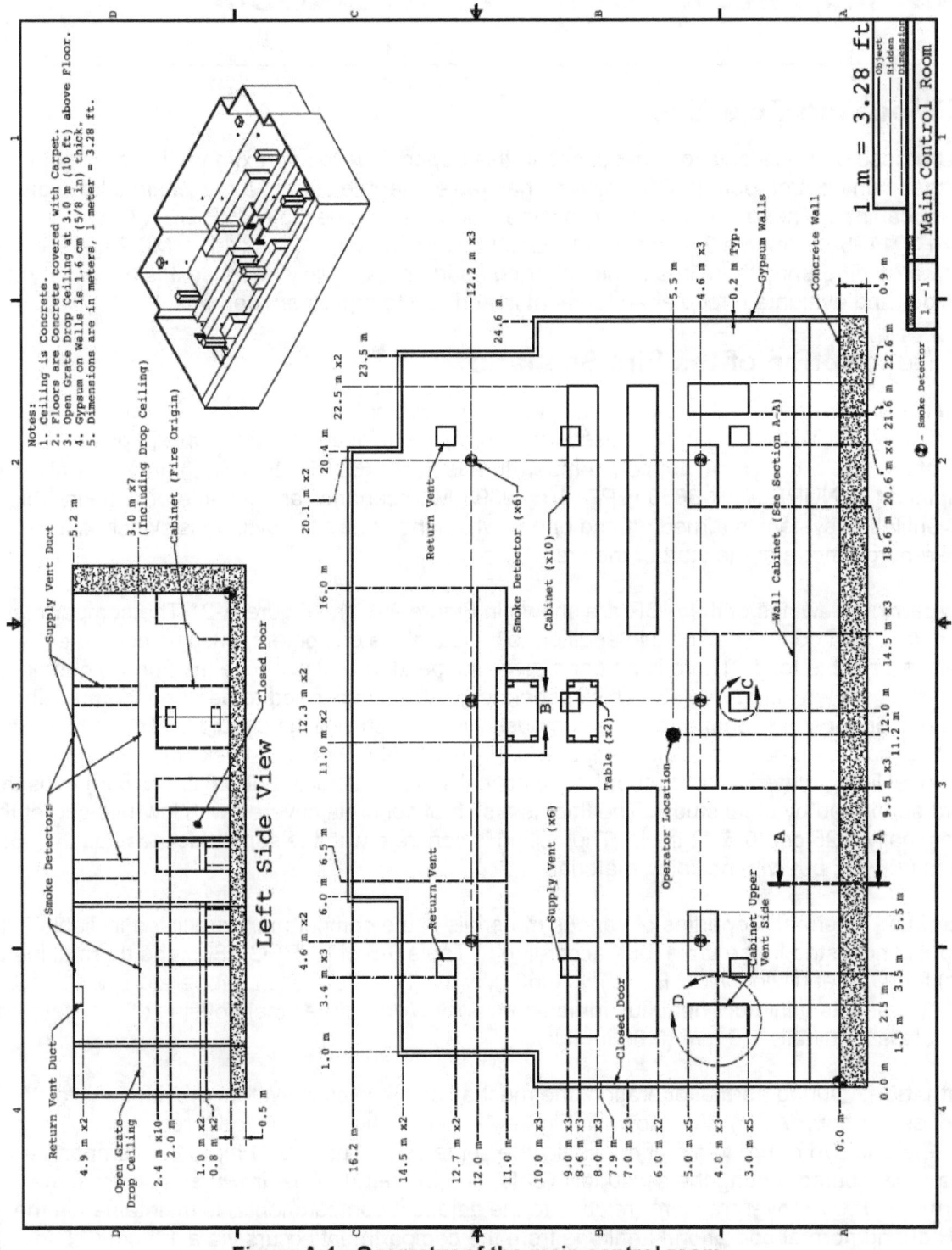

Figure A-1. Geometry of the main control room.

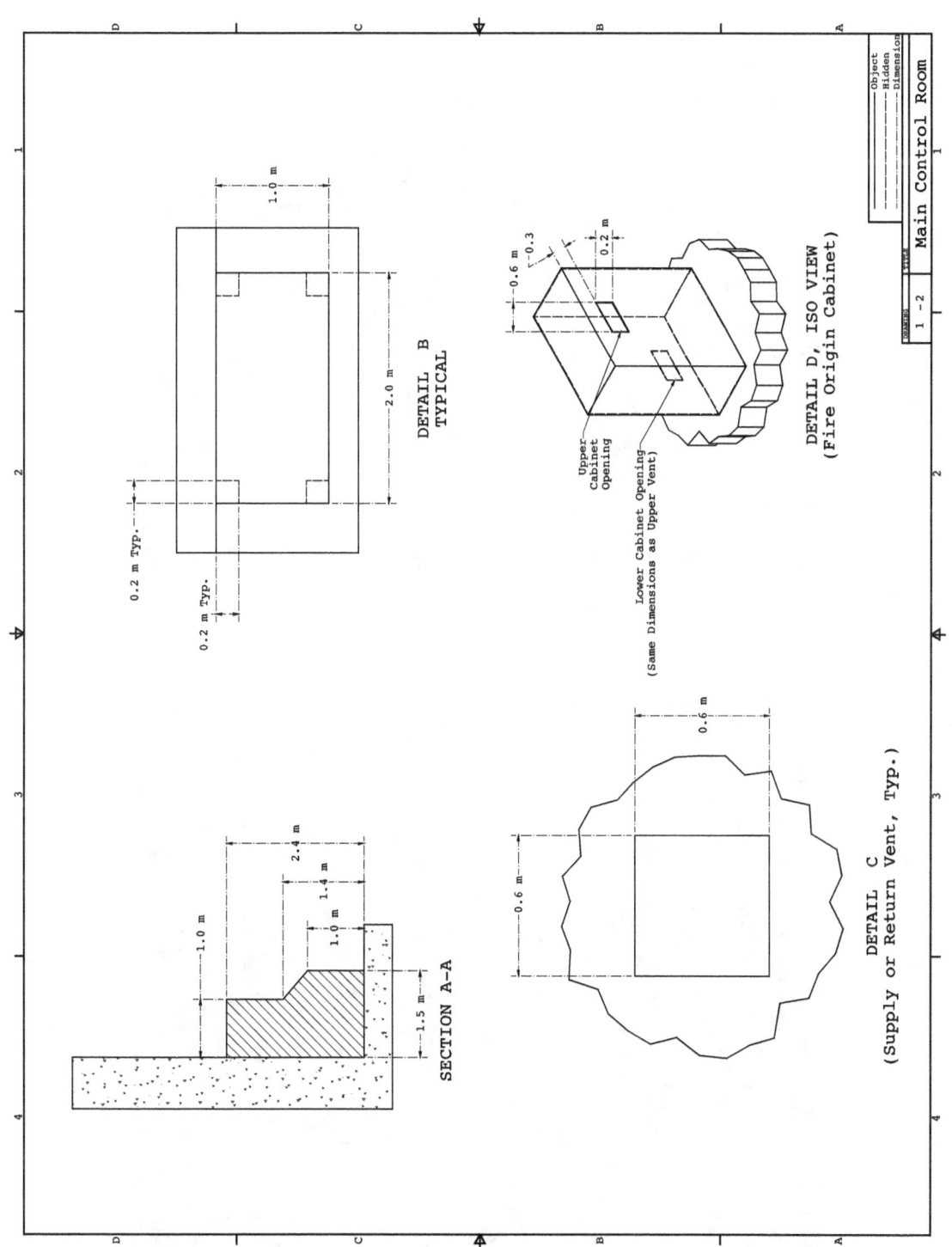

Figure A-2. Main control room details.

Figure A-3. Photograph of a typical open-grate ceiling.

Figure A-4. Photograph of a typical control cabinet.

Fire: The fire ignites due to an electrical malfunction in bundles of qualified XPE/neoprene cables inside an isolated control cabinet (Figure A-4), designated as the Fire Origin Cabinet in Figure A-2. The fire grows according to a "t-squared" curve to a maximum value of 702 kW in 12 min and remains steady for eight additional minutes, consistent with NUREG/CR-6850 (EPRI 1011989), Appendix G, for a low-voltage cabinet fire involving more than one bundle of qualified cable. After 20 min, the fire's heat release rate (HRR) decays linearly to zero in 19 min. A peak fire intensity of 702 kW represents the 98th percentile of the probability distribution for the HRR in cabinets of this general description. The HRR curve is shown in Figure A-5.

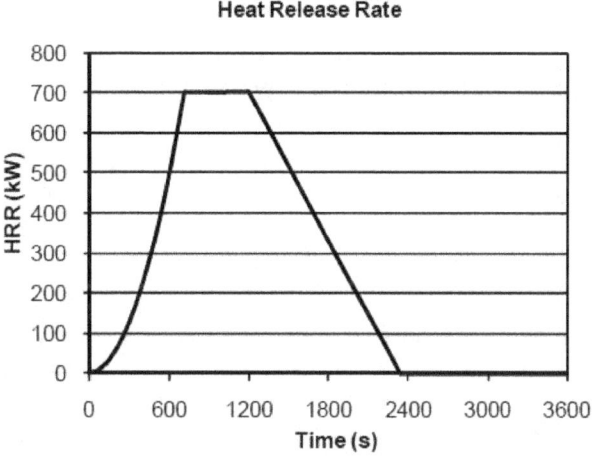

Figure A-5. Time history of the HRR used by all models in the MCR scenario.

Based on a physical assessment of the cabinet, it is determined that the exterior panels of the burning cabinet do not open before or during the fire. The smoke, heat, and flames are exhausted from an air vent in the side of the cabinet. The top of the air vent is 0.3 m (1 ft) below the top of the cabinet. The air vent is 0.6 m (2 ft) wide and 0.2 m (8 in) high. The cabinet is 2.4 m (8 ft) tall.

The heat of combustion and product yields for XPE/neoprene cable are taken from Table 3-4.16 of the *SFPE Handbook, 4th edition* and are listed in Table A-1. When estimating the composition of the fire's combustion products, the jacket and insulation material of the cable are taken as an equal-parts mixture of polyethylene (C_2H_4) and neoprene (C_4H_5Cl), with the effective chemical formula $C_3H_{4.5}Cl_{0.5}$.

Table A-1. Data for MCR fire based on XPE/neoprene electrical cable.

Parameter	Value	Source
Effective Fuel Formula	$C_3H_{4.5}Cl_{0.5}$	Combination of polyethylene and neoprene
Peak HRR	702 kW	NUREG/CR-6850 (EPRI 1011989), App. G
Time to reach peak HRR	720 s	NUREG/CR-6850 (EPRI 1011989), App. G
Heat of Combustion	10,300 kJ/kg	*SFPE Handbook*, 4th ed., Table 3-4.16
CO_2 Yield	0.63 kg/kg	*SFPE Handbook*, 4th ed., Table 3-4.16
Soot Yield	0.175 kg/kg	*SFPE Handbook*, 4th ed., Table 3-4.16
CO Yield	0.082 kg/kg	*SFPE Handbook*, 4th ed., Table 3-4.16
Radiative Fraction	0.53	*SFPE Handbook*, 4th ed., Table 3-4.16
Mass Extinction Coefficient	8700 m²/kg	Mulholland and Croarkin (2000)

Habitability: The MCR is manned 24 hours per day during normal plant operations. To assess habitability of the compartment, the operator position indicated in Figure A-1 is used. According to NUREG/CR-6850 (EPRI 1011989), Volume 2, Chapter 11, "Detailed Fire Modeling," a space is considered uninhabitable if at least one of the following occurs:

1. The incident heat flux at 1.8 m (6 ft) exceeds 1 kW/m^2. A smoke layer temperature of approximately 95 °C (200 °F) generates this level of heat flux.

2. The smoke layer descends below 1.8 m (6 ft) from the floor, and the optical density of the smoke is greater[15] than 3 m^{-1}.

A.3 Selection and Evaluation of Fire Models

This section discusses the overall modeling strategy. In particular, it describes the process of model selection, including a discussion of the validity of these models for the given fire scenario.

A.3.1 Temperature Criterion

The forced ventilation correlation of Foote, Pagni, and Alvares (FPA) that is incorporated in the Fire-Induced Vulnerability Evaluation (FIVE-Rev1) is used to estimate the compartment temperatures under smoke purge ventilation conditions. This correlation is not appropriate in the scenario without ventilation. The Consolidated Fire Growth and Smoke Transport Model (CFAST) and the Fire Dynamics Simulator (FDS) are used to estimate the hot gas layer (HGL) temperature for both the smoke purge and unventilated scenarios. One advantage of a Computational Fluid Dynamics (CFD) model like FDS for this fire scenario is that it can estimate habitability conditions at the specific location of the operator, whereas a zone model like CFAST can only estimate average conditions in the hot gas layer.

A.3.2 Heat Flux Criterion

The point source radiation heat flux model is first used to estimate the heat flux to the operator. FIVE-Rev1 and the Fire Dynamics Tools (FDT^s) both contain methods to estimate the heat flux from a fire to a target located a specified distance away. However, distant targets, including the operators, may not be exposed directly to the thermal radiation; it is also possible that the descending hot gas layer will be responsible for some fraction of the heat flux to which the operator is exposed. Neither FIVE-Rev1 nor the FDT^s include calculation methods to account for this source of thermal radiation. CFAST and FDS will be used to determine whether the thermal radiation from the HGL reaches the critical value for tenability.

A.3.3 Visibility Criterion

FIVE-Rev1 and the FDT^s do not include correlations to estimate visibility in a ventilated compartment, but correlations from the *SFPE Handbook* are applied to develop an estimate of smoke obscuration in the MCR under smoke purge conditions. With a smoke purge ventilation rate of 25 ACH, a well-mixed smoke environment is expected, and the average smoke concentration and visibility conditions can be calculated using these methods. CFAST and FDS are also used to address the visibility criterion. CFAST has an advantage over MAGIC for this

---

[15] The original edition of NUREG/CR-6850 (EPRI 1011989) contains an error in the specification of the optical density (NRC ADAMS Accession Number ML061630360). The value of 3 m^{-1} is correct.

analysis because it computes the smoke obscuration. FDS is particularly useful because it does not require that the smoke is confined to a uniform descending layer. It is expected that air currents in the room will mix the smoke to the floor. Zone models typically do not allow the smoke layer to descend below the height of the base of the fire.

The case of the unventilated enclosure fire scenario is addressed with CFAST and FDS to estimate the smoke layer interface position and smoke layer optical density as a function of time. One advantage of a CFD model like FDS for this fire scenario is that it can estimate habitability conditions at the specific location of the operator.

A.3.4 Validation

The principal source of validation data justifying the use of the fire models discussed above for this scenario is the U.S. Nuclear Regulatory Commission/Electric Power Research Institute (NRC/EPRI) verification and validation (V&V) study documented in NUREG-1824 (EPRI 1011999). The National Institute of Standards and Technology (NIST) has expanded the NRC/EPRI V&V to include the latest versions of CFAST (6.1.1) (Peacock, 2008) and FDS (5.5.3) (McGrattan, 2010). In particular, a Factory Mutual/Sandia National Laboratories (FM/SNL) test series was designed specifically as a mock-up of a control room in a nuclear power plant (NPP). One of these experiments (Test 21) involves a fire within a hollow steel cabinet.

Table A-2 lists the important non-dimensional parameters that characterize the fire scenario and the ranges for which the NRC/EPRI validation study is applicable. A few parameters fall outside the validation parameter space and are addressed individually:

- The Fire Froude Number falls outside the range. This parameter is essentially a measure of the fire's heat output relative to its base area. In this example, the fire is specified as emanating from the side of the cabinet, with the vent opening serving as its base. This leads to a higher value of $\dot{Q}^*$ than would be calculated if the fire were burning completely outside of the cabinet. Thus, the high value of $\dot{Q}^*$ is the represents more severe fire conditions than would be expected if the fire were specified as burning partially within the cabinet.

- The relatively low Equivalence Ratio for the compartment is a result of the relatively large amount of air forced into the room during the smoke purge mode. Twenty-five air changes per hour is a considerable flow rate, and no validation experiment in NUREG-1824 (EPRI 1011999) involved such a high ventilation rate. However, the results of all the model simulations indicate that the scenario in which the ventilation is turned off is most likely to compromise human habitability, and the presence of any level of ventilation reduces room temperature and heat flux and increases visibility.

For the scenario with no ventilation, the classic definition of the Equivalence Ratio does not apply because there is no supply of oxygen in the room. However, it can be shown that there is sufficient oxygen in the room to sustain the specified fire. The total mass of oxygen in the room is the product of the density of air, ρ, the volume of the room, V, and the mass fraction of oxygen in the air, Y_{O_2}:

$$m_{O_2,\text{tot}} = \rho V Y_{O_2} = 1.2 \text{ kg/m}^3 \times 1945 \text{ m}^3 \times 0.23 \cong 537 \text{ kg} \qquad \text{(A-1)}$$

CABINET FIRE IN MAIN CONTROL ROOM

The mass of oxygen required to sustain the fire is equal to the total energy produced by the fire divided by the energy released per unit mass oxygen consumed:

$$m_{O_2,req} = \frac{Q}{\Delta H_{O_2}} \cong \frac{702 \text{ kW} \times 60 \text{ s/min} \times \left(\frac{12}{3} + 8 + \frac{19}{2}\right) \text{ min}}{13{,}100 \text{ kJ/kg}} \cong 69 \text{ kg} \tag{A-2}$$

These calculations show that the quantity of oxygen in the room would be able to sustain the specified cabinet fire.

- The ratio of the Target Distance relative to the Fire Diameter, r/D, exceeds the range of the validation study. However, this parameter is only relevant to the point source radiation heat flux calculation, which is by definition more accurate, as the target moves further from the source. Thus, although the parameter is outside the validation range, it is not outside of the methodology's range of validity.

Table A-2. Normalized parameter calculations for the MCR fire scenario. See Table 2-5 for further details.

Quantity	Normalized Parameter Calculation		NUREG-1824 Validation Range	In Range?
Fire Froude Number	$\dot{Q}^* = \dfrac{\dot{Q}}{\rho_\infty c_p T_\infty D^{2.5}\sqrt{g}}$ $= \dfrac{702\ \text{kW}}{(1.2\ \text{kg/m}^3)(1.0\ \text{kJ/kg/K})(293\ \text{K})(0.4^{2.5}\ \text{m}^{2.5})\sqrt{9.8\ \text{m/s}^2}} \cong 6.2$		0.4 – 2.4	No
Flame Height, $H_f + L_f$, relative to the Ceiling Height, H_c	$\dfrac{H_f + L_f}{H_c} = \dfrac{2.1\ \text{m} + 2.7\ \text{m}}{5.2\ \text{m}} \cong 0.9$ $L_f = D\left(3.7\ \dot{Q}^{*\,2/5} - 1.02\right) = 0.4\ \text{m}\ (3.7 \times 6.2^{0.4} - 1.02) \cong 2.7\ \text{m}$		0.2 – 1.0	Yes
Ceiling Jet Radial Distance, r_{cj}, relative to the Ceiling Height, H_c	N/A – Ceiling jet targets are not included in simulation.		1.2 – 1.7	N/A
Equivalence Ratio, φ, of the room, based on Forced Ventilation of Purge Mode	$\varphi = \dfrac{\dot{Q}}{\Delta H_{O_2} \dot{m}_{O_2}} = \dfrac{702\ \text{kW}}{13{,}100\ \text{kJ/kg} \times 3.7\ \text{kg/s}} \cong 0.014$ $\dot{m}_{O_2} = Y_{O_2}\,\rho_\infty \dot{V} = 0.23 \times 1.2\ \text{kg/m}^3 \times 13.4\ \text{m}^3/\text{s} \cong 3.7\ \text{kg/s}$		0.04 – 0.6	No
Compartment Aspect Ratio	$\dfrac{L}{H_c} = \dfrac{24.6\ \text{m}}{5.2\ \text{m}} \cong 4.7$	$\dfrac{W}{H_c} = \dfrac{16.2\ \text{m}}{5.2\ \text{m}} \cong 3.1$	0.6 – 5.7	Yes
Target Distance, r, relative to the Fire Diameter, D	$\dfrac{r}{D} = \dfrac{8.8\ \text{m}}{0.4\ \text{m}} \cong 22$		2.2 – 5.7	No

Notes:

(1) The effective diameter of the base of the fire, D, is calculated using $D = \sqrt{4A/\pi}$, where A is the area of the cabinet vent.

(2) The Flame Height, $H_f + L_f$, is the sum of the height of the fire base above the floor and the fire's flame length.

A.4 Estimation of Fire-Generated Conditions

This section provides details as to how each of the models was set up and run.

A.4.1 Algebraic Models

General: The forced ventilation correlation of FPA is used to estimate the HGL temperature of the MCR for the smoke purge scenario. The point source method is used to estimate the heat flux from the fire to the operator for both the smoke purge and non-ventilated scenarios. The smoke concentration and visibility are calculated for the smoke purge scenario based on quasi-steady conditions, as described below.

Temperature in the Smoke Purge Scenario

Figure A-6 is a schematic diagram that illustrates the smoke purge scenario. The FPA correlation requires room dimensions to be specified in terms of length, width, and height. For this example, the selected compartment is not a rectangular parallelepiped, so it needs to be represented with an effective length, width, and height that provide the same volume and boundary surface area as the actual compartment. The effective compartment height is taken as the actual height of 5.2 m (17 ft) because it is important to maintain the same compartment height for smoke filling calculations. Next, the effective length and width are calculated to maintain the same volume and boundary surface area of the actual compartment. Because the actual and effective heights are the same, this is equivalent to maintaining the same floor area (A_{fl}) and wall perimeter (P) of the enclosure. From Figure A-1, the MCR floor area is 372 m^2 (4004 ft^2), and the perimeter is 81.6 m (274 ft). Maintaining the total floor area and perimeter yields effective compartment dimensions of 27.1 m (89 ft) by 13.8 m (45 ft). This is calculated by solving the following two equations for the effective length and width, L_e and W_e:

$$A_{fl} = L_e \times W_e \quad ; \quad P = 2 \times (L_e + W_e) \tag{A-3}$$

The parameters for the FIVE-Rev1 implementation of the FPA are listed in Table A-3. Note that the specified time-dependent HRR is used in the calculation, but that the calculation does not include either the fire's elevation above the floor or any other information about the fire. The walls, ceiling, and floor are all specified as gypsum board rather than concrete because the FPA correlation only accounts for one type of lining material. Gypsum board was chosen because it has a lower thermal conductivity than concrete, which results in a slightly higher HGL temperature.

Table A-3. Summary of input parameters for the FPA calculation of the MCR.

Parameter	Value	Source
Room height (H)	5.2 m	Figure A-1
Room effective length (L_e)	27.1 m	Equation (A-3)
Room effective width (W_e)	13.8 m	Equation (A-3)
Room boundary material	Gypsum board	Table 3-1
Mech. ventilation rate ($\dot{V}$)	13.4 m^3/s	Specified (25 ACH)
Ambient temperature (T_a)	20 °C	Specified
Fire parameters		Table A-1

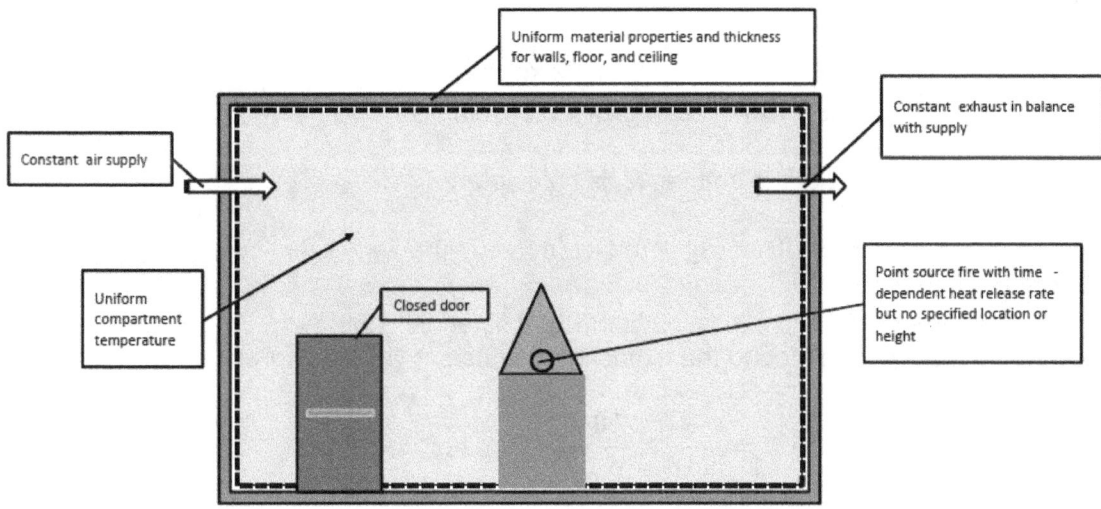

Figure A-6. Schematic diagram of the FPA calculation for the MCR smoke purge scenario.

Heat Flux

The point source model is used to estimate the heat flux from the flames to the operator when the fire is at its peak HRR. The peak HRR, $\dot{Q}$, is 702 kW, the radiative fraction, χ_r, is 0.53, and the distance from the cabinet vent to the operator is approximately 8.8 m (29 ft). The heat flux is calculated:

$$\dot{q}'' = \frac{\chi_r \, \dot{Q}}{4\pi \, r^2} = \frac{0.53 \times 702 \text{ kW}}{4\pi \times 8.8^2 \text{ m}^2} \cong 0.38 \text{ kW/m}^2 \tag{A-4}$$

While this heat flux prediction is well below the critical value of 1 kW/m^2, it does not account for the thermal radiation from the HGL. Thus, the point source method can be used as a screening tool, and further analysis can be performed by CFAST and FDS.

Smoke concentration and visibility

Neither the FDT^s nor FIVE-Rev1 include methods to calculate smoke concentrations or visibility in mechanically ventilated enclosure fires, but calculation methods provided in Section 3, Chapter 9, of the *SFPE Handbook, 4th ed.* are relatively simple to apply and are based on the same principles and concepts embodied in zone models. These hand calculations provide an estimate of the fire-generated smoke concentrations and visibility conditions for this scenario and will indicate whether more detailed modeling is warranted.

The soot mass generation rate, $\dot{m}_s$, is the product of the soot yield, y_s, and the mass burning rate of fuel, $\dot{m}_f$. The latter quantity is obtained by dividing the HRR, $\dot{Q}$, by the heat of combustion, ΔH:

$$\dot{m}_s = y_s \dot{m}_f = y_s \frac{\dot{Q}}{\Delta H} = 0.175 \times \frac{702 \text{ kW}}{10,300 \text{ kJ/kg}} \cong 0.012 \text{ kg/s} \tag{A-5}$$

CABINET FIRE IN MAIN CONTROL ROOM

The soot mass fraction in the smoke layer, Y_s, is then calculated:

$$Y_s = \frac{\dot{m}_s}{\dot{m}_{tot}} \cong \frac{\dot{m}_s}{\dot{m}_a} = \frac{\dot{m}_s}{\rho \dot{V}} = \frac{0.012 \text{ kg/s}}{1.2 \text{ kg/m}^3 \times 13.4 \text{ m}^3/\text{s}} \cong 0.00075 \text{ kg/kg} \qquad \text{(A-6)}$$

The extinction coefficient of the smoke, K, is calculated:

$$K = K_m \rho Y_s = 8700 \text{ m}^2/\text{kg} \times 1.2 \text{ kg/m}^3 \times 0.00075 \text{ kg/kg} \cong 7.8 \text{ m}^{-1} \qquad \text{(A-7)}$$

Here K_m is the mass-specific extinction coefficient listed in Table A-1. By definition, the optical density of the smoke is related to the extinction coefficient via the expression:

$$D = \frac{K}{\ln 10} \cong \frac{7.8 \text{ m}^{-1}}{2.3} \cong 3.4 \text{ m}^{-1} \qquad \text{(A-8)}$$

This calculated optical density is then compared with the tenability limit, 3 m^{-1}. This calculation indicates that the visibility criterion for tenability would be exceeded for the MCR smoke purge scenario based on the specified parameters and the MCR uniformly filling with smoke. However, because this analysis is based only on the peak rather than the time-dependent HRR, further modeling with zone and CFD models is warranted.

A.4.2 Zone Model

Geometry: CFAST divides the geometry into one or more compartments connected by vents. For this simulation, the entire compartment is modeled as a single compartment. As with the algebraic models, zone models simulate fires in compartments with rectangular floor areas. The strategy for selecting effective room dimensions is the same as for the FPA analysis described above. Figure A-7 shows the compartment geometry inputs to CFAST.

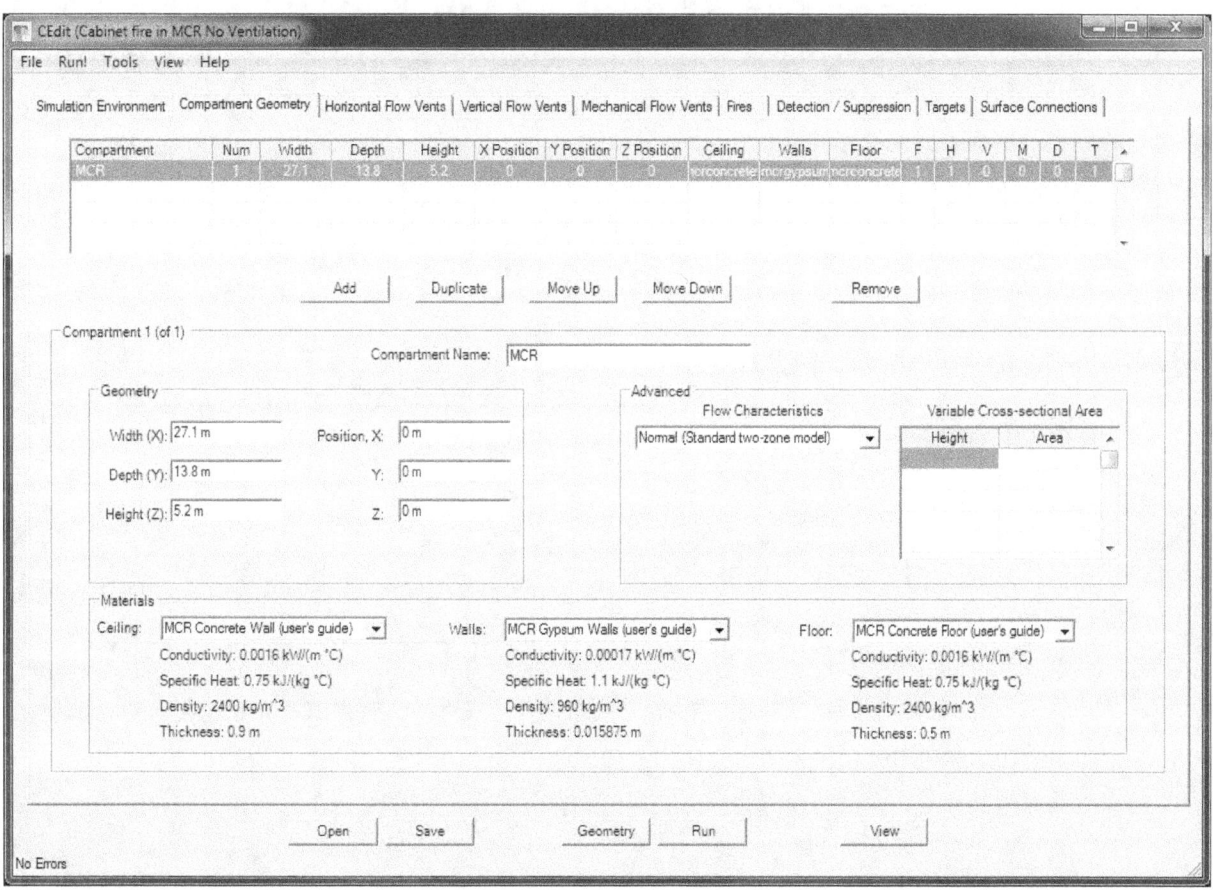

Figure A-7. Compartment geometry and surface material selection in CFAST for the MCR Fire.

While there are numerous cabinets and tables in the compartment, they do not significantly change the overall volume of the room and may be neglected as obstructions. There are no mechanisms within CFAST to account for the open-grate ceiling. This ceiling provides a negligible flow resistance to the heat and air that go through it, so it can be ignored. It is expected that neglecting it will lead to slightly higher HGL temperatures because there is less resistance for the rising smoke and hot gases.

Fire: In CFAST, a fire is described as a source of heat placed at a specific point within a compartment that generates combustion products according to user-specified combustion chemistry. Consistent with typical practice for the use of zone fire models for electrical cabinet fires, the fire is positioned at the top of the air vent, 0.3 m (1 ft) below the top of the cabinet, at the center of the cabinet. The air vent dimensions are 0.6 m (2 ft) wide and 0.2 m (0.7 ft) high. The effective diameter of the fire is 0.4 m (1.3 ft), as noted in Table A-2.

Combustion chemistry in CFAST is described, at a minimum, by the production rates of CO, CO_2, and soot. The yield of HCl is calculated by the model based on the specified molecular formula and the conversion of all of the chlorine into hydrochloric acid. Inputs for the fire specification in CFAST, taken directly from Table A-1, are shown in Figure A-8.

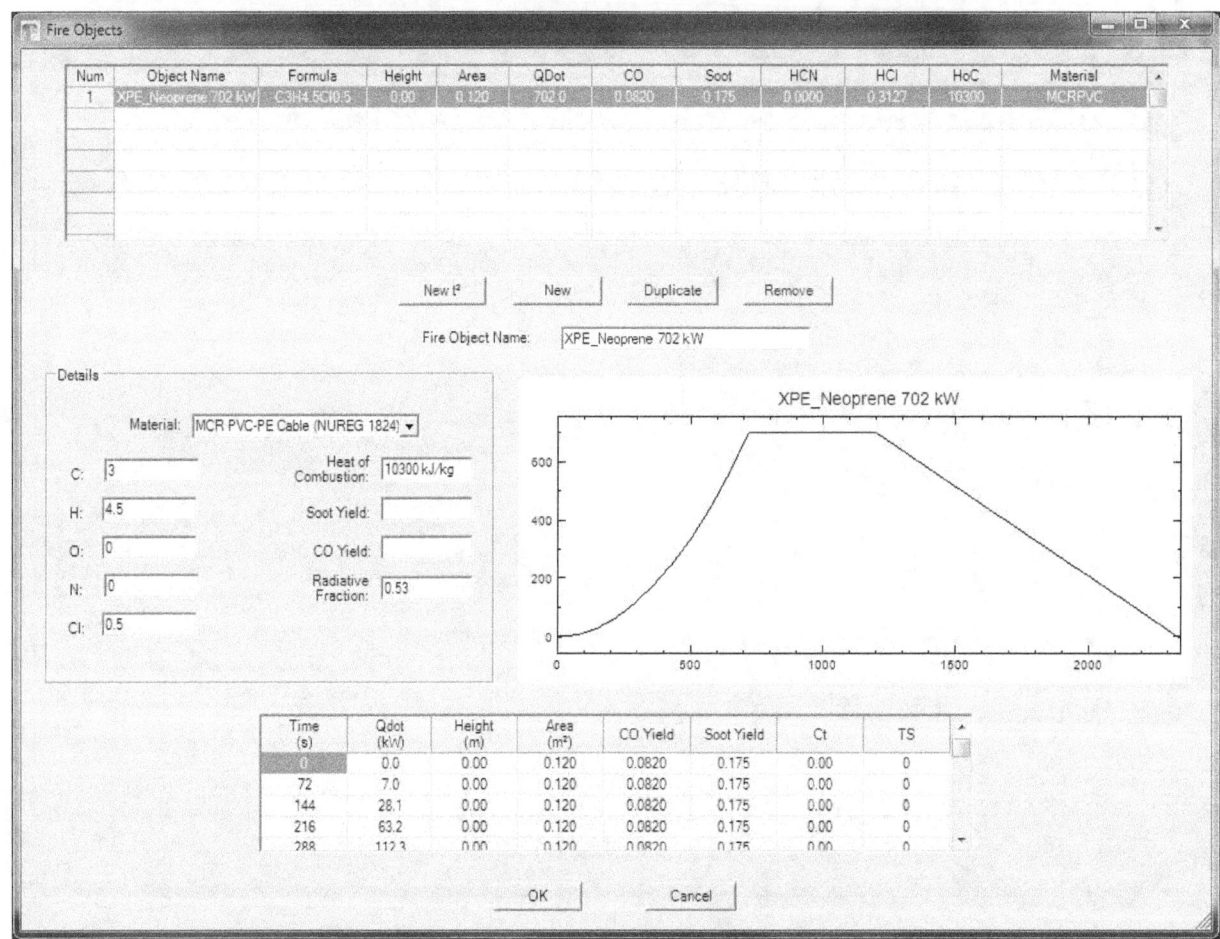

Note: Values for "Total Mass," "Heat of Gasification," and "Volatilization Temperature" are set at default values.

Figure A-8. Specification of the fire in CFAST for the MCR Fire.

Materials: CFAST does not include the ability to model individual walls of different materials, but it can model different materials for walls, floors and ceilings. For this example, the compartment walls are specified as being entirely made of gypsum wallboard which has a thermal conductivity lower than the other wall materials. The floor and ceiling are modeled as 0.5 m (1.6 ft) thick concrete.

Ventilation: For the smoke-purge calculation, air is supplied to the MCR via the six supply vents and exhausted through the two returns. The total ventilation rate is 25 air changes per hour, or 13.4 m^3/s. Mechanical ventilation inputs for CFAST are shown in Figure A-9. A snapshot of the CFAST simulation is shown in Figure A-10.

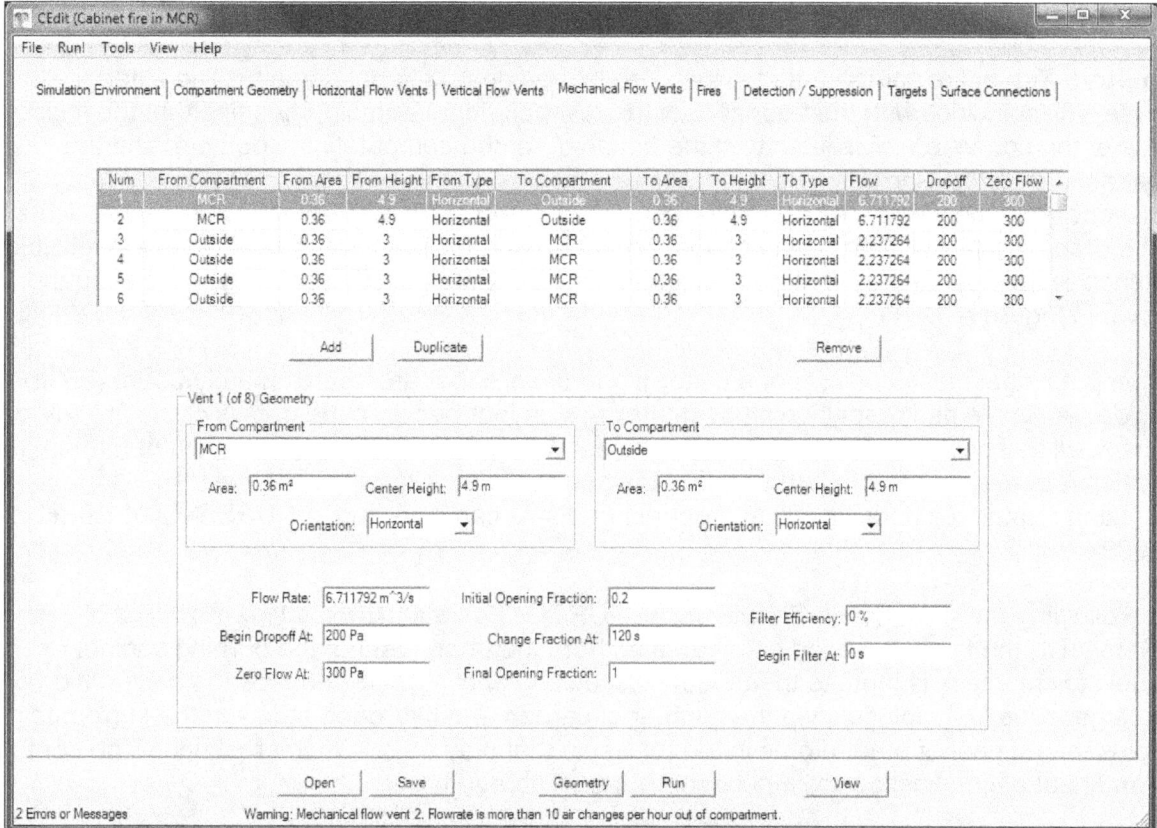

Note: Values for "Begin Dropoff At" and "Zero Flow At" are set at default values.

Figure A-9. Mechanical ventilation inputs in CFAST for the MCR fire.

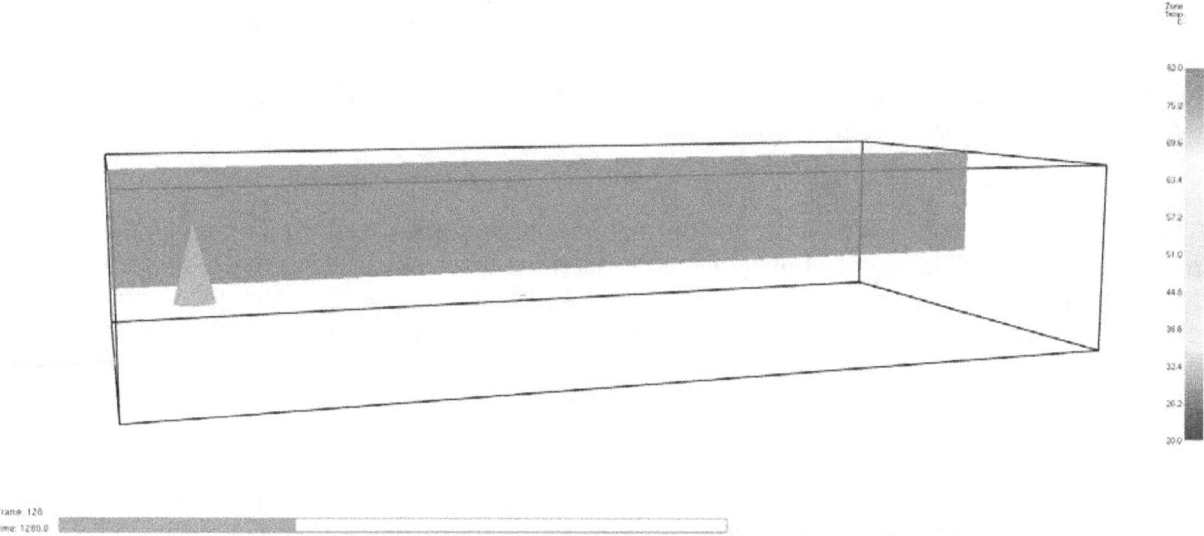

Figure A-10. Smokeview rendering of the CFAST simulation of the MCR fire with mechanical ventilation.

A.4.3 CFD Model

Geometry: The entire compartment is included in the computational domain. The exterior concrete wall coincides with the boundary of the computational domain, meaning that the inside surface of the concrete wall is flush with the boundary of the computational domain, and the properties of concrete (including its thickness) are applied to this boundary. The tables (made out of wood) and the electrical cabinets (made out of steel) are included in the simulation. Note that the drop ceiling is not modeled because it is open and for this example provides a negligible resistance to the heat and air that go through it. An FDS/Smokeview rendering of the scenario is shown in Figure A-12.

The computational mesh consists of a uniform grid of cells that are approximately 0.2 m (0.7 ft) on a side. A simple grid resolution study demonstrates that because the details of the fire (other than its specified heat and smoke production rates) within the cabinet are not relevant for this simulation, there is no need to further refine the grid in the vicinity of the cabinet. An explanation related to choosing the appropriate grid size can be found in NUREG-1824 (EPRI 1011999), Vol. 7.

Fire: Following the guidance in Chapter 12 of NUREG/CR-6850 (EPRI 1011989), Supplement 1, the fire is modeled as emanating from the upper vent of the burning cabinet. The fuel stoichiometry is input to the model as specified above. FDS requires the designation of a single gaseous fuel molecule via the number of carbon and hydrogen atoms in the surrogate fuel, plus the number of other atoms in the molecule that play no role in the reaction. The soot yield and heat of combustion are input directly from Table A-1.

Materials: The cabinets are represented by closed boxes with the specified properties of steel. The tables are assigned the properties of plywood that is 5 cm (2 in) thick. The table legs are not modeled because they play little role in the fire or heat transfer calculation to the solids. Concrete and gypsum properties are applied to the applicable walls and ceiling. The floor is modeled as a 1 cm (0.4 in) thick carpet over a 0.5 m (1.6 ft) thick concrete slab. The concrete properties are taken directly as specified. The carpet properties are obtained by choosing individual values of the thermal conductivity, k, density, ρ, and specific heat, c, so that their product equals the given value of $k\rho c$.

$$k\rho c = 0.0017 \text{ kW/m/K} \times 200 \text{ kg/m}^3 \times 2.0 \text{ kJ/kg/K} = 0.68 \text{ (kW/m}^2\text{/K)}^2 \text{ s}$$

The $k\rho c$ used to estimate the individual thermal parameters is derived from an empirical ignition model, and it may differ somewhat significantly from the true value. However, in this case, the heat losses to the carpet are not expected to play a significant role in the simulation and the choice of its thermal properties is not expected to affect the results. If the boundary is expected to be an important heat loss surface, then a more accurate determination of the thermal parameters is necessary.

Ventilation: Air is supplied to the MCR via the six supply vents and exhausted through the two returns. The supply rate is divided equally among the six supply vents, and the return rate is divided equally among the two returns. The leakage from the compartment is modeled by specifying a small vent located at the base of the door through which air escapes at a rate determined by the pressure difference between the MCR and ambient. Note that the door crack itself is not modeled explicitly, as the numerical grid is not fine enough. Rather, the leak is spread over a slightly larger area. The volume flow through the leakage area, $\dot{V}_L$, is estimated via the equation:

$$\dot{V}_L = A_L \sqrt{\frac{2\,\Delta p}{\rho_\infty}} \tag{A-9}$$

where $A_L A_L$ is the *actual* leakage area (0.9 m (3 ft) by 0.013 m (0.04 ft), or 0.0117 m^2 in this case), Δp is the pressure difference between the inside and outside of the compartment (Pa), and ρ_∞ is the ambient air density.

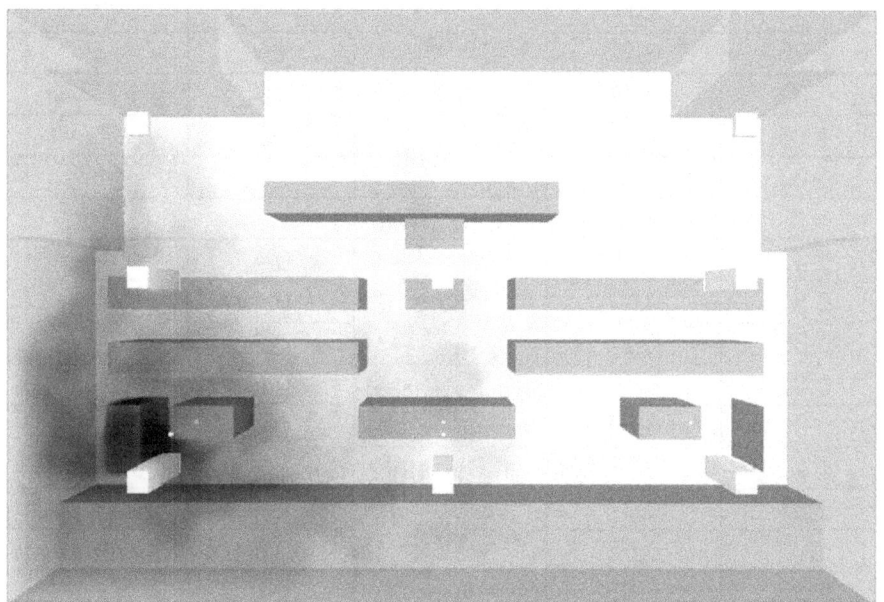

Figure A-11. FDS/Smokeview rendering of the MCR, as viewed from above.

A.5 Evaluation of Results

The habitability of the MCR depends on the temperature, heat flux, and smoke concentration to which the operators would be exposed. According to NUREG/CR-6850 (EPRI 1011989), Volume 2, Chapter 11, abandonment of the MCR is occurs if the gas temperature 1.8 m (6 ft) above the floor exceeds 95°C (200°F) or if the heat flux exceeds 1.0 kW/m^2 or if the optical density exceeds 3 m^{-1}.

Table A-4 summarizes the results of all the models for the three tenability criteria. For each predicted value, a calculation is performed to determine the Probability of Exceeding the Critical Value. The procedure for calculating this probability is given in Chapter 4, and it accounts for the model bias and standard deviation. The purpose of this table is to highlight the criterion that is most likely to be exceeded, so that further analysis can be focused on this criterion and the model or models that predict it. Each criterion is discussed in greater detail in the following subsections.

Table A-4. Summary of the model predictions of the MCR scenario.

Model	Bias Factor, δ	Standard Deviation, $\tilde{\sigma}_M$	Ventilation	Predicted Value	Critical Value	Probability of Exceeding
Temperature (°C), Initial Value = 20 °C						
FIVE-Rev1 (FPA)	1.56	0.32	Purge	70	95	0.000
CFAST	1.06	0.12		61	95	0.000
FDS	1.03	0.07		48	95	0.000
CFAST	1.06	0.12	No Vent.	82	95	0.009
FDS	1.03	0.07		70	95	0.000
Heat Flux (kW/m^2)						
FIVE-Rev1	1.42	0.55	Purge	0.4	1	0.000
CFAST	0.81	0.47		0.1	1	0.000
FDS	0.85	0.22		0.2	1	0.000
CFAST	0.81	0.47	No Vent.	0.6	1	0.228
FDS	0.85	0.22		0.4	1	0.000
Optical Density (m^{-1})						
CFAST	2.65	0.63	Purge	7.6	3	0.471
FDS	2.7	0.55		0.5	3	0.000
CFAST	2.65	0.63	No Vent.	54	3	0.912
FDS	2.7	0.55		31	3	0.909

A.5.1 Temperature Criterion

The HGL temperature and height predictions are summarized in Table A-4 and are shown in detail in Figure A-12. None of the analyses imply that the temperature tenability limit would be exceeded by a fire of this type, regardless of the ventilation system. It is important to note that neither the FPA correlation nor CFAST estimate the temperature at the operator location specifically. For the purpose of assessing habitability, the HGL temperature is used to approximate the flux condition to which the operator would be exposed, regardless of whether the HGL descends to the operator's height. This means that the FPA and CFAST analyses have an extra level of conservatism built in for this particular case.

The FPA correlation predicts a peak HGL temperature of 70 °C (158 °F) when the smoke purge system is on, but, based on the CFAST calculation, it is expected that the layer height would not descend to the operator level if the purge system were in operation.

CFAST predicts that the HGL temperature reaches just above 80 °C (176 °F) in 20 min when the smoke purge system is off. The HGL descends to 2 m (6.6 ft) above the floor in approximately the same amount of time and thus remains above the head of the operator. When the smoke purge system is on, CFAST predicts that the peak HGL temperature reaches approximately 60 °C (140 °F), but that the smoke layer does not descend beyond a meter below the ceiling due to the operation of the smoke exhaust system.

The FDS predictions of HGL temperature are lower than those of the other models because FDS accounts for the mixing of heat and smoke with ambient air due to the high purging flow, since it models flow within the compartment in detail.

The HGL temperature is largely a function of the amount of energy from the fire that is carried aloft in the smoke plume. The model simulations have all used, based on data from the *SFPE Handbook, 4th edition*, a convective fraction of the HRR that is relatively low for the kind of cables under consideration. Typically, approximately 65% of the fire's energy is lofted upwards in the plume, whereas in this case the models have all used a convective fraction of only 47% (one minus the radiative fraction). As a result, the HGL temperature might be lower than one would expect from a typical fire because a higher percentage of its energy is radiated away. Referring to Table 4-3, the HGL temperature rise is proportional to the HRR to the two-thirds power. For the purpose of this analysis, the HRR can be regarded as the convective HRR. If the convective HRR were increased by 38%, the HGL temperature rise would increase by approximately two-thirds of 38%, or 25%. This would have the effect of increasing the CFAST-predicted temperature rise from 60 °C (140 °F) to 75 °C (167 °F). Given an ambient temperature of 20 °C (68 °F), this means that if CFAST were to use the conventional 35% radiative fraction (65% convective fraction), its prediction[16] of the HGL in the no-ventilation case would be approximately 95 °C (203 °F), the critical value for abandonment. A similar argument can be made for the other HGL predictions.

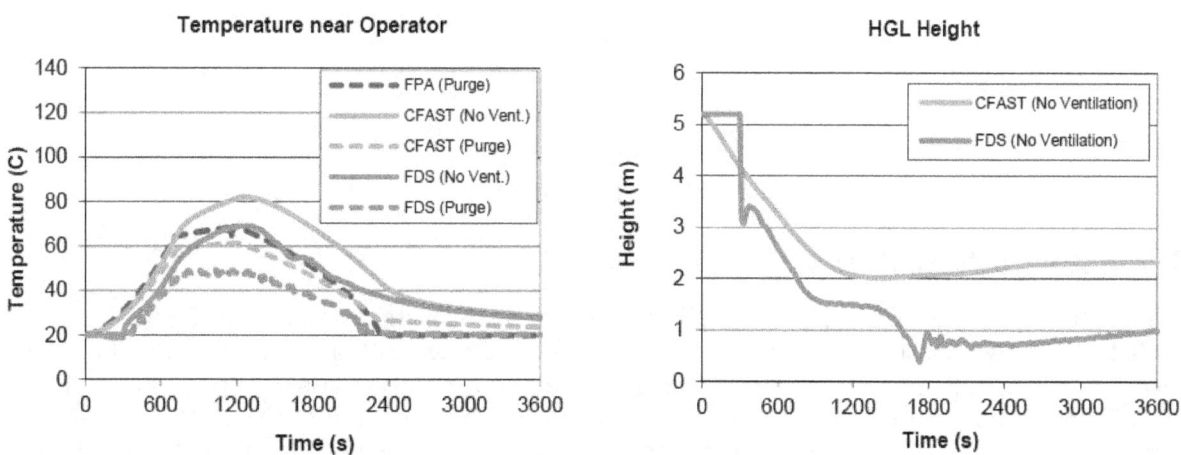

Figure A-12. Hot Gas Layer Temperature and Height for the MCR scenario.

A.5.2 Heat Flux Criterion

In the fire scenario that includes the operation of the smoke purge system, none of the models predict that the heat flux to the operator exceeds the tenability criterion (see Figure A-13). In fact, CFAST and FDS estimate a peak flux of approximately 0.1 kW/m^2, a value that is one-tenth the critical value. With the smoke purge system turned off, FDS predicts a peak heat flux of

---

[16] The bias factor for the CFAST predictions of HGL temperature is 1.06. Taking this into account, the CFAST prediction of HGL temperature for the 35% radiative fraction case would be 90 °C (194 °F) rather than 95 °C (203 °F). This is close enough to the critical value to warrant further modeling.

0.4 kW/m^2 and CFAST predicts 0.6 kW/m^2. However, as in the case of the HGL temperature criterion, it is important to consider the ramifications of the decision to use a radiative fraction of 53% rather than a value more typical of most fires, 35%. Table 4-3 suggests that the heat flux is proportional to the HRR to the four-thirds power. If the models were to use a radiative fraction of 35% rather than 53%, the convective HRR would be 38% greater, in which case the heat flux from the HGL layer onto the operator could increase by as much as 4/3 times 38%, or 50%. Referring to Figure A-13, this would have the effect of increasing the CFAST prediction from 0.6 kW/m^2 to 0.9 kW/m^2, close to the critical value of 1 kW/m^2. In fact, the validation study documented in NUREG-1824 (EPRI 1011999) indicates that CFAST tends to underpredict the total heat flux by 19%, on average. Given this fact and the discussion above on the HGL temperature, it should be noted that in the unventilated case, CFAST predicts that the HGL would descend to a level comparable in height to the operator (approximately 2 m), and it is reasonable to conclude that the operator would be exposed to a heat flux comparable to the habitability threshold.

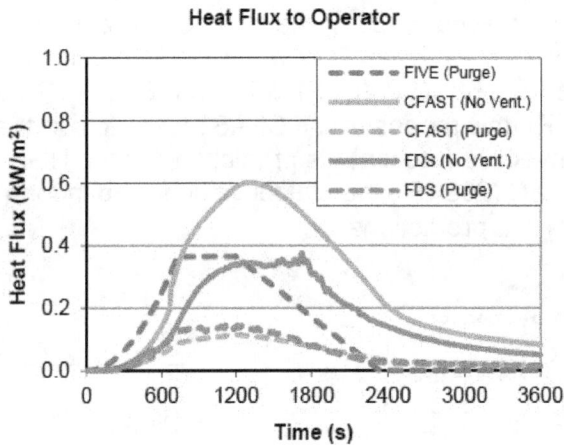

Figure A-13. Predicted heat flux at the location of the operator.

A.5.3 Visibility Criterion

The optical density results are shown in Figure A-14. As with temperature, the CFAST prediction is based on its upper layer smoke concentration calculation, whereas that of FDS is based on the actual operator location. The simple algebraic techniques described in A.4.1 and CFAST both predict that the optical density will exceed the critical threshold, even when the purge system is on. FDS, however, predicts a much lower optical density in the purge mode scenario for two reasons. First, FDS does not limit the transport of smoke to a descending layer like CFAST; and, second, FDS does not uniformly mix the smoke over the entire compartment volume like the simple algebraic model. As the operator stands relatively close to two supply vents, the supplied fresh air keeps this vicinity clearer than other areas.

When the smoke purge system is off, FDS predicts that the visibility tenability criterion will be exceeded at the operator position in about 12 min. Such conditions would force abandonment of the MCR. CFAST predicts that the visibility tenability criterion would be exceeded in the smoke layer in approximately 7 min, but it also predicts that the smoke layer remains above the operator's head throughout the fire simulation, which suggests that the MCR would not need to be abandoned. The fact that the HGL remains above 2 m (6.6 ft) is partially an artifact of the zone model. There is no mechanism in CFAST for the smoke layer to descend below the base of the fire; a fire with a lower base height could result in a lower HGL elevation.

There is considerable uncertainty in the smoke yield of real fires, especially in cases where the fire might be under-ventilated inside of a cabinet. A value of 0.175 (kg soot per kg fuel consumed) was chosen for the smoke yield in the models, even though literature values range from 0.01 to 0.2 (kg soot per kg fuel consumed), depending on the fuel. In addition to the uncertainty in the specified input value of the smoke yield, the NRC/EPRI V&V study (NUREG-1824 (EPRI 1011999)) indicates that both CFAST and FDS overestimate measured smoke concentrations, on average, by factors of 2.65 and 2.70, respectively. In light of these uncertainties in both models and in the input parameters, it is prudent to consider the sensitivity of the simulation results to the selected value of the smoke yield. Table 4-3 indicates that the optical density is directly proportional to the smoke yield. This means that if the smoke yield is doubled, the predicted optical density is doubled as well. The curves in Figure A-14 can easily be adjusted to show the effect of a variation in the smoke yield, but the predicted abandonment times in the unventilated scenario do not change significantly with changes in the smoke yield.

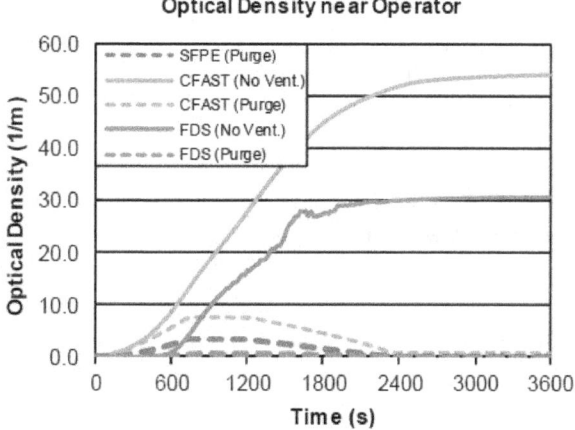

Figure A-14. Optical density predictions for the MCR scenario.

A.6 Conclusion

A fire modeling analysis has been performed to assess the habitability of the MCR in the event of a fire within an isolated electrical cabinet. The fire is not expected to spread to other cabinets. Of the three MCR abandonment criteria, it is most likely that the operators would be forced to abandon the MCR because the optical density would surpass 3 m^{-1} approximately 12 minutes after the fire ignites if the smoke purge system is not activated before this time, according to the FDS analysis. A simple analytical method and the zone model CFAST indicate that the optical density would exceed the critical value with the smoke purge system on and with the ventilation system turned off. However, these analyses are based on the use of several important conservative parameters. For the smoke purge case, the analytical method predicts that the smoke fills the entire compartment uniformly, even though the FDS analysis shows that the supply vents maintain visibility in the vicinity of the operator location. CFAST reports the optical density of the upper layer, but does not predict that the upper layer would descend to the level of the operator in either the purge or no-ventilation scenario based on the conservative specifications, at least for a fire having a base height of 2 m (6.6 ft).

A.7 References

1. Mulholland, G.W., and C. Croarkin, "Specific Extinction Coefficient of Flame Generated Smoke," *Fire and Materials*, 24:227–230, 2000.
2. NIST SP 1018-5, *Fire Dynamics Simulator (Version 5), Technical Reference Guide, Volume 3, Experimental Validation*, 2010.
3. NIST SP 1030. *CFAST: An Engineering Tool for Estimating Fire Growth and Smoke Transport, Version 5 - Technical Reference Guide*, National Institute of Standards and Technology, Gaithersburg, Maryland, 2004.
4. NUREG-1805, *Fire Dynamics Tools (FDT^s) Quantitative Fire Hazard Analysis Methods for the U.S. Nuclear Regulatory Commission Fire Protection Inspection Program*, 2004.
5. NUREG-1824 (EPRI 1011999), *Verification and Validation of Selected Fire Models for Nuclear Power Plant Applications*, 2007.
6. *NUREG/CR-6850 (EPRI 1011989), EPRI/NRC-RES Fire PRA Methodology for Nuclear Power Facilities*, 2005.
7. *SFPE Handbook of Fire Protection Engineering*, 4th edition, 2008.

A.8 Attachments (on CD)

1. FDS input files:

    a. Main_Control_Room_No_Purge.fds
    b. Main_Control_Room_Purge.fds

2. CFAST input files:

    a. Cabinet fire in MCR No Ventilation.in
    b. Cabinet fire in MCR.in

3. Algebraic calculation input files:

    a. FPA_AppA.xlsx

APPENDIX B
CABINET FIRE IN SWITCHGEAR ROOM

B.1 Modeling Objective

The calculations described in this appendix estimate the effects of fire in a switchgear room electrical cabinet on overhead cables and other nearby cabinets (targets). These calculations are part of a larger fire analysis described in Chapter 11 of NUREG/CR-6850 (EPRI 1011989), Volume 2, "Detailed Fire Modeling (Task 11)." The switchgear room (SWGR) contains both Train A and Train B safety-related equipment that is not separated as required by 10 CFR 50, Appendix R. The lack of separation between the two has been identified as an unanalyzed condition. The purpose of the calculation is to analyze this condition and determine whether these targets fail, and, if so, at what time failure occurs.

B.2 Description of the Fire Scenario

General Description: The 4160 V SWGR is located in the auxiliary building and contains three banks of cabinets (labeled A, B, and C in Figure B-1). Cabinet banks A and C contain safety-related equipment. In addition to the cabinets, there are nine cable trays, with three stacks of three trays each located, directly above each of the cabinet banks. The lower two trays above the middle bank of cabinets contain control cables for safety-related equipment. The compartment is not normally occupied. Ambient temperature in the room is 20 °C (68 °F).

Construction: Plan and section views of the SWGR are shown in Figure B-1. The compartment floor, ceiling, and walls are concrete, nominally 0.5 m (1.6 ft) thick. The cabinets, ventilation ducts, and cable trays are made of steel. The cabinet housings are 1.5 mm (0.06 in) thick. The ducts are 2 mm (0.08 in) thick. The cable tray steel is 3 mm (0.12 in) thick. The trays are stacked 0.5 m (20 in) apart, bottom to bottom. The tray width is 0.8 m (31 in).

Materials: The thermal properties of the concrete and steel are listed in Table 3-1. The cable trays are filled with PE-insulated, PVC-jacketed control cables, which have a diameter of approximately 15 mm (0.6 in), a jacket thickness of approximately 1.5 mm (0.06 in), and seven conductors. The cable mass per unit length is 0.38 kg/m (0.25 lb/ft), and the mass fraction of copper is 0.68. The cables are contained in nine stacked cable trays. There are 50 cables in each tray.

Thermoplastic (TP) cables are considered damaged when their interior[17] temperature reaches 205 °C (400 °F) or the exposure heat flux reaches 6 kW/m^2 (NUREG-1805, Appendix A). The damage criteria for the adjacent cabinet is taken to be equal to that for PE/PVC cable since the cables inside the cabinet are unqualified.

Ventilation: There are three supply and three return vents located near the side walls, as shown in Figure B-1. Each vent is 0.5 m (1.6 ft) by 0.6 m (2.0 ft), and each has an airflow rate of 0.47 m^3/s (1000 cfm). The mechanical ventilation is normally on, and normal operations continue during the fire. The supply air to the compartment is equally distributed among the supply vents, and the return air is drawn equally from the returns. The compartment has only

---

[17] The cable interior temperature is taken to be that of the insulation material surrounding the conductors.

one door, which is normally closed. The room temperature is maintained at 20 °C (68 °F), and the pressure is comparable to adjacent compartments. Leakage from the compartment occurs via a 2.5 cm (1 in) high crack under the 0.91 m (3 ft) wide door indicated on the drawing. All other penetrations are sealed.

Fire: The fire ignites in one cabinet in the middle bank, as specified in Figure B-1. The cabinet door is closed, but there are vents on the top of the cabinet for air circulation. The cabinet contains more than one bundle of unqualified cable. The fire grows following a "t-squared" curve to a maximum value of 464 kW in 12 min and remains steady for eight additional min, consistent with NUREG/CR-6850 (EPRI 1011989), page G-5, for a cabinet with more than one bundle of unqualified cable. After 20 min, the fire heat release rate (HRR) decays linearly to zero in 12 min. A peak fire intensity of 464 kW represents the 98th percentile of the probability distribution for HRRs in cabinets with unqualified cable in scenarios where flames propagate through cable bundles. From a cabinet configuration perspective, this selection is appropriate for control cables where cable loading is typically higher than in other types of cabinets. From an applications perspective, the use of the 98th percentile is consistent with the guidance provided in NUREG/CR-6850 (EPRI 1011989) for evaluating fire conditions with different fire intensities (including the 98th percentile) within the probability distribution range.

There is an air vent on the top the cabinet. The air vent is 0.6 m (2 ft) wide and 0.3 m (1 ft) long. The cabinet is 2.4 m (8 ft) tall. Consistent with NUREG/CR-6850 (EPRI 1011989), the fire burns within the interior of the cabinet, and the smoke, heat, and possibly flames exhaust from the air vent at the top of the cabinet.

In this scenario, the cables above Cabinet B are not only targets, they are also "intervening combustibles" that may add to the overall HRR of the fire. Predicting the ignition and growth of a cable fire is challenging for all the models, which is why empirical models are used instead. In this case, the 464 kW cabinet fire described above is supplemented by additional heat from the cable fire. In this scenario, the cables are ignited by the cabinet fire, after which a relatively simple model for predicting the growth and spread of a fire within a vertical stack of horizontal cable trays is applied. The model is referred to as FLASH-CAT, short for Flame Spread over Horizontal Cable Trays. This scenario follows Appendix R of NUREG/CR-6850 (EPRI 1011989), with some additional information provided by the small- and intermediate-scale experiments described in NUREG/CR-7010. The FLASH-CAT model makes use of the following information:

- The cable trays are horizontal and stacked vertically.
- There are no barriers separating the trays, and the tray tops and bottoms are open.
- The cables are not protected with coatings, shielding, or thermal blankets.
- There is a fire beneath the lowest tray.
- Each tray has at least a single row of cables.

In this scenario, the fire propagates upward through the array of cable trays according to an empirically determined timing sequence. First, ignition of the cables in the lowest tray occurs when the internal temperature of a target cable within that tray reaches the failure temperature of 205 °C (400 °F). This analysis is based on guidance provided in NUREG/CR-6850 (EPRI 1011989), Appendix R. The calculation of the cable's internal temperature is based on the Thermally-Induced Electrical Failure (THIEF) methodology (NUREG/CR-6931, Volume 3). Following ignition, the cables in the lowest tray burn at a rate of 250 kW/m^2, a value appropriate for TP cables (NUREG/CR-7010, Volume 1). The width of the burning cable is the same as the width of the trays (0.8 m). The lateral extent of burning cable in the lowest tray, before the onset of lateral spread, is equal to that of the ignition source. The lateral extent of the burning cable in upper trays, before the onset of lateral spread, is given by the formula:

$$L_i = L_{i-1} + 2\,h_i \tan(35°) \tag{B-1}$$

where L_i is the length of tray i and h_i is the distance (bottom to bottom) between tray $i-1$ and tray i. The 35° upward spread angle is described in NUREG/CR-6850 (EPRI 1011989), Appendix R. In this example, L_1 is equal to the length of the vent in the cabinet (0.6 m (2 ft)) and h_i is the distance between trays (bottom to bottom).

Following ignition, the fire in the first tray spreads laterally at a rate of 3.2 m/h (NUREG/CR-6850 (EPRI 1011989), Appendix R). The fire in the second tray ignites 4 min after the first, and the lateral extent of the initial fire in the second tray is widened based on Equation (B-1) to 1.3 m. The burning and spread rates of the fire in the second tray are the same as the first. The fire in the third tray ignites 3 min after the fire in the second, and the initial lateral extent of the fire is widened yet again following the 35° spread angle to 2.0 m.

Local burnout of the fire occurs when the cable plastic is consumed. The time to burnout is calculated as follows. First, determine the combustible mass per unit area of tray, m_c'':

$$m_c'' = \frac{n\,Y_p\,(1-v)m'}{W} = \frac{50 \times 0.32 \times (1-0) \times 0.38\ \text{kg/m}}{0.8\ \text{m}} = 7.6\ \text{kg/m}^2 \tag{B-2}$$

where n is the number of cables per tray, Y_p is the mass fraction of combustible (i.e., non-metallic or plastic) material in the cable, v is the residue yield, m' is the total mass per unit length of a single cable, and W is the tray width. Next, calculate the burnout time, Δt:

$$\Delta t = \frac{m_c''\,\Delta H}{5\,\dot{q}_{\text{avg}}''/6} = \frac{7.6\ \text{kg/m}^2\ \times 20{,}900\ \text{kJ/kg}}{5/6 \times 250\ \text{kW/m}^2} \cong 762\ \text{s} \tag{B-3}$$

where ΔH is the heat of combustion[18], and $\dot{q}_{\text{avg}}''$ is the average HRR per unit area of tray. The FLASH-CAT model asserts that the HRR per unit area ramps linearly to its average value over a time period of $\Delta t/6$, remains steady for a time period of $2\,\Delta t/3$, and then decreases linearly to zero over a time period of $\Delta t/6$. The linear ramp-up and ramp-down are typical ways of approximating the time history of an item's HRR. Further details of the FLASH-CAT model are provided in NUREG/CR-7010, Volume 1.

The heat of combustion and product yields for PE/PVC cables are taken from Table 3-4.16 of the *SFPE Handbook, 4th edition*. Note that five different types of PE/PVC cables are listed in the chapter. The values listed in Table B-1 are for a cable with relatively high soot and CO yields, typical of an under-ventilated fire burning within a closed cabinet. Note also that the non-metallic components of the cables are a mixture of PE (C_2H_4) and PVC (C_2H_3Cl). Because the mixture consists of approximately the same mass of each, the cable materials with an effective chemical formula of $C_2H_{3.5}Cl_{0.5}$ have been selected. Table B-1 summarizes the fuel and combustion parameters for this scenario.

---

[18] By default, the FLASH-CAT model uses a heat of combustion of 16,000 kJ/kg unless there is experimental data that is more appropriate. In this case, the chosen heat of combustion is based on a measurement of PE/PVC cable.

CABINET FIRE IN SWITCHGEAR ROOM

Table B-1. Products of combustion for switchgear room cabinet and cable fire.

Parameter	Value	Source
Effective Fuel Formula	$C_2H_{3.5}Cl_{0.5}$	Combination of polyethylene and PVC
Peak HRR	464 kW	NUREG/CR-6850 (EPRI 1011989), App. G
Heat of Combustion	20,900 kJ/kg	*SFPE Handbook,* 4th Ed., Table 3-4.16
CO_2 Yield	1.29 kg/kg	*SFPE Handbook,* 4th Ed., Table 3-4.16
Soot Yield	0.136 kg/kg	*SFPE Handbook,* 4th Ed., Table 3-4.16
CO Yield	0.147 kg/kg	*SFPE Handbook,* 4th Ed., Table 3-4.16
Radiative Fraction	0.49	*SFPE Handbook,* 4th Ed., Table 3-4.16

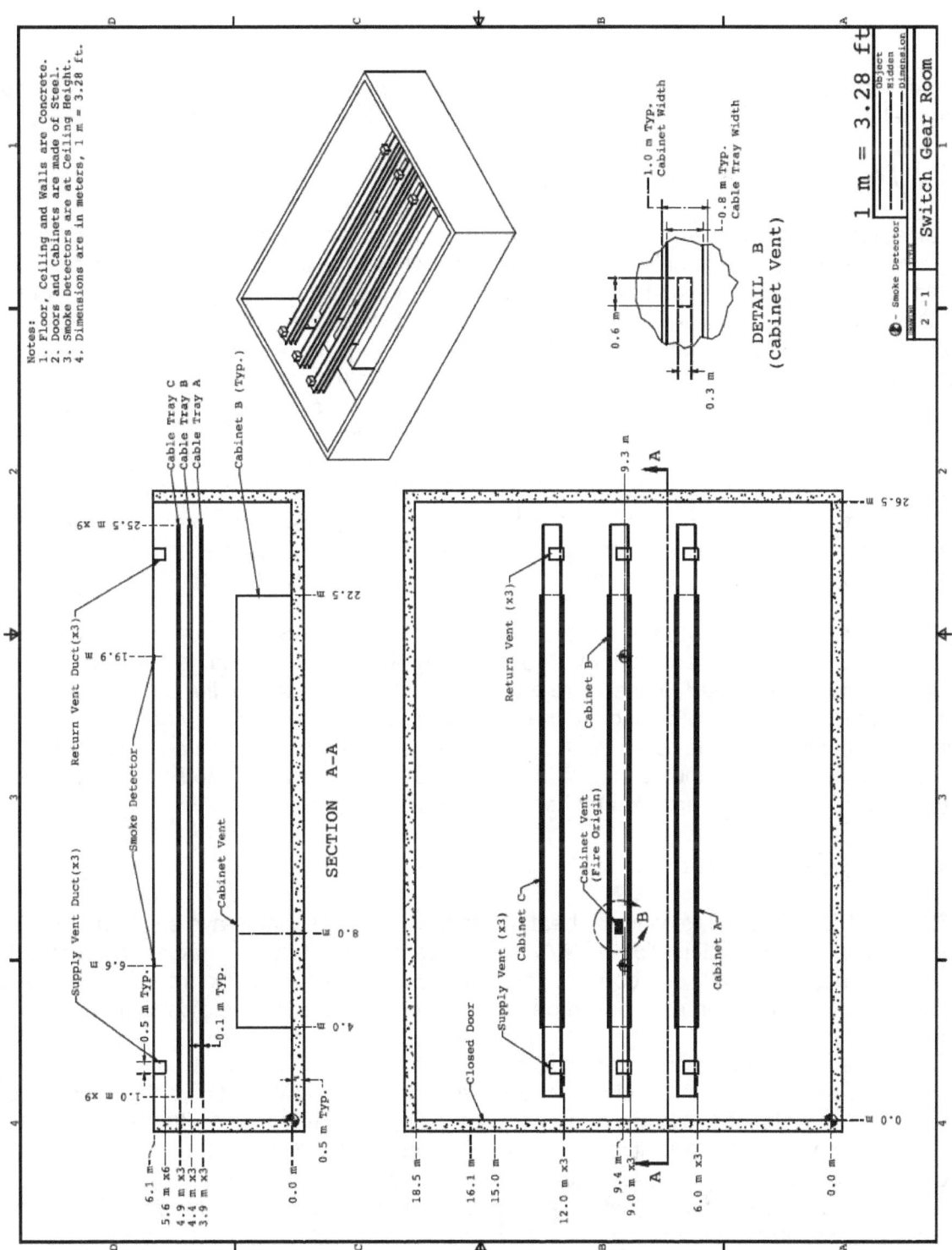

Figure B-1. Geometry of the switchgear room.

B.3 Selection and Evaluation of Fire Models

This section discusses the overall modeling strategy. In particular, it describes the process of model selection, including a discussion of the validity of these models for the given fire scenario.

B.3.1 Temperature Criterion

The algebraic models are used in this scenario to estimate the flame height and plume temperatures from the cabinet fire, first to determine whether the overhead cable trays would be damaged by the cabinet fire and then to calculate the average hot gas layer (HGL) temperature using the Foote, Pagni, and Alvares (FPA) correlation to determine whether cables in the adjacent cabinets would be damaged by the cabinet fire only. Note that the Fire Dynamics Tools (FDT^s) do not allow the HRR to be input as a function of time. With a constant HRR, the FDT^s are use a conservative approximation of an instantaneous, fully developed fire that remains at peak HRR for the duration of the fire scenario. Although this approach would be more likely to estimate a conservative shorter time to failure than tools that utilize a time-dependent HRR, it is useful as a screening tool to determine whether failure is likely to occur at any time. The FPA correlation is used to estimate the average HGL temperature that would result from the initial cabinet fire alone; this analysis using the FPA correlation does not consider the potential ignition of cable trays, or their potential contribution to the fire's HRR.

The Consolidated Fire Growth and Smoke Transport Model (CFAST) and the Fire Dynamics Simulator (FDS) estimate damage to and ignition time of the overhead cable trays, and also use the development of elevated temperature and heat flux from the cabinet fire and any ignited overhead cables to estimate damage to the adjacent cabinets. CFAST provides time-dependent conditions for the cable trays and cabinet, and uses a point source for radiation from the fire sources. FDS distributes the fire in a more realistic way, which should provide more accurate estimates of temperature and heat flux to adjacent cabinets.

B.3.2 Heat Flux Criterion

CFAST and FDS are used to estimate heat flux to the overhead cable trays and to the adjacent cabinets. The heat flux calculation in CFAST includes the contribution of radiation from the fire, upper and lower gas layers, and bounding surfaces, as well as convection from nearby gases. Radiation from the fire sources is based on a point source for the radiation calculation.

The relative position of the cabinet fire and cable trays may provide a challenge because the algorithms used by the zone models to assess target damage are based on a fire radiation point source. FDS models the fire in much the same way as the zone models, with the fire on top of the cabinet. However, because it is a CFD model, FDS can estimate local conditions at the specific location of the target cables and adjacent cabinet. Thus, instead of locating the fire at a single point, FDS more naturally distributes the fire amongst the trays, and the resulting heat flux calculation is more realistic.

B.3.3 Validation

NUREG-1824 (EPRI 1011999) contains experimental validation results for CFAST and FDS that are appropriate for this scenario. The National Institute of Standards and Technology (NIST) has expanded the U.S. Nuclear Regulatory Commission/Electric Power Research Institute

(NRC/EPRI) V&V to include the latest versions of CFAST (6.1.1) (Peacock, 2008) and FDS (5.5.3) (McGrattan, 2010). In particular, the International Collaborative Fire Model Project (ICFMP) Benchmark Exercise #3 test series was designed specifically as a mock-up of a real SWGR. These experiments include ventilation effects on, heat fluxes to, and temperatures of various targets, particularly cables. Fire sizes in these experiments bound those used in this scenario. Also, the cable failure and cable fire spread algorithms, THIEF and FLASH-CAT, are developed and validated in NUREG/CR-6931 (Vol. 3) and NUREG/CR-7010 (Vol. 1).

Table B-2 below lists the important non-dimensionalized parameters that characterize this fire scenario. With the exception of the Fire Froude Number, all of the scenario parameters fall within the range of the NRC/EPRI fire model validation study (NUREG-1824 (EPRI 1011999)). The Fire Froude Number is essentially a measure of the fire's heat output relative to its base area. In this example, the fire is located at the cabinet's top vent, with the vent opening serving as the area of the base of the fire. This method leads to a higher value of $\dot{Q}^*$ than would be calculated if the fire burns completely outside of the cabinet. Thus, the high value of $\dot{Q}^*$ is the result of this method that will lead to more severe fire conditions than would be expected if the fire were to burn partially within the cabinet. Therefore, the model predictions should be valid for this scenario. Also, the Fire Froude Number is calculated to be within the validation range when the heat release rate for the initial cabinet fire and the burning cables are considered together with an average effective diameter of 1 m (3.2 ft).

Table B-2. Normalized parameters and their ranges of applicability to NUREG-1824 (EPRI 1011999).

Quantity	Normalized Parameter Calculation	NUREG-1824 Validation Range	In Range?
Fire Froude Number	$\dot{Q}^* = \dfrac{\dot{Q}}{\rho_\infty c_p T_\infty D^{2.5}\sqrt{g}}$ $= \dfrac{464\text{ kW}}{(1.2\text{ kg/m}^3)(1.0\text{ kJ/kg/K})(293\text{ K})(0.48^{2.5}\text{ m}^{2.5})\sqrt{9.8\text{ m/s}^2}} \cong 2.6$ $\dot{Q}^* = \dfrac{\dot{Q}}{\rho_\infty c_p T_\infty D^{2.5}\sqrt{g}}$ $= \dfrac{1600\text{ kW}}{(1.2\text{ kg/m}^3)(1.0\text{ kJ/kg/K})(293\text{ K})(1^{2.5}\text{ m}^{2.5})\sqrt{9.8\text{ m/s}^2}} \cong 1.4$	0.4 – 2.4	No
Flame Length, L_f, relative to the Ceiling Height, H_c	$\dfrac{H_f + L_f}{H_c} = \dfrac{2.4\text{ m} + 2.1\text{ m}}{6.1\text{ m}} \cong 0.7$ $L_f = D\left(3.7\,\dot{Q}^{*2/5} - 1.02\right) = 0.48\text{ m }(3.7 \times 2.6^{0.4} - 1.02) \cong 2.1\text{ m}$	0.2 – 1.0	Yes
Ceiling Jet Radial Distance, r_{cj}, relative to the Ceiling Height, H_c	N/A – Ceiling jet targets are not included in simulation.	1.2 – 1.7	N/A
Equivalence Ratio, φ, as an indicator of the Ventilation Rate	$\varphi = \dfrac{\dot{Q}}{\Delta H_{O_2}\,\dot{m}_{O_2}} = \dfrac{1{,}600\text{ kW}}{13{,}100\text{ kJ/kg} \times 0.4\text{ kg/s}} \cong 0.31$ (based on peak fire size) $\dot{m}_{O_2} = 0.23\,\rho_\infty \dot{V} = 0.23 \times 1.2\text{ kg/m}^3 \times 1.4\text{ m}^3/\text{s} \cong 0.4\text{ kg/s}$	0.04 – 0.6	Yes
Compartment Aspect Ratio	$\dfrac{L}{H_c} = \dfrac{26.5\text{ m}}{6.1\text{ m}} \cong 4.3 \qquad \dfrac{W}{H_c} = \dfrac{18.5\text{ m}}{6.1\text{ m}} \cong 3.0$	0.6 – 5.7	Yes
Target Distance, r, relative to the Fire Diameter, D	$\dfrac{r}{D} = \dfrac{1.5\text{ m}}{0.48\text{ m}} \cong 3.1$	2.2 – 5.7	Yes

Notes:

(1) The effective diameter of the fire is determined from the formula, $D = \sqrt{4A/\pi}$, where A is the area of the vent on the cabinet. Fire area varies for the cable fire; an area of 1 m^2 is typical at the peak heat release rate.

(2) The Fire Height, $H_f + L_f$, is the sum of the height of the fire from the floor, plus the fire's flame length.

B.4 Estimation of Fire-Generated Conditions

This section provides details specific to each model.

B.4.1 Algebraic Models

General: The general approach to using the algebraic models for this scenario is to first calculate the flame height of the cabinet fire to determine whether the flame reaches one or more of the overhead cable trays, then to calculate the plume temperatures from the cabinet fire to determine which of the overhead cable trays would be damaged by the cabinet fire, and finally to calculate the average hot gas layer temperature using the FPA correlation to determine whether the cables in the adjacent cable trays would be damaged by the cabinet fire only. The general scenario is depicted schematically in Figure B-2.

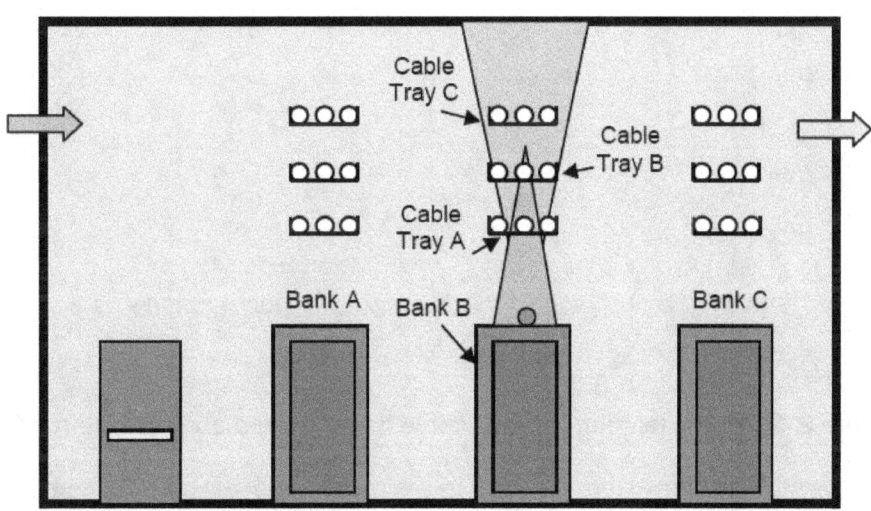

Figure B-2. Schematic diagram of cabinet fire in switchgear room.

The first step in using algebraic models is to determine whether the cables in the cable trays located directly above the cabinet fire are likely to be damaged and potentially ignited by the cabinet fire. As shown in Figure B-1, the top of the cabinet is located at an elevation of 2.4 m (7.9 ft), and the lowest overhead cable tray is located at an elevation of 3.9 m (12.8 ft), which is 1.5 m (4.9 ft) above the top of the cabinet. As shown in Table B-2, the flame length of the cabinet fire is calculated to be 2.1 m (6.9 ft) at the cabinet peak heat release rate of 464 kW, so this empirical correlation for flame length can be used to confirm that the overhead cables would likely be damaged in this scenario. This calculation supports the scenario that the overhead cable trays would be ignited by the cabinet fire, as discussed in Section B.2.

The next step is to calculate the fire plume temperatures that develop from the cabinet fire to determine which of the three cable trays located above the cabinet would be damaged by the cabinet fire. The Heskestad plume temperature correlation included in the FDT[s] and the Fire-Induced Vulnerability Evaluation (FIVE-Rev1) was used to calculate the plume centerline temperature above the cabinet fire. The results of this calculation are shown in Figure B-3. These results show that the plume temperature at all three cable trays would exceed the cable damage threshold temperature of 205 °C (400 °F). However, the Heskestad plume temperature correlation is based on an unobstructed plume. The obstruction caused by the position of the cable trays within the fire plume would alter the actual fire plume entrainment and temperatures. Nonetheless, these results demonstrate that the potential for damage and ignition of the cable trays as a result of the cabinet fire warrants more detailed analysis.

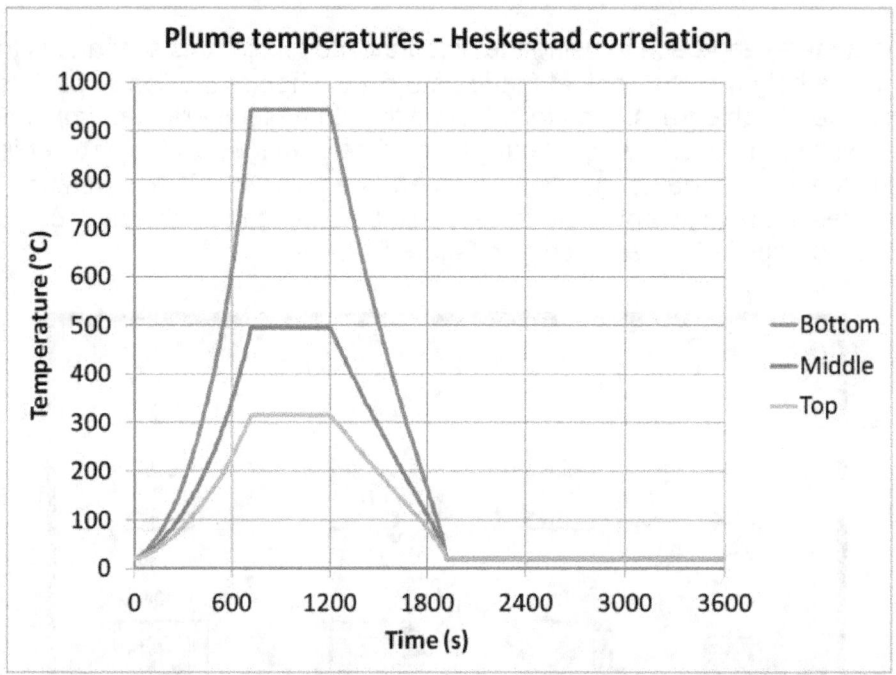

Figure B-3. Plume temperatures at cable trays located above a cabinet fire.

The FPA forced ventilation correlation is used to estimate the average HGL temperature of the SWGR resulting only from the cabinet fire only, based on the parameters described in the following subsections and summarized in Table B-3.

Geometry: The FPA correlation requires room dimensions to be specified in terms of length, width, and height. For this example, the selected compartment is a rectangular parallelepiped, so its length, width, and height are specified directly from dimensions shown in Figure B-1.

Fire: As applied to this scenario, the FPA correlation is used with the time-dependent HRR specified for the cabinet fire only. This HRR history is shown in Figure B-4.

Materials: The walls, ceiling, and floor are all specified as concrete, with the thermal properties specified in Table 3-1.

Ventilation: The ventilation rate of the smoke purge mode is 1.42 m³/s (3,000 cfm). This value is used as a direct input parameter in the FPA correlation.

Table B-3. **Summary of input parameters for FPA analysis of switchgear room scenario.**

Parameter	Value	Source
Room height (H)	5.2 m	Figure B-1
Room length (L)	26.5 m	Figure B-1
Room effective width (W_e)	18.5 m	Calculation
Room boundary material	Concrete	Figure B-1. See Table 3-1 for properties.
Mech. Ventilation rate ($\dot{V}$)	1.42 m³/s	From scenario description
Fire elevation (H_f)	2.4 m	From scenario description of cabinet height and vent location.
Ambient temperature (T_a)	20°C	Specified
Fire parameters	See Table B-1	

Temperature: The FPA HGL temperature correlation for mechanically ventilated spaces is expressed in non-dimensional terms as:

$$\frac{\Delta T_g}{T_\infty} = 0.63 \left(\frac{\dot{Q}}{\dot{m} c_p T_\infty} \right)^{0.72} \left(\frac{h_k A_T}{\dot{m} c_p} \right)^{-0.36} \tag{B-4}$$

ΔT_g is the HGL temperature rise above ambient, T_∞ is the ambient air absolute temperature, $\dot{Q}$ is the HRR of the fire compartment, $\dot{m}$ is the mass ventilation flow rate, c_p is the specific heat of air, h_k is the heat transfer coefficient, and A_T is the total area of the compartment enclosing surfaces. The results for the cabinet fire are shown in Figure B-4, based on the input parameters specified in Table B-3 and the HRR history shown in Figure B-4. These results show that, for the specified parameters, the average HGL temperature reaches a maximum of approximately 65 °C (149 °F) at 20 minutes, based on the peak cabinet fire HRR of 464 kW. These results show that cables in the two adjacent cabinets would not be damaged by the initial cabinet fire alone. However, further analysis is required to determine the potential impact of overhead cable ignition on the potential for damage to cables in the adjacent cabinets. CFAST and FDS are used to perform this more detailed analysis.

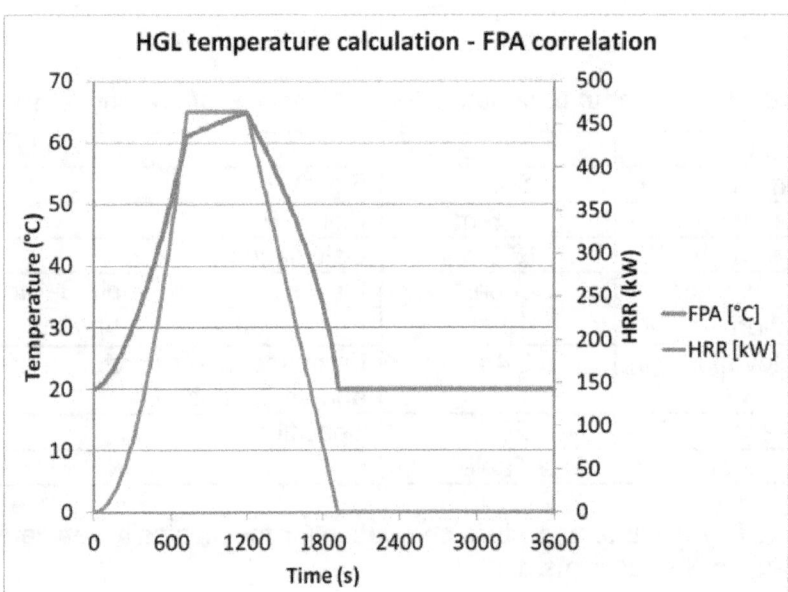

Figure B-4. Average HGL temperature (red line) from FPA correlation and HRR (blue line) for SWGR cabinet fire scenario.

B.4.2 Zone Model

Geometry: The CFAST analysis defines the compartment as a single rectangular parallelepiped with the specified dimensions. While there are a number of cable trays in the compartment, the compartment is large enough that the cable trays do not occupy a significant fraction of the total volume, so the compartment dimensions are taken directly from the scenario description and Figure B-1. Figure B-5 illustrates the scenario as modeled by CFAST. Figure B-6 shows the CFAST inputs for the geometry.

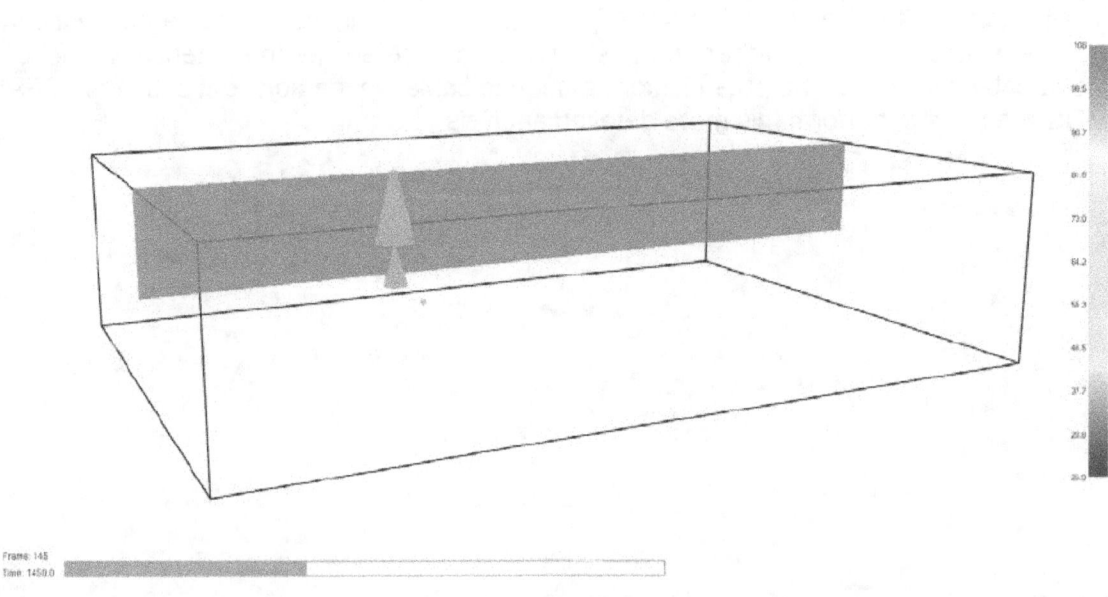

Figure B-5. Average CFAST/Smokeview rendering of SWGR.

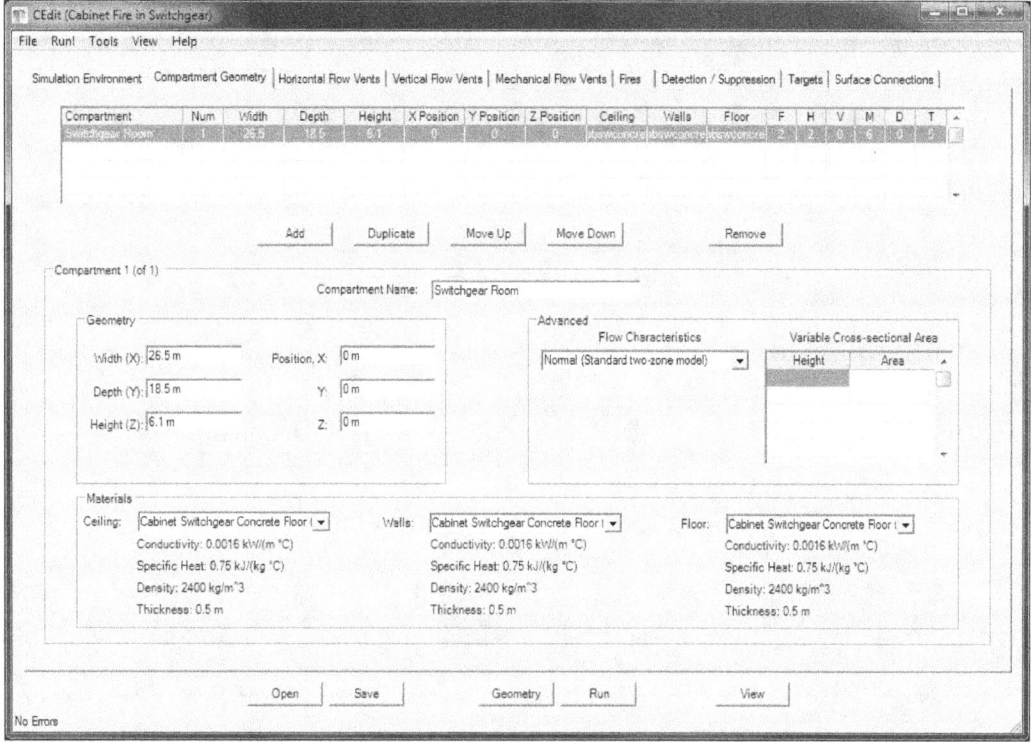

Figure B-6. CFAST inputs for compartment geometry in SWGR scenario.

Fire: CFAST requires a user-specified, time-dependent HRR and stoichiometry for the combustion of fuel and oxygen. The HRR is the combined cabinet/cable fire described above. Figure B-7 shows the CFAST inputs for the fire taken directly from Table B-1.

CABINET FIRE IN SWITCHGEAR ROOM

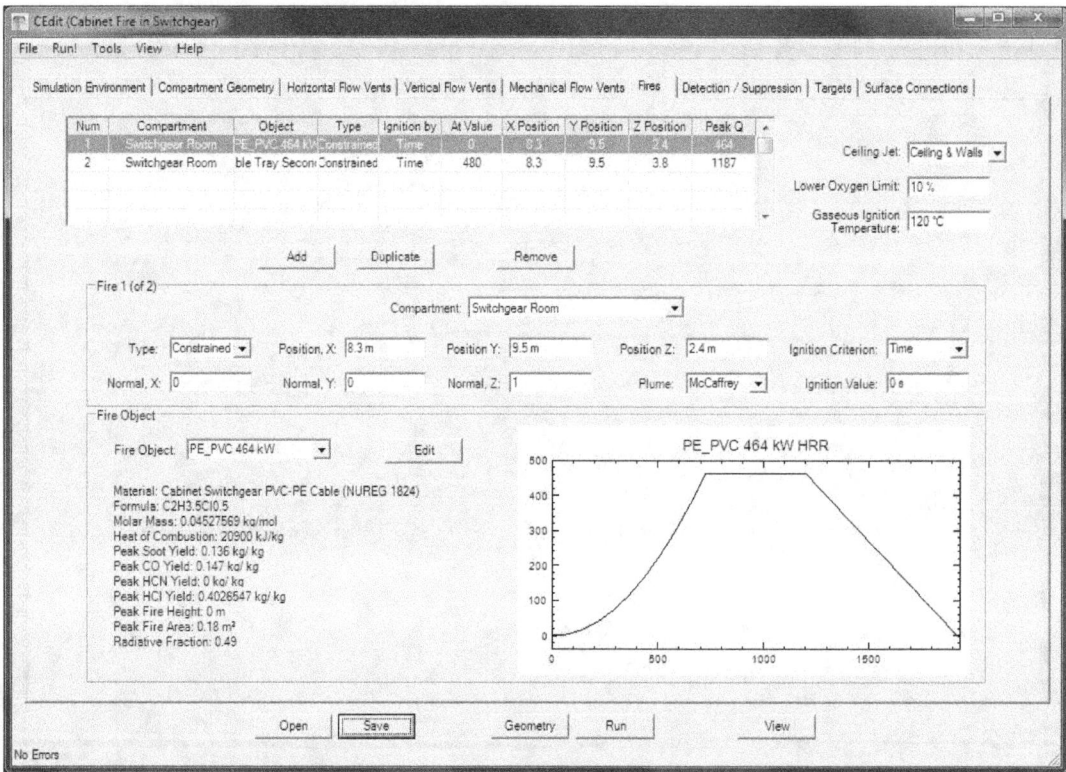

Note: Values for "Lower Oxygen Limit" and "Gaseous Ignition Temperature are set at default values.

Figure B-7. CFAST fire specification inputs for the SWGR scenario.

Materials: The walls, floor, and ceiling are specified as concrete in CFAST with properties as previously described. The target properties are provided directly as input.

Ventilation: Mechanical ventilation and leakage are specified as input to CFAST directly from the scenario description. CFAST uses three inlet and three outlet vents for the mechanical ventilation at the heights specified in the scenario description. Horizontal placement of the mechanical ventilation within the compartment does not affect the zone model calculation and is not part of the input. Figure B-8 shows the CFAST inputs for the mechanical ventilation. Since pressure for this scenario peaks approximately at the lower fan cutoff pressure threshold, default values for these pressures are used as they should not impact the calculations.

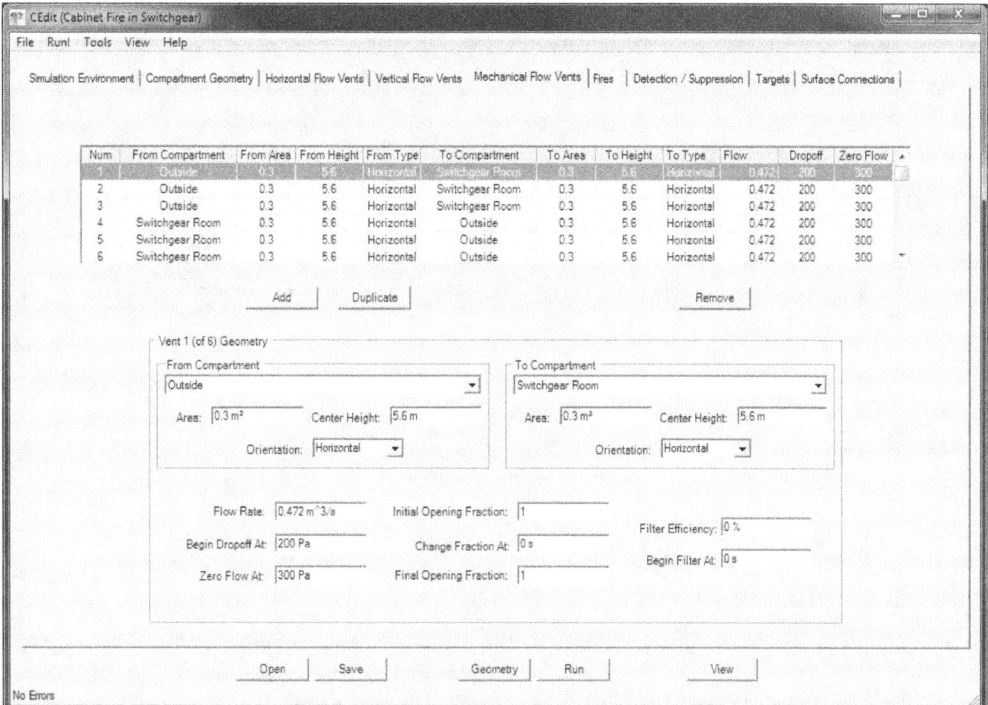

Note: Values for "Begin Dropoff At" and "Zero Flow At" are set at default values.

Figure B-8. CFAST mechanical ventilation inputs for the SWGR scenario.

Cable Targets: In CFAST, target temperatures are calculated with a one-dimensional cylindrical heat transfer calculation based on the material properties and cable diameter, as specified in the scenario description. CFAST uses the THIEF methodology developed as part of the Cable Response to Live Fire (CAROLFIRE) program (NUREG/CR-6931, Vol. 3). Each of the target cables is specified directly in the model. Thermal properties are taken directly from NUREG/CR-6931. Cable density is calculated from the specified mass per unit length and the cross-sectional area of the cable as 0.38 kg/m /$(\pi \times 0.0075$ m$)^2$ = 2150 kg/m^3. Figure B-9 shows the CFAST inputs for targets in the scenario. Ignition time for the target cables was calculated with an initial CFAST run with only the cabinet fire. This run estimated this ignition time (as the time flames reached the secondary cable target) as 480 s.

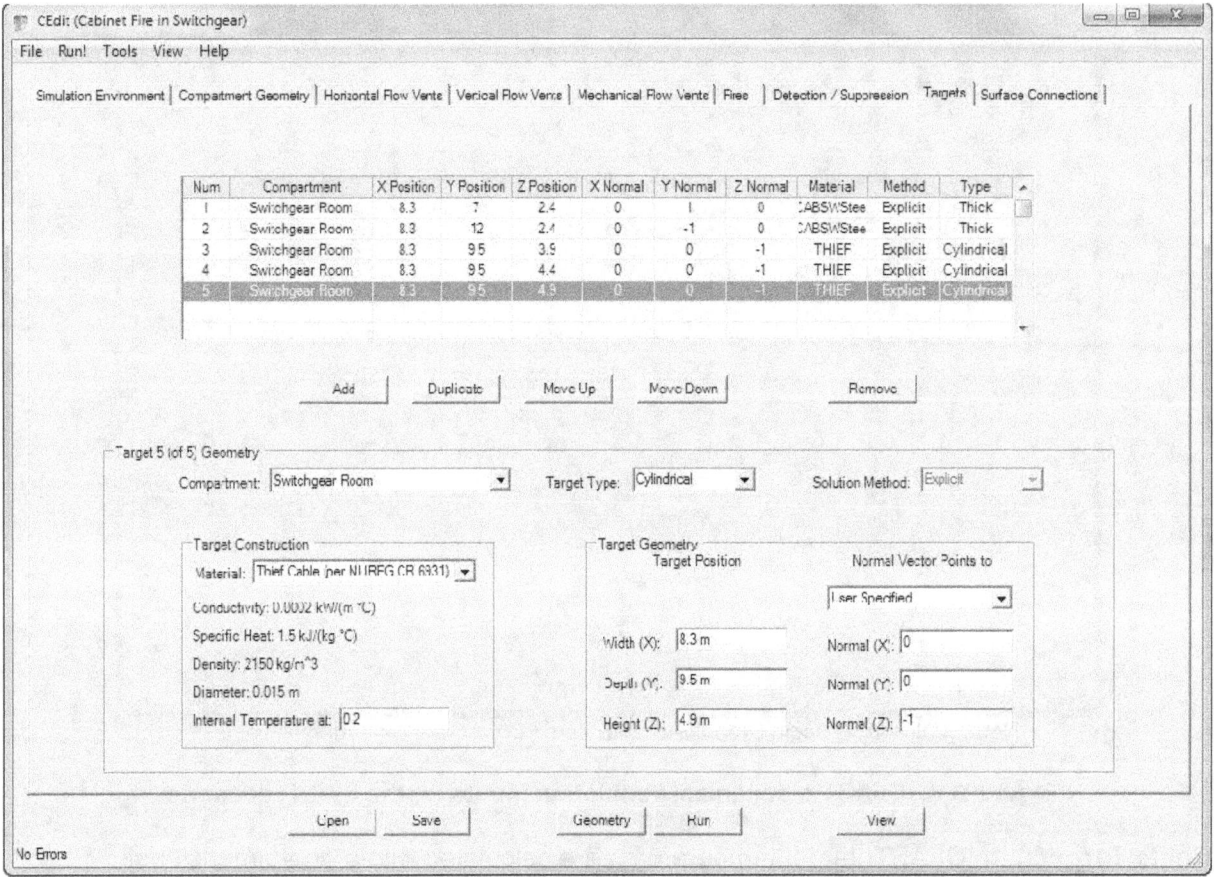

Figure B-9. CFAST inputs for cabinet and cable targets for the SWGR scenario.

B.4.3 CFD Model

Geometry: The compartment has a simple rectangular geometry that coincides with the external boundary of the computational domain. In other words, the exterior walls are not explicitly declared, but are defined by default to be the external boundaries of the domain with the surface properties of concrete, given above. The cabinets are modeled simply as boxes constructed of steel, whose properties are specified above. No attempt is made to model the interior of the cabinets because the fire has been specified as originating at or near the top of one of the cabinets. Figure B-10 shows the compartment geometry used in FDS.

The numerical mesh away from the fire consists of uniform grid cells, approximately 0.2 m (8 in) on a side. In the vicinity of the fire, the mesh cells are approximately 0.1 m (4 in) on a side. This refinement of the mesh is needed to better locate the actual tray locations. Further refinement of the mesh is unnecessary because the fire's growth and spread are to be specified based on the empirical model FLASH-CAT.

Materials: The material properties are applied directly as specified to the walls, floor, ceiling, and cabinet. The cabinets adjacent to the center bank are modeled as hollow steel boxes whose interiors remain at ambient temperature. For the cables, the thermal properties are superseded by the algebraic ignition and flame spread model called FLASH-CAT. In FDS, the model is applied as follows. First, each cable tray is declared an obstruction, the top and bottom of which are given the thermal properties of PVC cables from NUREG/CR-6850 (EPRI 1011989), Volume 2, Appendix R.4.1.1 (k=0.192 W/m/K, ρ=1,380 kg/m^3, c=1.289 kJ/kg/K). These properties are not particularly important because the cable failure, ignition, and fire spread are not dependent on them, but merely provide approximate thermal boundary conditions for those cables that are not burning. Next, the top surface of each tray is divided into three sections. The middle section ignites and burns according to the timing sequence of the FLASH-CAT model. The two adjacent sections ignite at the same time, but only at the edge abutting the middle section. The fire then spreads laterally at a rate specified by the FLASH-CAT model, in this case, 0.9 mm/s.

Fire: The initial fire source is modeled as a 0.6 m (2 ft) x 0.3 m (1 ft) "gas burner" atop the central cabinet with the specified HRR. This is meant to represent a fire that burns near the top of the cabinet and exhausts through the vent. The ignition and growth of the cable fire is based on the empirical FLASH-CAT model described above. Figure B-11 shows a snapshot of the burning cable during the simulation.

Ventilation: The door is included in the calculation merely as a surface with different properties than the default concrete wall. The supply and return vents are specified according to the drawing and given volume flow rates. Note that because of the relative coarseness of the underlying numerical grid, the ventilation rate is input directly in terms of the volume flow rate (m^3/s) rather than as a separate vent area (m^2) and velocity (m/s). The model automatically adjusts the dimensions of all objects to conform to the numerical mesh, and it also adjusts the velocity of the air stream to properly reflect the desired volume flow rate.

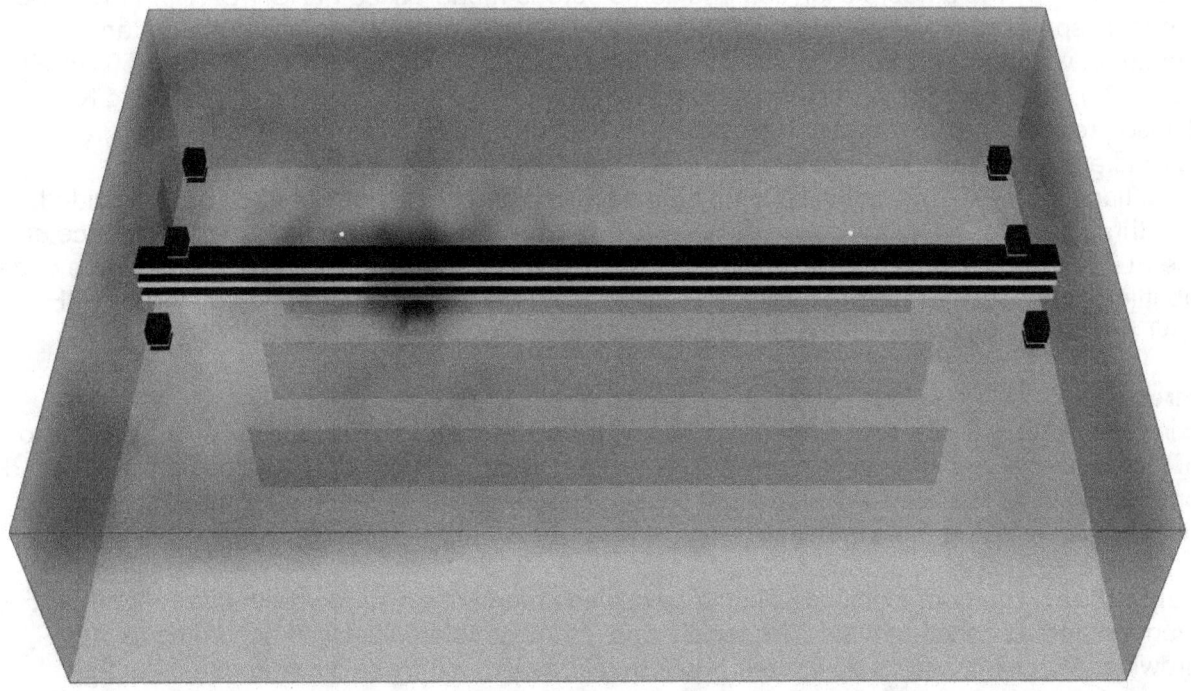

Figure B-10. FDS/Smokeview rendering of the SWGR.

Figure B-11. FDS/Smokeview rendering of the SWGR fire showing localized ignition of extinction of secondary cable fires resulting from initial cabinet fire.

B.5 Evaluation of Results

In this example, the objective of the calculations is to estimate the effects of fire in a cabinet in a SWGR on nearby cable and cabinet targets; that is, to determine whether and when temperatures and/or heat flux to the cable and adjacent cabinets exceed established critical values. Cables are considered damaged when the temperature of the cable reaches 205 °C (400 °F) or the exposure heat flux reaches 6 kW/m^2 (NUREG/CR 6850 (EPRI 1011989), Appendix H, Table H-1). These criteria are intended to be indicative of electrical failure, but, based on experimental observation, are routinely also used as ignition criteria.

Table B-4 summarizes the results of all the models for the chosen failure criteria. For each Predicted Value, a calculation is performed to determine the Probability of Exceeding the Critical Value. The procedure for calculating this probability is given in Chapter 4, and it accounts for the model bias and scatter. The purpose of this table is to highlight the criterion that is most likely to be exceeded, so that further analysis can be focused on this criterion and the model or models that predict it. Each criterion is discussed in greater detail in the following subsections.

Table B-4. Summary of the model predictions of the cabinet fire scenario.

Model	Bias Factor, δ	Standard Deviation, $\tilde{\sigma}_M$	Location	Predicted Value	Critical Value	Probability of Exceeding
Temperature (°C), Initial Value = 20 °C						
CFAST	1.00	0.27	Cable Tray A	335	205	0.937
FDS	1.02	0.13		755	205	1.000
CFAST	1.00	0.27	Cabinet A	168	205	0.177
FDS	1.02	0.13		136	205	0.000
Heat Flux (kW/m^2)						
CFAST	0.81	0.47	Cabinet A	5.3	6	0.576
FDS	0.85	0.22		4.2	6	0.159

B.5.1 Cable Ignition and Damage

The algebraic models cannot be used in this case to accurately assess the damage to cables. FIVE-Rev1 does not have an algorithm that considers the thermal inertia of the cables. FDT^s does, but the model is only applicable when the exposing temperature is constant, which is not the case for this example. However, the algebraic models can be used for screening purposes, to show that the plume temperatures resulting from the specified cabinet fire are high enough to potentially cause damage to and ignition of cables located in cable trays directly above the cabinet.

CFAST and FDS estimate conditions resulting from the ignition (based on initial model runs of the initial cabinet fire) and burning of the cables (based on calculations from the FLASH-CAT model). NUREG/CR-6850 (EPRI 1011989) contains some guidance on modeling cable ignition, flame spread, and the fire's resulting heat release based on a limited set of fire test data. The differences in HRR between the models (Figure B-12) result from variations in the implementation of this guidance. Figure B-12(a) shows the HRR from the initial cabinet fire source only. Figure B-12(b) shows the overall HRR, including the initial cabinet fire source and the cables ignited by this initial fire.

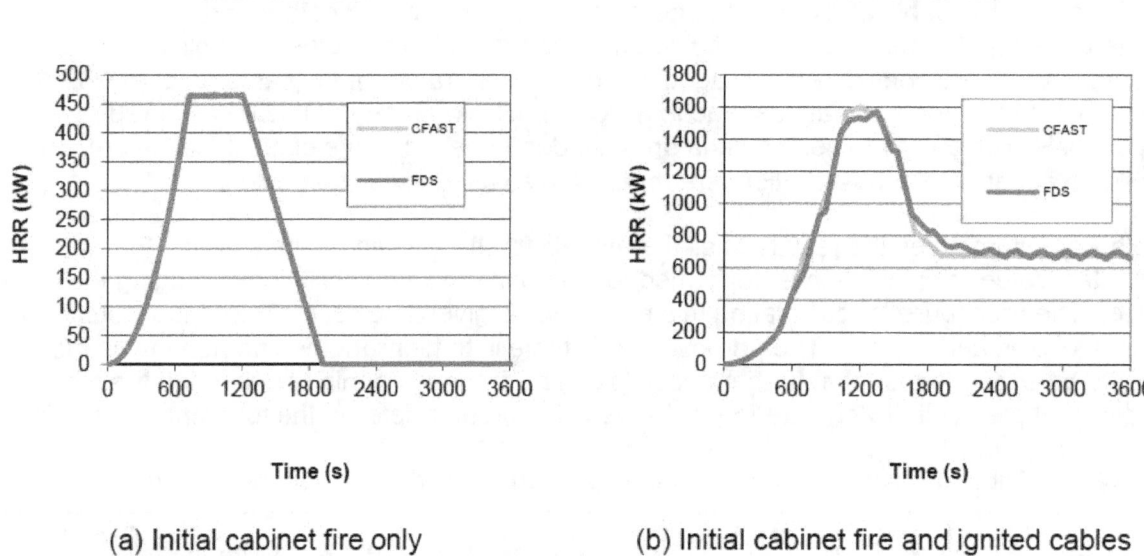

(a) Initial cabinet fire only   (b) Initial cabinet fire and ignited cables

Figure B-12. Heat release rate inputs to CFAST and FDS for a SWGR cabinet fire scenario.

There are several ways to assess the initial fire source's potential to ignite the lowest cable tray (Cable Tray A). Cables are considered damaged when the temperature reaches 205 °C (400 °F) or the exposure heat flux reaches 6 kW/m^2 (NUREG-1805, Appendix A). These criteria are intended to indicate electrical failure, but are routinely also applied as ignition criteria. In newer studies in NUREG/CR-7010, cable ignition was not observed at fluxes below 25 kW/m^2, and most often only with direct flame impingement. Handbook values for minimum ignition flux for power and communication cables are reported in the range of 15 kW/m^2 to 35 kW/m^2 (SFPE Handbook, Table 3-4.2). For this scenario, CFAST predicts that the flame height reaches the cable tray in approximately 490 s, quite similar to the temperature-based prediction. Table B-5 shows the lowest cable tray's estimated time to ignition for a variety of ignition criteria. For this simulation, 490 s was chosen.

Table B-5. Estimated time to ignition of lowest cable tray by CFAST for the SWGR cabinet fire.

Ignition Criterion	Time
Gas temperature ≥ 205 °C	270 s
Cable temperature ≥ 205 °C	860 s
Heat flux ≥ 6 kW/m^2	490 s
Heat flux ≥ 15 kW/m^2	740 s
Flame impingement	490 s

The CFAST and FDS temperature predictions for the Cable Tray A cables are shown in Figure B-13. FDS predicts cable failure in Tray A at about 495 s, CFAST in about 600 s. Peak temperatures from both models are well above the failure criteria for the cables, so it can be expected that the cables will ignite and provide an additional source of fire.

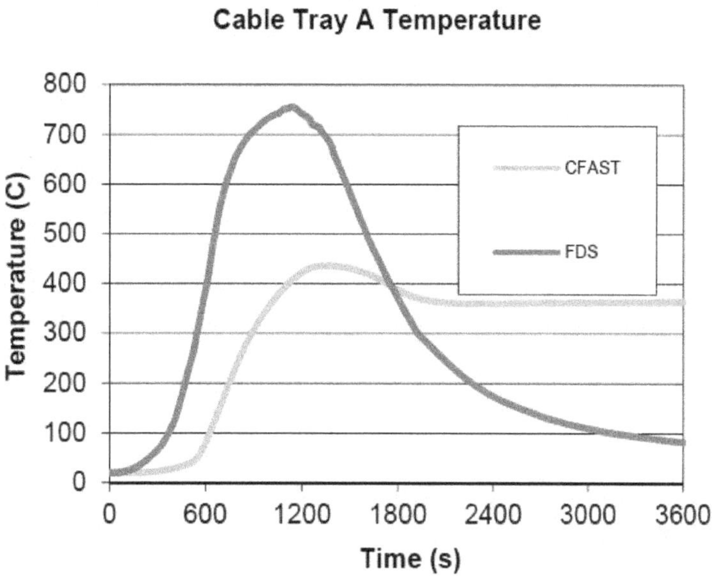

Cable Tray A Temperature

Figure B-13. Estimated temperatures for Cable Tray A directly above the fire source for a SWGR cabinet fire scenario.

Qualitatively, the results of the CFAST and FDS predictions are quite different. The radiation from the fire source in CFAST is calculated based on a point source fire positioned at the base of the fire. Thus, once the fire grows and the flame height approaches the target cable tray, CFAST can be expected to underestimate the local cable temperature and heat flux, since the cable would actually be immersed within the flames. CFAST does include an estimate of the flame height, which can also be used as an indicator of damage to the cable. For this scenario, CFAST predicts that the flame height will reach the cable tray in approximately 490 s, quite similar to the temperature-based prediction. Past this point, CFAST estimates of the local target temperature are expected to be under-predictions. FDS predictions include the impact of direct flame impingement and immersion of the target in flames. Thus, the higher temperatures predicted by FDS are expected.

Upon ignition of the bottom cable tray (Cable Tray A), the higher cable trays are ignited, consistent with the FLASH-CAT model.

B.5.2 Cabinet Damage

To assess potential damage to adjacent cabinets, both the predicted temperatures and heat fluxes are evaluated. Because the two adjacent cabinets are equidistant from the fire and have similar properties, only one is considered here. The critical damage thresholds are the same for these cabinets as for the cables in trays. Figure B-14 shows estimated temperature and heat flux on the cabinet surface.

The algebraic models are not capable of estimating the temperature of a target such as an electrical cabinet, whereas the other models do provide this capability. CFAST and FDS estimate peak temperatures below 165 °C (330 °F), which is well below the threshold of 205 °C

(400 °F). The somewhat higher cabinet temperature and heat flux predicted by CFAST is consistent with the point source radiation calculations for the CFAST simulation. CFAST and FDS both estimate an incident heat flux below about 5 kW/m^2, with the difference caused by the more simple radiation calculation in CFAST.

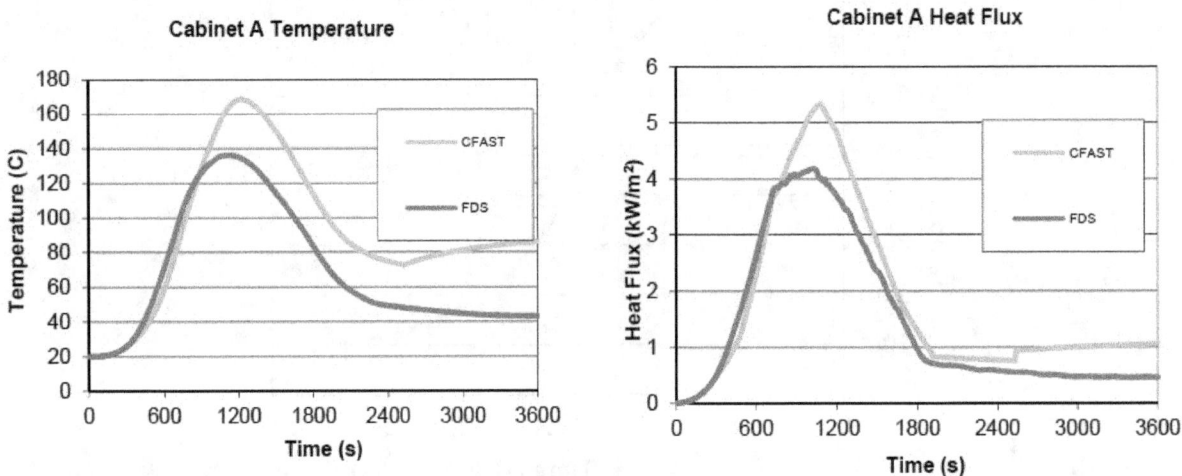

Figure B-14. Estimated temperature and heat flux to a cabinet adjacent to the fire source in a SWGR cabinet fire scenario.

B.5.3 Parameter Uncertainty Propagation

The analysis above has shown that a 98th percentile cabinet fire is very likely to ignite cables in the trays above the cabinet and could potentially damage adjacent cabinets. Within the context of a PRA, the next step is to calculate the probability of cable damage for *any* fire within the cabinet, not just the 98th percentile fire.

Figure B-15 displays the distribution[19] of peak heat release rates for cabinets with more than one bundle of unqualified cable (NUREG/CR-6850, Appendix G). The analysis described above made use of the 98th percentile fire from this distribution, with a peak of 464 kW.

---

[19] NUREG/CR-6850 specifies gamma distributions for the various types of combustibles found within an NPP. Microsoft Excel® provides a built-in function (GAMMA.DIST) that calculates the probability density function given the parameters α and β. In this case, these parameters are 2.6 and 67.8, respectively.

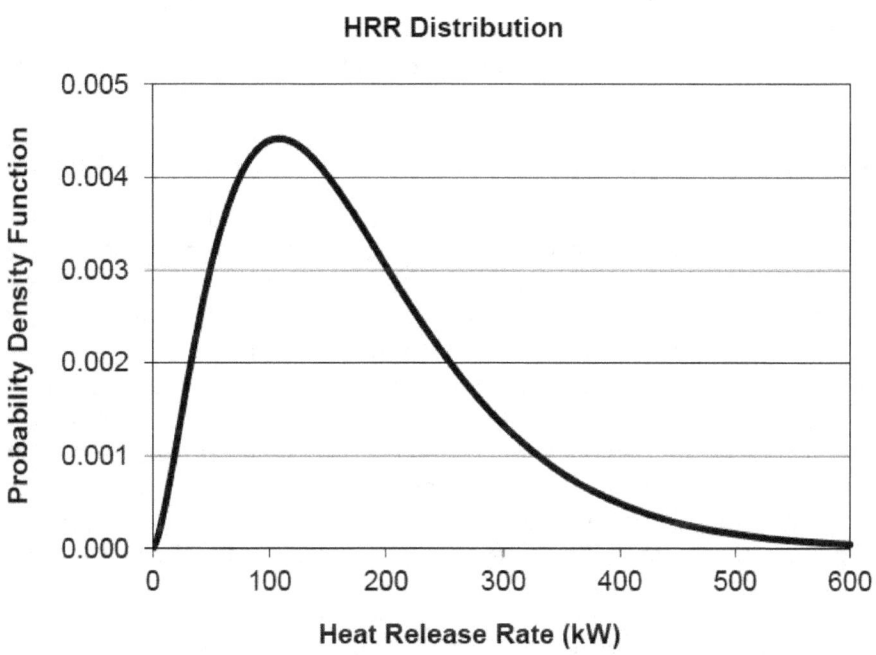

Figure B-15. Distribution of HRR for an electrical cabinet fire.

Applying Heskestad's flame height correlation to the entire range of HRR, now taken as a random variable, leads to a distribution of flame height shown in Figure B-16.

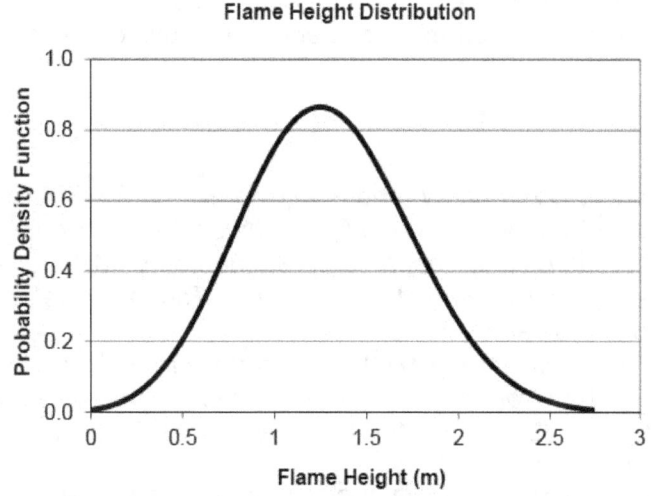

Figure B-16. Distribution of flame heights for the entire range of cabinet fires.

The cable tray is 1.5 m (4.9 ft) above the top of the cabinet. The probability that the flames from a randomly chosen fire will reach the cables is equal to the area beneath the curve in Figure B-16 for flame heights greater than 1.5 m (4.9 ft), or approximately 0.31. Consistent with the guidance in NUREG/CR-6850, this resulting probability can be used as the severity factor for the quantification of corresponding fire ignition frequencies. It is interesting to note that an analysis that uses the 50th percentile fire would lead to the conclusion that the flame from the cabinet fire does not impinge on the cables and that ignition of the cables is not likely to occur. This changes the outcome of the analysis significantly because an analysis that uses only the mean HRR without explicit quantification of the uncertainty would result in an overly optimistic modeling conclusion and an inappropriate evaluation of the scenario within the integrated fire analysis.

B.6 Conclusion

This analysis has considered the potential that a fire in an electrical cabinet in a 4160 V SWGR will damage overhead cables and adjacent electrical cabinets. Algebraic equations from the FDTs and FIVE-Rev1, including the Heskestad flame height correlation and the Heskestad plume temperature correlation, were used for screening purposes, to evaluate the potential for damage as well as to determine whether more detailed analysis with CFAST and FDS was warranted. The algebraic equations demonstrate that the calculated flame height from the cabinet fire would be high enough to potentially ignite the lowest of the three horizontal cable trays located directly above the cabinet fire. They also demonstrate that the calculated fire plume temperatures are high enough at all three horizontal cable trays located directly above the cabinet fire to potentially damage cables in all three trays. As applied in this scenario, the algebraic equations demonstrate that a more detailed analysis with CFAST and FDS is warranted.

The more detailed analyses with CFAST and FDS demonstrate that the cabinet fire is likely to ignite the electrical cables in the lowest cable tray directly above the cabinet fire in approximately 10 min. The additional cable trays directly above the lowest tray would then ignite in turn. In addition, the models indicate that it is possible that the combined cabinet and cable fire could damage adjacent cabinets. These conclusions are based on an analysis of a severe (i.e., 98th percentile) fire. A subsequent analysis involving the propagation of the entire HRR distribution through a simple flame height model indicates that 31 % of all possible fires within the cabinet would ignite the overhead trays and the resulting fire could damage adjacent cabinets.

B.7 References

1. Jones, W., R.D. Peacock, G.P. Forney, and P.A. Reneke, *CFAST - Consolidated Model of Fire Growth and Smoke Transport (Version 6), Technical Reference Guide*, SP 1026, National Institute of Standards and Technology, Gaithersburg, MD, 2009.
2. NIST SP 1018-5, *Fire Dynamics Simulator (Version 5), Technical Reference Guide, Volume 3, Experimental Validation*, 2010.
3. NUREG-1805, *Fire Dynamics Tools (FDT[s]) Quantitative Fire Hazard Analysis Methods for the U.S. Nuclear Regulatory Commission Fire Protection Inspection Program*, 2004.
4. NUREG-1824 (EPRI 1011999), *Verification and Validation of Selected Fire Models for Nuclear Power Plant Applications*, 2007.
5. *NUREG/CR-6850 (EPRI 1011989), EPRI/NRC-RES Fire PRA Methodology for Nuclear Power Facilities*, 2005.
6. *SFPE Handbook of Fire Protection Engineering*, 4th edition, 2008.

B.8 Attachments (on CD)

1. FDS input file: Switchgear_Room_Cabinet.fds

2.   CFAST input files:

    a.   Initial Fire Only.in
    b.   Cabinet Fire in Switchgear.in

3.   Algebraic calculation input files:

    a.   FPA_AppB.xlsx

APPENDIX C
LUBRICATING OIL FIRE IN PUMP COMPARTMENT

C.1 Modeling Objective

The purpose of the calculations described in this appendix is to determine whether important safe-shutdown cables within a pump room will fail in the event of a lubricating oil fire, and, if so, at what time failure occurs. These cables are protected by an electrical raceway fire barrier system (ERFBS), but there is a concern that the existing barrier system will not provide the required protection. The impact of opening a door during the fire is also investigated.

C.2 Description of the Fire Scenario

General Description: The compartment is of fire-resistive construction and contains an emergency core cooling system pump and a single tray containing safe-shutdown cables that are protected by an ERFBS. The pump is surrounded by a dike designed to contain any lubricating oil that may leak or spill, with a maximum capacity of 190 L (50 gal). The compartment contains one smoke detector and one sprinkler. The compartment is mechanically ventilated. The fire occurs when pump oil leaks into the dike area and ignites. Large oil fires are likely to cause flashover conditions. Flashover refers to the rapid transition from the growth period of a fire (pre-flashover) to the fully developed fire (post-flashover). A flashover condition is typically expected when the hot gas layer (HGL) temperature reaches 500 °C (932 °F) or greater. Post-flashover conditions are expected in this scenario.

Geometry: The pump room is relatively small and has only one door. As shown in Figure C-1, the walls are 0.1 m (0.3 ft) thick. The floor and ceiling are 0.9 m (3 ft) thick.

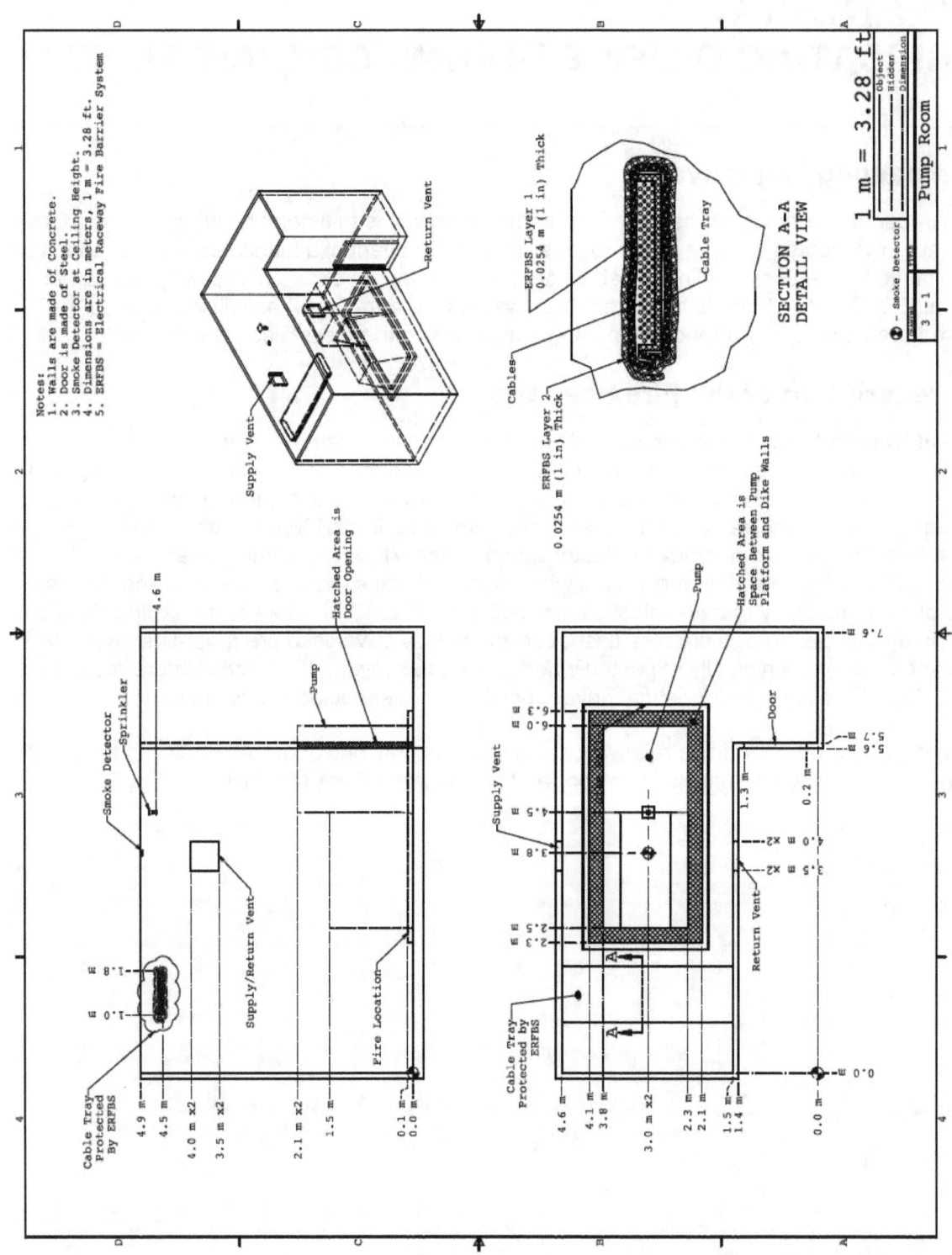

Figure C-1. Geometry of the pump room.

Materials: The walls, ceiling, and floor are all constructed of concrete. Nominal values for the thermal properties of various materials in the compartment are listed in Table 3-1. The single cable tray in this compartment is filled with PE/PVC cables with copper conductors. The properties of the cables are listed in Table C-1. The damage criterion is taken to be the point at which the cable temperature reaches 205 °C (400 °F) (NUREG-1805, Appendix A). The cable tray is protected by an ERFBS, which is two layers of ceramic fiber insulation blankets, covered by 0.0254 mm (1 mil) foil. The properties of the insulation material and cables are listed in Table C-1. Note that the properties of the ceramic fiber material are typically dependent on temperature, and that this material undergoes a series of chemical reactions that are designed to absorb heat and protect the underlying cables. Detailed thermophysical properties for this particular material are not known. However, the ERFBS has undergone a fire endurance furnace test in which the average temperature of the electrical raceway was maintained below 121 °C (250 °F) and the maximum temperature below 163 °C (325 °F) for an hour when exposed to the standard ASTM E 119 temperature curve.

The damage criterion is taken to be when the cable surface temperature reaches 205 °C (400 °C) (NUREG-1805, Appendix A). Using the temperature on the cable surface rather than inside the cable jacket is conservative, but is chosen in this case to allow a comparison of results between MAGIC and FDS at equivalent positions in the two models.

Table C-1. Data for ERFBS and cable insulation.

Material	Parameter	Value*
Ceramic Fiber Insulation	Thickness (2 layers)	5 cm
	Thermal conductivity	0.06 W/m/K
	Density	128 kg/m^3
	Specific heat	1.07 kJ/kg/K
	Emissivity	0.9
Cable	Diameter	15 mm
	Jacket thickness	2 mm
	Insulation/jacket conductivity	0.192 W/m/K
	Insulation/jacket density	1380 kg/m^3
	Insulation/jacket specific heat	1.289 kJ/kg/K
	Mass per unit length	0.4 kg/m
	Conductor mass fractions	33% PE/PVC, 67% copper

*Source: Product literature (ERFBS) and NUREG/CR-6850 (EPRI 1011989), Volume 2, Appendix R (PVC cable insulation).

Fire Protection Systems: As shown in Figure C-1, a smoke detector and a sprinkler are located in the pump room. However, to determine whether the barrier system alone will provide the required protection, the fire detection and suppression systems are not credited in the fire scenario under consideration.

Ventilation: There is one supply and one return air vent, each with an area of 0.25 m^2 (2.7 ft^2), providing a volume flow rate of 0.25 m^3/s (530 cfm). The locations are shown in Figure C-1. The ventilation system continues to operate during the fire, with no changes brought about by fire-related pressure effects. This does not imply that the fire does not impact the ventilation system, but rather that there is typically limited information about the ventilation network that connects to a given compartment. The pump compartment has one door; it is 1.1 m (3.6 ft) wide and 2.1 m (6.9 ft) tall. The door is normally closed, but it is opened 10 min after ignition by the fire brigade. Before the door opens, leakage from the doorway occurs via a 1.3 cm (0.5 in) gap under the door.

Fire: The fire starts following an accidental release of 190 L (50 gal) of lubricating oil. The spill is contained by the dike. Lubricating oil is a mixture of hydrocarbons, mostly alkanes, which have the chemical formula C_nH_{2n+2} (with n ranging from 12 to 15). For the purpose of modeling, the fuel is specified to be $C_{14}H_{30}$. Fuel properties for the lubricating oil are summarized in Table C-2. The properties obtained from NUREG-1805 correspond to those for transformer oil, based on the statement in Table 3-4 in NUREG-1805 that lubricating and transformer oils are similar.

Table C-2. Data for lubricating oil fire.

Parameter	Value	Source
Effective Fuel Formula	C_nH_{2n+2}	Specified as $C_{14}H_{30}$
Mass burning rate	0.039 kg/s.m^2	NUREG-1805 Table 3-4
Fuel volume	190 L	Specified
Fuel density	760 kg/m^3	NUREG-1805 Table 3-4
Heat of Combustion	46,000 kJ/kg	NUREG-1805 Table 3-4
Heat of Combustion per unit mass of oxygen consumed	13,100 kJ/kg	Huggett 1980, Average value
CO_2 Yield	2.64 kg/kg	*SFPE Handbook*, 4th ed., Table 3-4.16*
Soot Yield	0.059 kg/kg	*SFPE Handbook*, 4th ed., Table 3-4.16*
CO Yield	0.019 kg/kg	*SFPE Handbook*, 4th ed., Table 3-4.16*
Radiative Fraction	0.34	*SFPE Handbook*, 4th ed., Table 3-4.16*
Mass Extinction Coefficient	8700 m^2/kg	Mulholland and Croarkin (2000)

*Material identified as "Hydrocarbon" in SFPE Handbook was used to derive the properties.

C.3 Selection and Evaluation of Models

This section describes the applicability of the models to this scenario.

C.3.1 Fire Sustainability

This fire scenario involves a large fire in a ventilated room that may not have sufficient oxygen to sustain the fire. An algebraic model is used to determine the oxygen availability within the room with the door closed.

C.3.2 Temperature Criterion

Algebraic Models: Neither the Fire-Induced Vulnerability Evaluation (FIVE-Rev1) nor the Fire Dynamics Tools (FDT^s) contain correlations to estimate the hot gas layer (HGL) temperatures within a flashed-over, under-ventilated compartment. Also, the point source radiation heat flux calculation included within FIVE-Rev1 and the FDT^s cannot account for the attenuation of thermal radiation by the smoke that fills the compartment. Consequently, neither model is used for this scenario. However, an algebraic calculation is used to examine the availability of oxygen.

Zone Models: This fire scenario is not a typical application of a zone model because it involves post-flashover conditions, where the two layers essentially become one. Nevertheless, zone models can transition to this state when the HGL essentially descends all the way to the floor and the room becomes a well-stirred reactor. Conservation laws of mass and energy still apply within the single layer; additionally, the processes governing the heating of a target immersed in the HGL still apply, even when the HGL fills the entire compartment. The zone model MAGIC has been selected for this application.

CFD Models: This fire scenario is a challenging application, even for a computational fluid dynamics (CFD) model. It involves relatively high temperatures, under-ventilated conditions, and flashover. The primary advantage of a CFD model for this fire scenario is that CFD models typically include combustion algorithms that estimate near- and post-flashover conditions.

C.3.3 Validation

Table C-3 lists various important non-dimensional parameters and the ranges for which the validation study NUREG-1824 (EPRI 1011999) is applicable. The only parameters that fall outside the validation parameter space are the equivalence ratios for the mechanically ventilated portion of the scenario (first 10 minutes, while the door is closed) and for the natural ventilation portion of the scenario (after 10 minutes when the door is opened). In both cases, the high equivalence ratios for the compartment are a result of the relatively large fire and low airflows.

For MAGIC, a sensitivity case is run, in which double doors, rather than a single door, are opened after 10 minutes to determine whether an increase in airflow would cause higher temperatures in the room. With the enlarged opening, the equivalence ratio is 0.5, putting it within the verification and validation (V&V) range. The results of the sensitivity case, presented in Section C.5.2, show that the HGL temperatures predicted in the room are not sensitive to the size of the door opening.

As part of the work performed at the National Institute of Standards and Technology (NIST) for the investigation of the World Trade Center disaster, the Fire Dynamics Simulator (FDS) has been validated against large-scale fire experiments. The experiments involved fairly large fires in a relatively small compartment, limited ventilation, a liquid fuel spray fire, and the measurement of the heat flux to and temperatures of insulated steel (similar to the cables protected by ceramic fiber blankets). The large-scale tests with furniture had equivalence ratios of approximately 1.4, which provides a validation basis for FDS under conditions similar to the natural ventilation portion of the scenario (after the door has been opened). The NIST experiments and the FDS simulations are described in NIST NCSTAR 1-5F.

Table C-3. Normalized parameter calculations for the pump room fire scenario.

Quantity	Normalized Parameter Calculation	NUREG-1824 Validation Range	In Range?
Fire Froude Number	$$\dot{Q}^* = \frac{\dot{Q}}{\rho_\infty c_p T_\infty D^{2.5}\sqrt{g}}$$ $$= \frac{4934 \text{ kW}}{(1.2 \text{ kg/m}^3)(1.0 \text{ kJ/kg/K})(293 \text{ K})(1.9^{2.5} \text{ m}^{2.5})\sqrt{9.8 \text{ m/s}^2}} \cong 0.9$$	0.4 – 2.4	Yes
Flame Length, L_f, relative to the Ceiling Height, H_c	$$\frac{L_f}{H_c} = \frac{4.8 \text{ m}}{4.9 \text{ m}} \cong 0.99$$ $$L_f = D\left(3.7\,\dot{Q}^{*\,2/5} - 1.02\right) = 1.9 \text{ m}\,(3.7 \times 0.93^{0.4} - 1.02) \cong 4.8 \text{ m}$$	0.2 – 1.0	Yes
Ceiling Jet Radial Distance, r_{cj}, relative to the Ceiling Height, H_c	N/A	1.2 – 1.7	N/A
Equivalence Ratio, φ, based on the mechanical ventilation rate	$$\varphi = \frac{\dot{Q}}{\Delta H_{O_2}\,\dot{m}_{O_2}} = \frac{4934 \text{ kW}}{13{,}100 \text{ kJ/kg} \times 0.07 \text{ kg/s}} \cong 5.5$$ $$\dot{m}_{O_2} = 0.23\,\rho_\infty \dot{V} = 0.23 \times 1.2 \text{ kg/m}^3 \times 0.25 \text{ m}^3/\text{s} \cong 0.07 \text{ kg/s}$$	0.04 – 0.6	No
Equivalence Ratio, φ, based on natural ventilation	$$\varphi = \frac{\dot{Q}}{\Delta H_{O_2}\,\dot{m}_{O_2}} = \frac{4934 \text{ kW}}{13{,}100 \text{ kJ/kg} \times 0.38 \text{ kg/s}} \cong 0.99$$ $$\dot{m}_{O_2} = 0.23 \cdot 0.5 A_o \sqrt{h_o} = 0.23 \times 0.5 \times 2.31 \text{ m}^2\,\sqrt{2.1 \text{ m}} \cong 0.38 \text{ kg/s}$$	0.04 – 0.6	No
Compartment Aspect Ratios	$$\frac{L}{H_c} = \frac{9.4 \text{ m}}{4.9 \text{ m}} \cong 1.9 \qquad \frac{W}{H_c} = \frac{2.8 \text{ m}}{4.9 \text{ m}} \cong 0.6$$	0.6 – 5.7	Yes
Target Distance, r, relative to the Fire Diameter, D	N/A	2.2 – 5.7	N/A

Notes:
    (1)   The non-dimensional parameters are explained in Table 2-5.

(2)   The equivalent fire diameter, $D = \sqrt{4A/\pi}$, where A is the area of the spilled lubricating oil.

(3)   The compartment aspect ratios are calculated using the equivalent length and width.

C.4 Estimation of Fire-Generated Conditions

This scenario is modeled using algebraic calculations, the zone model MAGIC, and the CFD model FDS.

C.4.1 Calculation of Oxygen Availability

At the start of the scenario, the mechanical ventilation is operational, the door is closed, and the fire output immediately jumps to the peak heat release rate (HRR) with a total spill area of approximately 2.75 m^2 (29.6 ft^2), as shown in the hatched area of Figure C-1. The peak HRR, $\dot{Q}$, is computed from the fuel mass burning rate, $\dot{m}''$, the heat of combustion, ΔH, and the specified area of the spill, A:

$$\dot{Q} = \dot{m}'' \, \Delta H \, A = 0.039 \text{ kg/m}^2/\text{s} \times 46{,}000 \text{ kJ/kg} \times 2.75 \text{ m}^2 \cong 4{,}934 \text{ kW} \qquad \text{(C-1)}$$

The oxygen needed to sustain the fire is calculated from the following equation:

$$\frac{\dot{Q}}{\Delta H_{O_2}} = \frac{4934 \text{ kW}}{13{,}100 \text{ kJ/kg}} = 0.377 \text{ kg/s} \qquad \text{(C-2)}$$

where ΔH_{O_2} is the heat of combustion per unit mass of oxygen consumed. The quantity of oxygen provided by the ventilation system is calculated by multiplying the oxygen content (0.23) by the density and the ventilation rate of the air:

$$0.23 \, \rho_\infty \dot{V} = 0.23 \times 1.2 \text{ kg/m}^3 \times 0.25 \text{ m}^3/\text{s} = 0.069 \text{ kg/s} \qquad \text{(C-3)}$$

The oxygen provided by the ventilation system is much lower than the amount needed to sustain the fire. The oxygen initially in the room can provide the additional oxygen needed for combustion for a short time. The available oxygen in the room, calculated from the room dimensions (Table C-4), is:

$$0.23 \, \rho_\infty LWH_c = 0.23 \times 1.2 \text{ kg/m}^3 \times (2.81 \times 9.39 \times 4.9) \text{ m}^3 = 35.7 \text{ kg} \qquad \text{(C-4)}$$

The oxygen initially in the room can sustain the fire for an amount of time equal to the oxygen quantity in the room divided by the consumption rate minus the ventilation supply rate, as shown below:

$$\frac{35.7 \text{ kg}}{(0.377 \text{ kg/s} - 0.069 \text{ kg/s})} = 116 \text{ s} \qquad \text{(C-5)}$$

Equation C-4 requires that all the oxygen within the room be consumed by the fire. This establishes an upper limit to the burning duration before the fire becomes ventilation-limited. After 116 s, the size of the fire is maintained only by the ventilation system and is limited to:

$$0.069 \text{ kg/s} \times 13{,}100 \text{ kJ/kg} = 904 \text{ kW} \tag{C-6}$$

These results show that the oxygen supply available to the room will only allow a fire of reduced size to burn until the door is opened (under-ventilated condition).

C.4.2 Zone Model

The following paragraphs outline the data utilized to model the scenario using MAGIC. Figure C-2 provides an illustration of the scenario, as rendered by MAGIC.

Geometry: To model this scenario with MAGIC, the pump compartment is modeled as a single compartment having the same total volume and surface area as the actual enclosure. This allows the volume in which the HGL develops and the surface area through which energy is transferred from the compartment to be maintained. Maintaining the total volume and surface area while leaving the ceiling height unchanged at 4.9 m (16 ft) yields an effective compartment size of 9.39 m (30.8 ft) by 2.81 m (9.2 ft). The modification to the geometry can be seen by comparing Figure C-1 and Figure C-2. All other aspects of the geometry are relatively unchanged.

Table C-4. Calculated input for lubricating oil fire.

Parameter	Value
Effective Length	2.81 m
Effective Width	9.39 m
Fire Diameter	1.87 m
Peak Heat Release Rate	4,934 kW
Fire Duration	1345 s
Mass of Fuel (kg)	144.4
Stoichiometric Mass-Oxygen	3.5 g/g
Specific Area	513 m^2/g

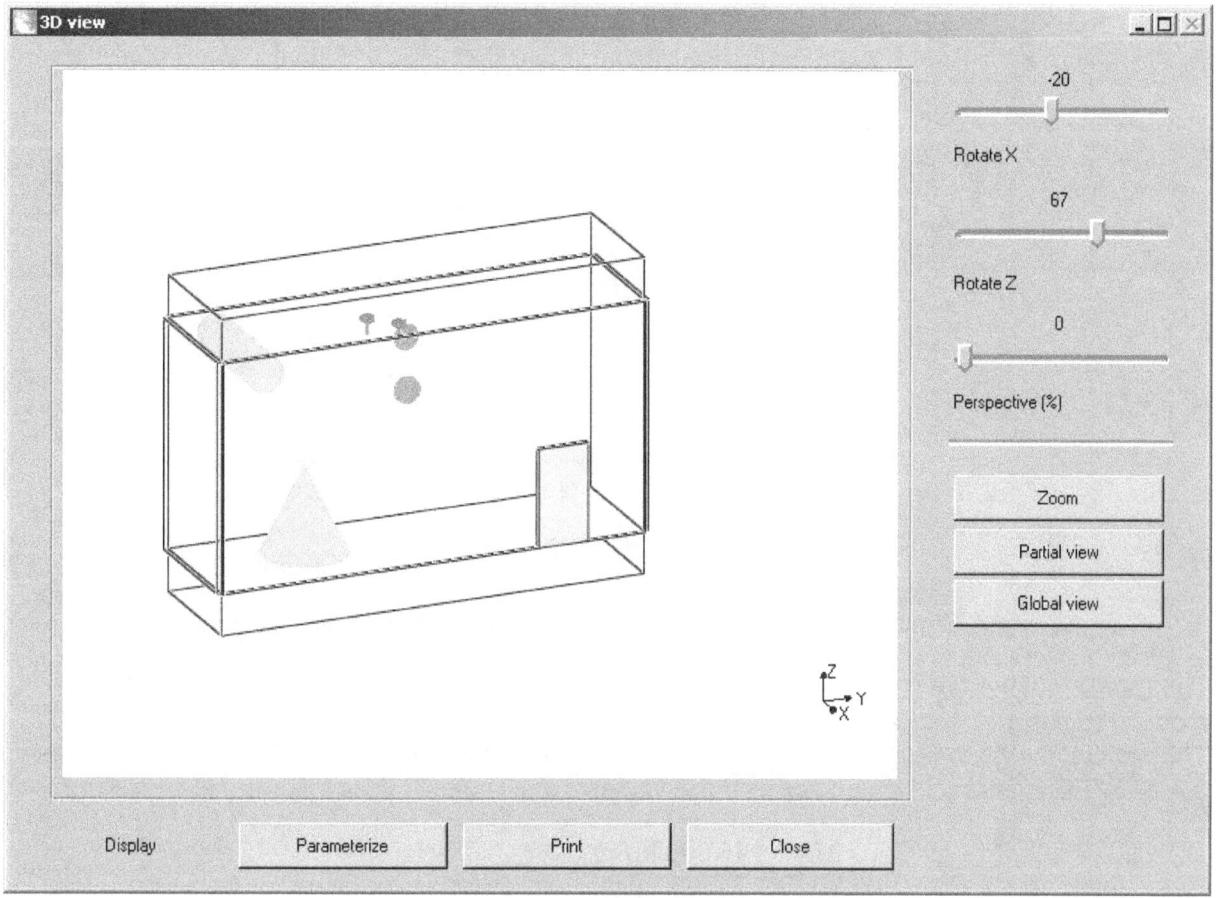

Figure C-2. MAGIC view of the pump room.

Fire: For this scenario, the lubricating oil is preheated prior to the spill, such that the HRR reaches the peak immediately upon fire initiation, as shown in the HRR curve plotted in Figure C-3. The lower oxygen level is specified to be 10%. Using the specified spill area and volume, the spill depth is calculated as 0.069 m (0.23 ft).

The fire is modeled as a single circular area of equivalent diameter. The actual entrainment for the pool fire is proportional to the perimeter of the fire, which is significantly greater than the perimeter of the circular area. However, the enclosure is small and the smoke filling rates are expected to be short regardless of the fire shape.

The fire duration, Δt, is determined from the pool depth, δ, density, ρ, and burning rate, $\dot{m}''$:

$$\Delta t = \frac{\delta\,\rho}{\dot{m}''} = \frac{0.069 \text{ m} \times 760 \text{ kg/m}^3}{0.039 \text{ kg/m}^2/\text{s}} \cong 1345 \text{ s} \quad (22.4 \text{ min}) \tag{C-7}$$

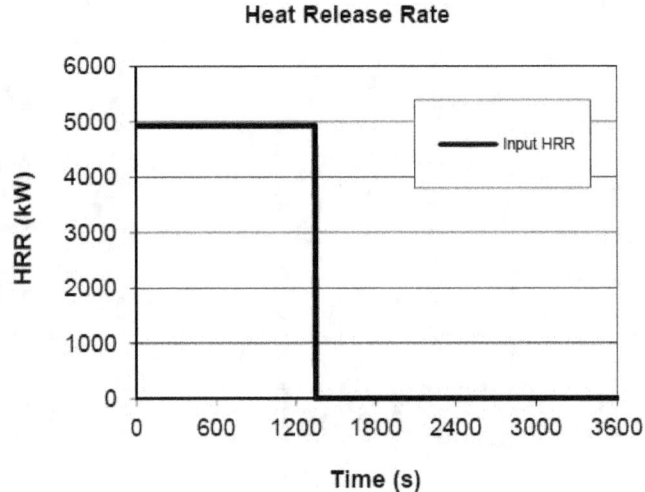

Figure C-3. Heat release rate curve for lubricating oil fire.

The location of the fire is placed at the edge of the dike, closest to the target cable. The total mass of fuel is 144.4 kg, calculated from the volume multiplied by the density from Table C-2. The stoichiometric mass-oxygen-to-fuel ratio, ψ, is calculated using Equation 22 from Chapter 3-4 of *SFPE Handbook, 4th Edition* and the values from Table C-1, as follows:

$$\psi = \frac{\Delta H_T}{\Delta H_{O_2}} = \frac{46,000 \text{ kJ/kg}}{13,100 \text{ kJ/kg}} = 3.5 \text{ kg/kg} \tag{C-8}$$

where ΔH_T is heat of combustion per unit mass of fuel consumed and ΔH_{O_2} is the heat of combustion per unit mass of oxygen consumed. One of the inputs required by MAGIC is the specific area, s, a measure of the smoke generation, which Is calculated as described in NUREG-1824 (EPRI 1011999), Volume 6, Section 3.2.7:

$$s = y_s k_m = 0.059 \times 8,700 \text{ m}^2/\text{kg} \cong 513 \text{ m}^2/\text{kg} \tag{C-9}$$

where k_m is the mass extinction coefficient and y_s is the soot yield, as listed in Table C-2. The pyrolysis rate (g/s) is calculated for input to MAGIC by dividing the HRR values (4934 kW) at each time step by the heat of combustion (46,000 kJ/kg). Other inputs needed for MAGIC are listed in Tables 3-1, C-1, C-2, and C-4. Figure C-4 is a screenshot of the source fire in the MAGIC input file.

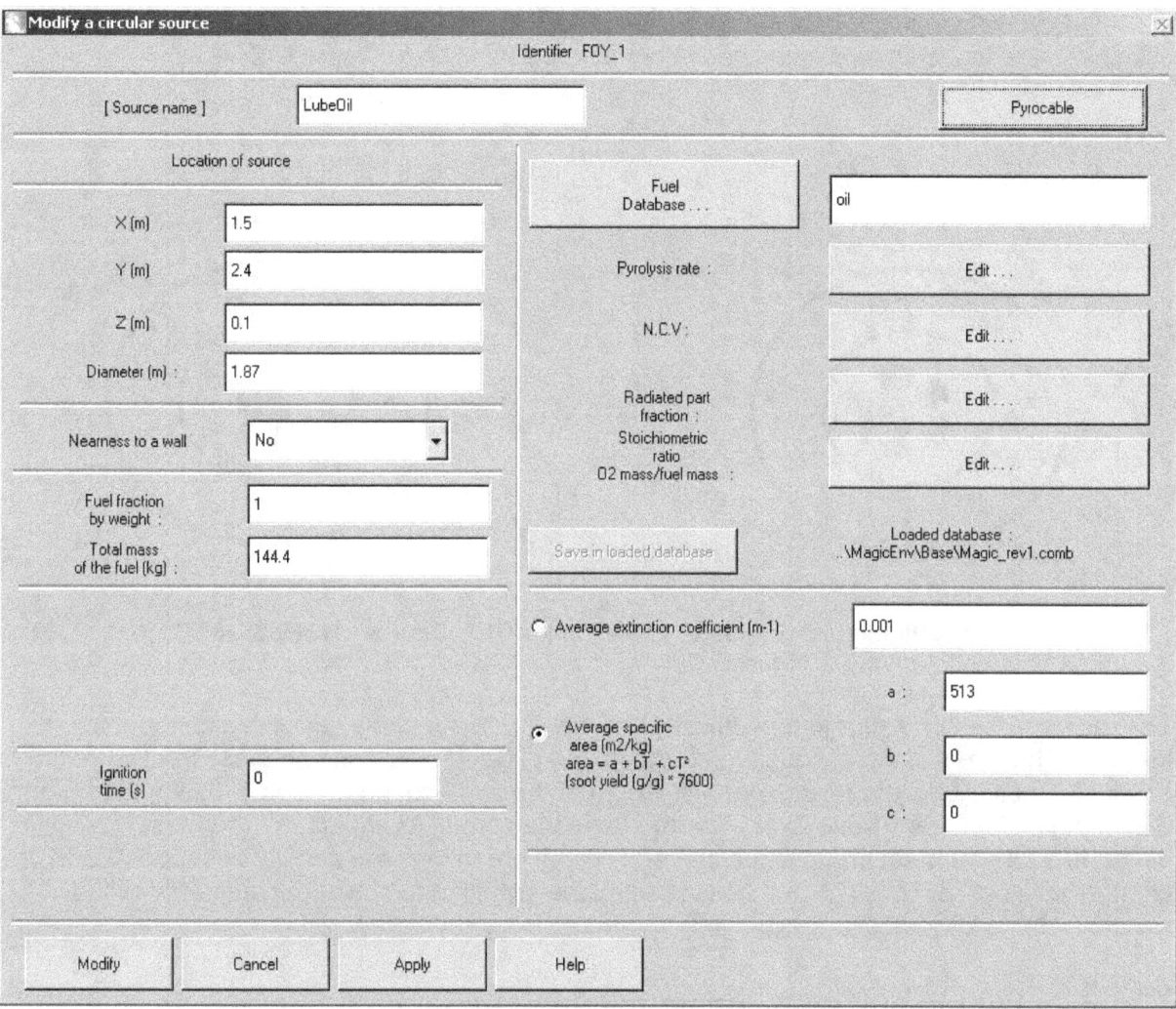

Figure C-4. MAGIC fire input screen for pump fire.

In MAGIC, multi-conductor cables, composed of jacket, insulation, and conductor (copper, in this case), are modeled as single-conductor cables, as shown in Figure C-5.

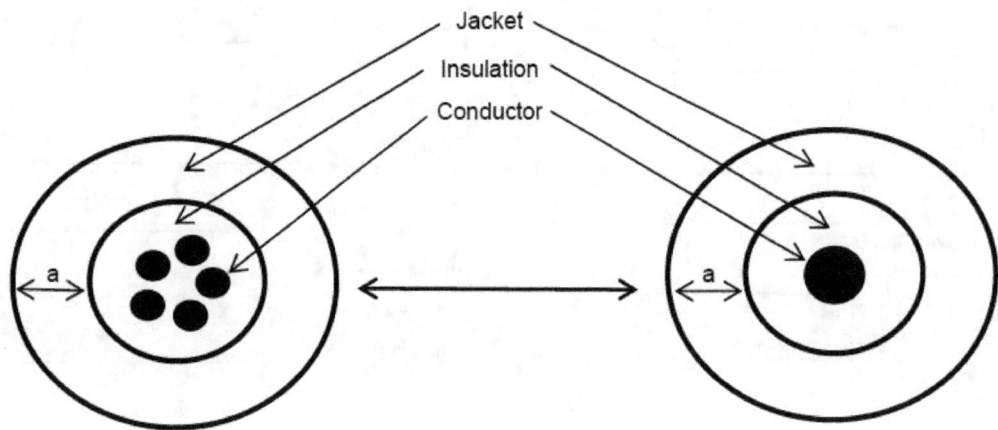

Figure C-5. Modeling multi-conductor cables in MAGIC. Source: NUREG-1824 (EPRI 1011999), Volume 6, Figure 3-3.

The radius of the conductor in an equivalent single-conductor cable is needed for input to MAGIC. The mass of the conductor per unit length is calculated from the mass fraction of the conductor multiplied by the mass per unit length of the multi-conductor cable: 0.67 x 0.4 kg/m = 0.27 kg/m (values from Table C-1). To determine the conductor radius, r_c, the mass per unit length (0.27 kg/m) is set equal to the cross-sectional area times the density of copper, 8954 kg/m^3 (Table 3-1) or $0.27 = \pi r_c^2 \rho$. Rearranging the equation to solve for conductor radius results in the following:

$$r_c = \sqrt{\frac{0.27}{\pi \times 8954}} = 3.1 \text{ mm} \qquad (C-10)$$

The insulation thickness is calculated by cable radius – jacket thickness – copper thickness = 7.5 mm – 2 mm – 3.1 mm = 2.4 mm. Since the jacket and the insulation are both composed of PE/PVC, the thicknesses are added together for a total thickness of 4.4 mm. As a result, the ERFBS protected cable raceway is modeled with three layers (Figure C-6): ceramic fiber blanket (5 cm), PE/PVC (4.4 mm), and copper (3.1 mm). The input screen for the layers of the ERFBS and the cables is shown in Figure C-7.

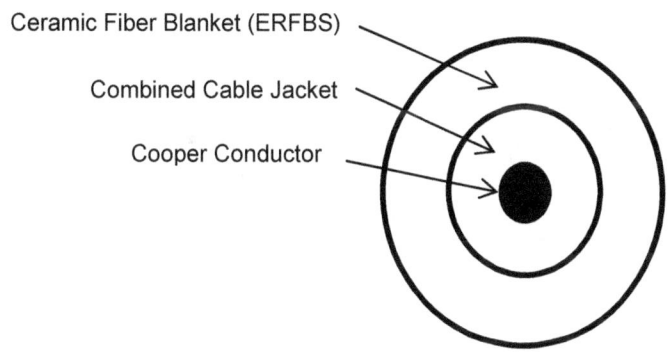

Figure C-6. Representation of the ERFBS protected cable raceway for MAGIC.

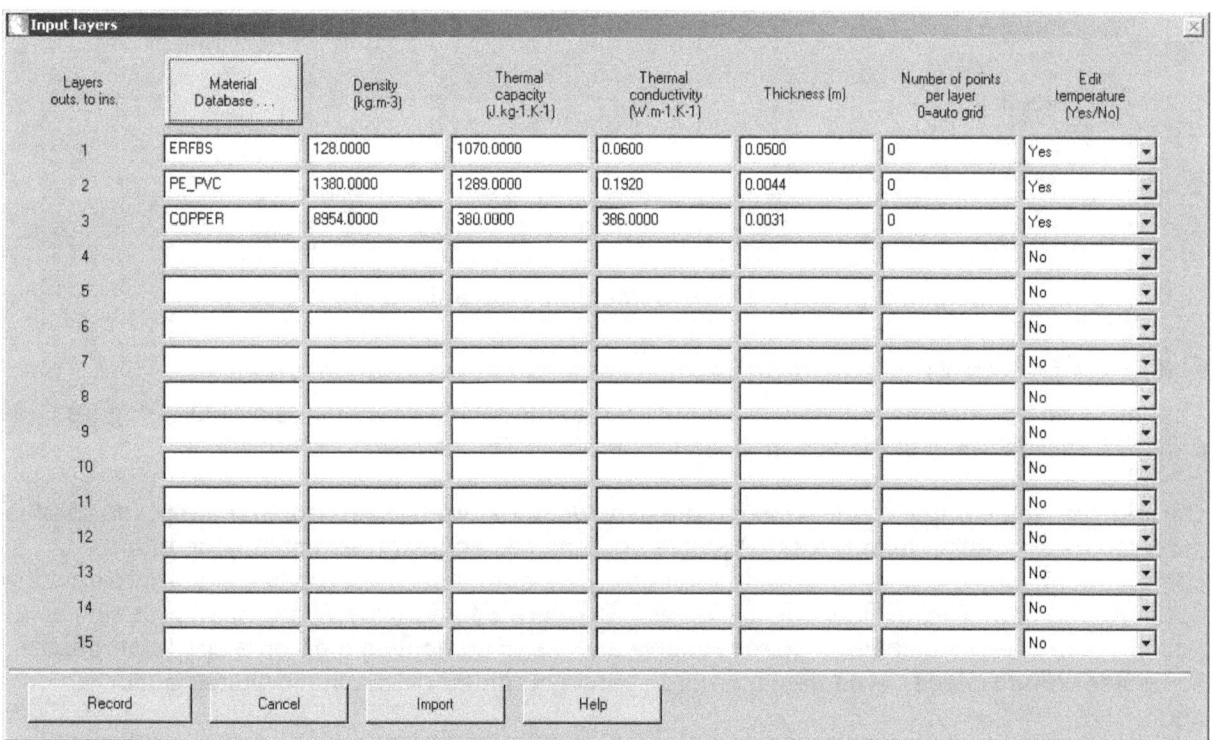

Figure C-7. MAGIC input screen for ERFBS layers.

Ventilation: Mechanical ventilation is maintained constantly during the simulation, using the values provided above. MAGIC uses circular ducts, so the rectangular ducts seen in Figure C-1 are modeled as circular areas with equivalent diameter of 0.56 m (1.8 ft). As noted above, the door is normally closed, but it is opened 10 minutes after ignition by the arriving fire brigade. Before the door opens, leakage due to the doorway occurs via a 1.3 cm (0.5 in) gap under the door. The MAGIC input screen for the doorway is shown in Figure C-8.

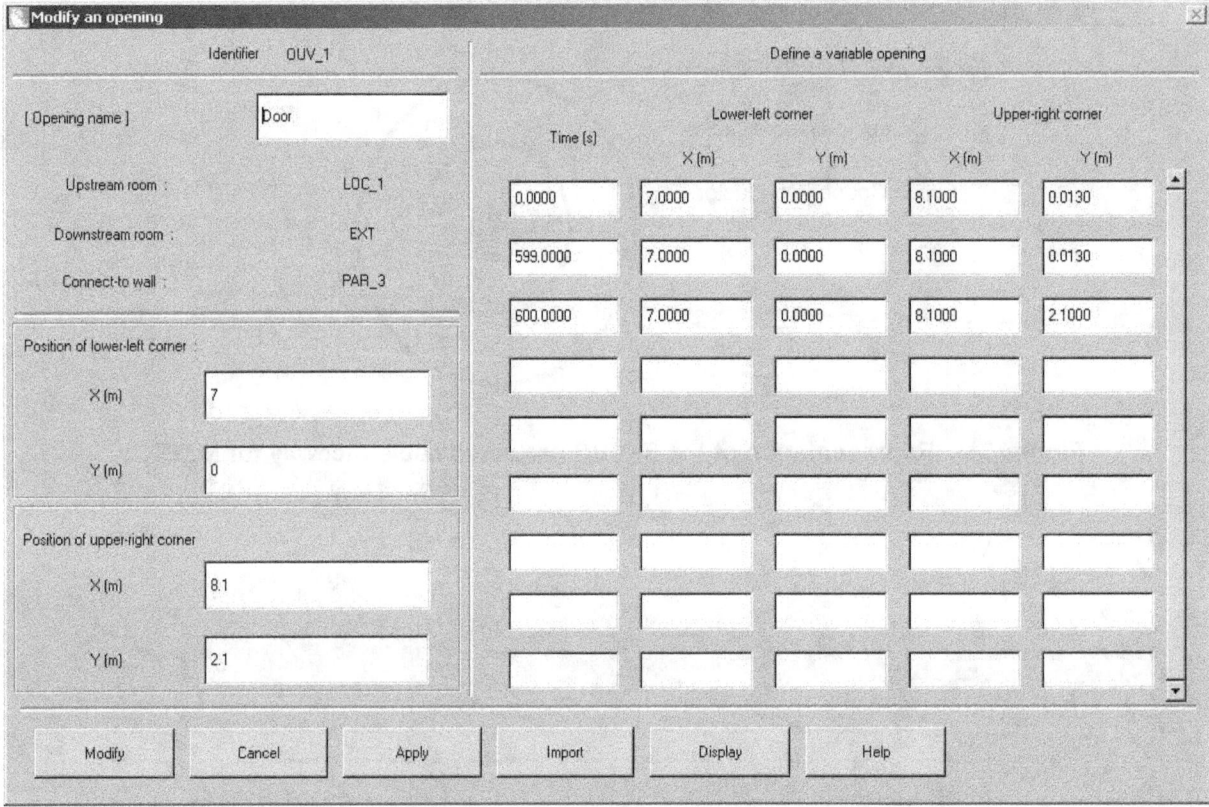

Figure C-8. MAGIC input screen for ventilation through door.

C.4.3 CFD Model

The following paragraphs outline the data utilized to model the scenario using FDS. Figure C-9 provides an illustration of the scenario, as rendered by Smokeview.

Geometry: The compartment is modeled as shown in Figure C-1, except that the pump itself is modeled as two rectangular boxes. A single uniform, rectangular mesh spans the entire compartment, plus the hallway outside the door. The numerical mesh consists of approximately 0.2 m (0.7 ft) grid cells. A finer calculation with 0.1 m (0.3 ft) cells was performed with similar results. The latter calculation requires roughly a week of computing time on a single processor computer (2008 vintage), whereas the more coarsely gridded calculation requires about 10 hours.

Materials: All material properties are as specified above. The protected cable tray is modeled as a rectangular box with the same dimensions as the tray wrapped in a blanket. The box is made solely of 5 cm (2 in) of ceramic fiber insulation. The tray is neglected. A cable target is positioned within the box pointing downwards, as this is the hottest surface of the box. The exact dimensions of the box are not an issue; what matters is that the cable within the box is exposed to the heat that penetrates the thermal blanket. The cable temperature is computed using the Thermally-Induced Electrical Failure (THIEF) methodology (NUREG/CR-6931).

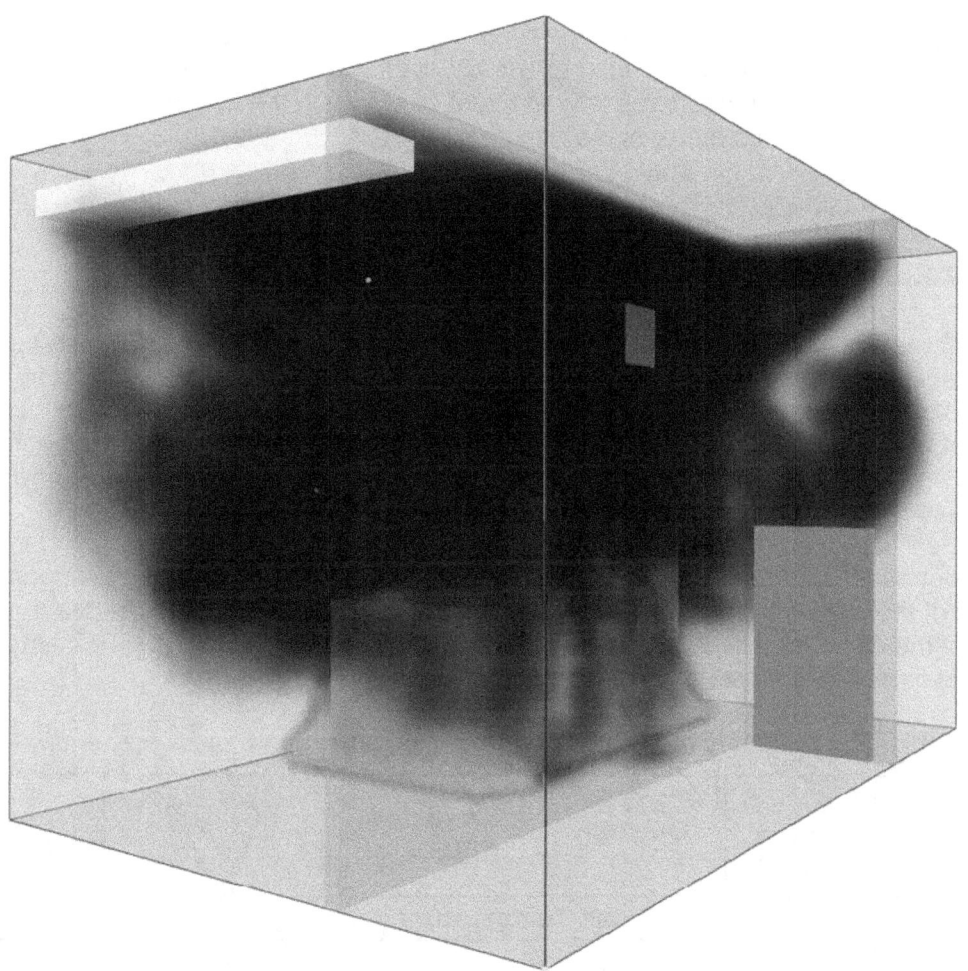

Figure C-9. FDS/Smokeview rendering of the pump room scenario at the early stage of the fire, before the compartment becomes under-ventilated.

Fire: Due to the limited amount of validation data available for scenarios of this type and the considerable uncertainties involved, the approach taken is to *specify*, rather than attempt to *predict*, the burning rate of the fuel, even though the FDS model does provide the physical mechanisms to estimate burning rates. The fire is specified in the diked area surrounding the pump. Although FDS has a liquid fuel burning model, it is not used here because there is not enough information about the fuel, and, more importantly, it lacks the exact geometry of the pump and diked area. Typically, FDS would expect that the oil has formed a relatively deep pool with relatively little influence by the surrounding solids. This is not the case here. Instead, the specified burning rate, 0.039 kg/m^2/s, is applied directly to the model over an area of 2.75 m^2 (29.6 ft^2), yielding a burning rate of 0.107 kg/s. The density of the oil is 0.76 kg/L, which means that the oil burns at a rate of 0.141 L/s. At this rate, 190 L (50 gal) will require 1,348 s to burn out. Note that this is slightly different from the burning duration of 1345 s computed for the MAGIC input. The fire duration computation for FDS converts the mass data to volumetric data, thus introducing an additional step and some rounding. The slight difference in the burning durations is due to this additional rounding and is not significant. The vaporized fuel is a mixture of various hydrocarbons, but FDS uses only one fuel molecule in the combustion sub-model. For this calculation the fuel molecule is modeled as $C_{14}H_{30}$.

Ventilation: The volume flow rates are applied as specified.

C.5 Evaluation of Results

The primary purpose of these calculations is to assess whether the Kaowool ERFBS applied to the critical cables within the pump room would be damaged in the event of a lubricating oil fire. The results of the zone model MAGIC and the CFD model FDS are consistent in their HRR and compartment temperatures. This is expected because the models use the same specified burning rate, the same fuel stoichiometry, and the same basic rules of gas phase flame extinction based on oxygen and temperature levels in the vicinity of the fire.

Table C-5 summarizes the predicted cable temperatures from MAGIC and FDS, including an assessment of the model uncertainty. Note that the results are based only on a direct calculation of the cable temperature and do not include an assessment of the sensitivity studies that are discussed in the next section.

Table C-5. Summary of the model predictions of the pump room scenario.

Model	Bias Factor, δ	Standard Deviation, $\tilde{\sigma}_M$	Predicted Value	Critical Value	Probability of Exceeding
Cable Temperature (°C)					
MAGIC	1.19	0.27	135	205	0.000
FDS	1.02	0.13	145	205	0.000

C.5.1 The Fire

The HRR curves predicted by the hand calculations and the MAGIC and FDS models are shown in Figure C-10 for the entire simulation (first plot) and the first five minutes (second plot). The figures show the pronounced drop in the HRR soon after the start of the fire, which demonstrates that there is insufficient air (i.e., oxygen) within the compartment to sustain the postulated fire. Based on the calculations, the drop in HRR does not occur until after about 120 s, which is later than the predictions by MAGIC and FDS because the algebraic calculations allow all of the oxygen in the room to be consumed. MAGIC uses a lower oxygen limit of 10%, and FDS uses a lower oxygen limit that depends on temperature. At high temperatures, FDS expects all of the oxygen is consumed. After 120 s, but before the door is opened at 600 s, the value of the HRR calculated by the algebraic calculations (approximately 900 kW) is higher than the values predicted by MAGIC and FDS (approximately 350 kW) due to the differences in the lower oxygen level. The sudden jump in the HRR, predicted by FDS at 600 seconds, is caused by the unburned fuel igniting as the door is opened. Note that none of the models has an algorithm capable of determining whether or not the fire would be sustained at this reduced burning rate until the time when the door is opened.

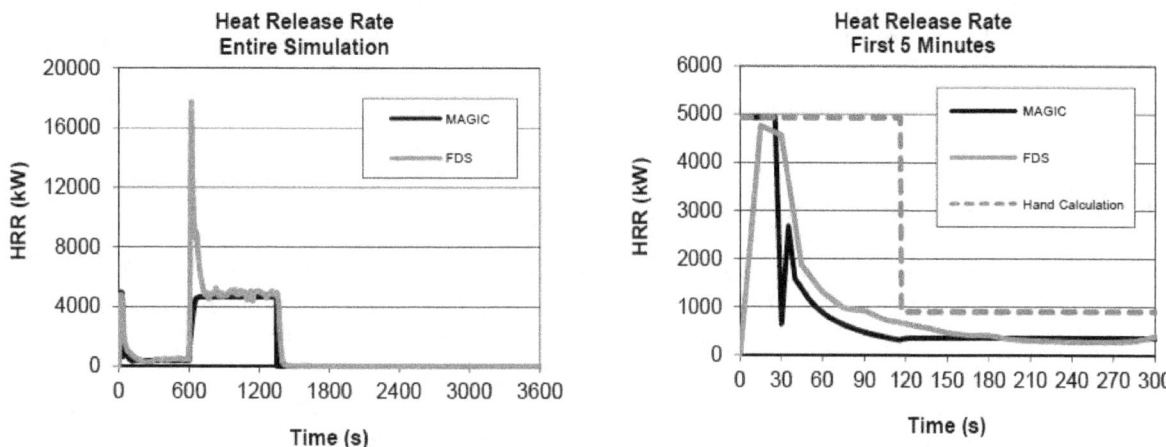

Figure C-10. HRR predicted by algebraic methods, MAGIC, and FDS for the pump room fire scenario.

C.5.2 Temperature Criterion

MAGIC and FDS estimate the temperature of the hot gas layer as a function of time, as shown in Figure C-11. As expected, the HGL temperature changes in accordance with the altered (oxygen-starved) HRR. Once the door opens at 600 seconds, the increased HRR causes the HGL temperature to rapidly increase until the fire consumes the available fuel. After the fire burns out, the HGL temperature slowly drops as heat leaves the HGL through the bounding surfaces and open door.

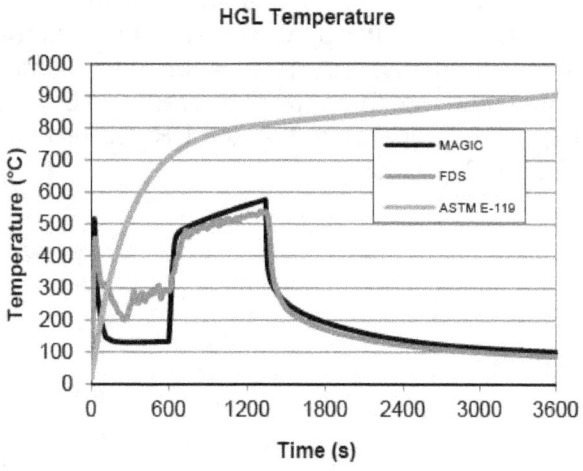

Figure C-11. HGL Temperature Predicted by MAGIC and FDS for the pump room fire scenario.

There are two strategies for assessing the viability of the protected cables. The first is to compare the predicted HGL temperature to the standard fire endurance curve under which the ERFBS was tested to determine whether the predicted thermal exposure is comparable to the qualification test. The second is to calculate the temperature of the cable directly, using the nominal thermal properties of the fiber insulating blanket. Each strategy will be assessed in turn below.

Comparison to the Standard Fire Endurance Temperature Curve

Figure C-11 includes the standard ASTM E119 temperature curve to which the ERFBS was subjected during its qualification test. The predicted HGL temperatures of both MAGIC and FDS fall below this curve during most of the hour-long simulation, but there is a period near the beginning of the fire where the models' predicted temperatures exceed the standard curve. In order to compare the relative exposure of the ERFBS, it is necessary to consider the integrated incident heat flux corresponding to the model HGL predictions and the ASTM E 119 temperature curve. The integrated heat flux is given by the following formula:

$$q'' = \int_{t_0}^{t_1} \dot{q}''(t)\, dt = \int_{0}^{3600} \sigma(T^4 - T_0^4) + h(T - T_0)\ dt \qquad \text{(C-11)}$$

Here, q'' is the integrated heat flux received by the ERFBS, T is the HGL temperature, T_0 is the ambient temperature (20 °C), h is the convective heat transfer coefficient (about 0.025 kW/m²/K in fully developed fires), and σ is the Stefan-Boltzmann constant (5.67 × 10⁻¹¹ kW/m²/K⁴). Note that Equation C-11 is the total energy transferred from a thermally thick (emissivity of 1) hot gas to a cold target. It is intended only to compare the different temperature curves. In reality, the net heat transferred to a target in the compartment decreases as the target heats up.

Applying Equation (C-11) to each of the HGL temperature curves in Figure C-11 yields values of 346 MJ/m² for the ASTM E119 curve and approximately 40 MJ/m² for both FDS and MAGIC. This 40 MJ/m² exposure corresponds to an approximately 14 min exposure within the standard test furnace. Table C-6 lists the thermal exposure as a function of time in the standard test furnace. It is also significant to note that the maximum predicted exposure temperature remains lower than the maximum exposure temperature that the ERFBS protected raceway was exposed to during the ASTM E119 fire test.

Table C-6. Integrated thermal exposure of an object subjected to the ASTM E119 temperature curve.

Time (min)	Thermal Exposure (MJ/m^2)
5	6
10	23
15	47
20	75
25	104
30	135
35	167
40	200
45	235
50	270
55	307
60	346

Direct Calculation of Cable Temperature

MAGIC and FDS have heat conduction algorithms to account for the multiple layers of insulation and cable materials. The surface temperature predictions of the cables protected by the ERFBS (ceramic fiber insulation in this case) are shown in Figure C-12. MAGIC predicts a maximum cable surface temperature of approximately 135 °C (275 °F). FDS predicts a maximum cable surface temperature of approximately 145 °C (293 °F). Note that although the HGL temperature drops and then increases dramatically when the door opens, as shown in Figure C-11, the cable temperature slowly rises. This is due to the thermal lag caused by the ERFBS.

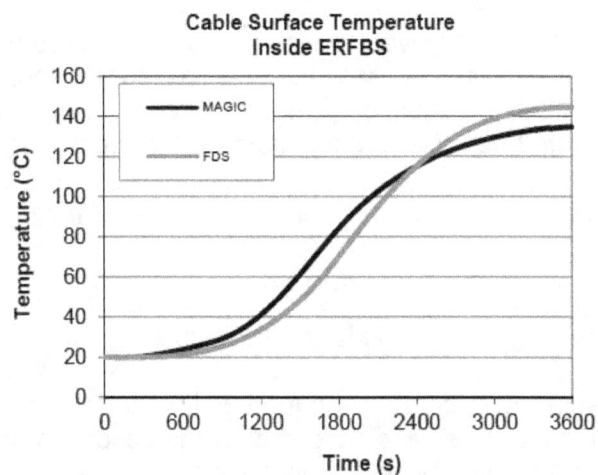

Figure C-12. Cable surface temperature predicted by MAGIC and FDS for the pump room fire scenario.

Sensitivity of the ERFBS Construction

Comparing Figure C-11 and Figure C-12 shows that the ERFBS has a large impact on the temperature of the target cable. To determine the sensitivity of the target cable temperature to the insulation installation technique, two additional MAGIC cases are run. In the first case (file: Pump_Room_thinner_wrapping.cas.), the thickness of the ceramic insulation blanket is reduced by 25% to 0.0375 m. In the second case (file: Pump_Room_tighter_wrapping.cas.), the thickness of the ceramic insulation blanket is reduced by 25% while the density is increased to 171 kg/m^3, such that the mass per area remains constant, which simulates a tighter installation of the insulation. The results, plotted in Figure C-13, show that both cases led to a higher cable temperature.

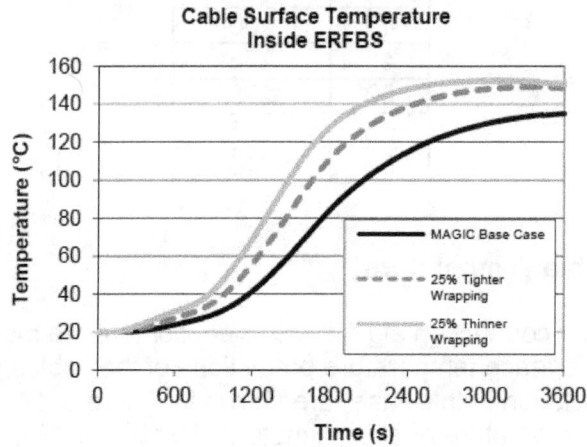

Figure C-13. Cable surface temperature predicted by MAGIC for changes to insulation wrapping.

Sensitivity of the Door Size

As mentioned in Section C.3.3, the equivalence ratio for the pump room scenario falls outside of the validation range. As a sensitivity test, MAGIC was run with the door area doubled, such that the equivalence ratio falls within the applicable validation range (0.04 – 0.6) for the portion of the simulation when the doors are open (file: Pump_Room_2Doors.cas), as calculated below:

$$\dot{m}_{O_2} = 0.23 \cdot 0.5 A_o \sqrt{h_o} = 0.23 \times 0.5 \times 4.62 \text{ m}^2 \sqrt{2.1 \text{ m}} \cong 0.77 \text{ kg/s} \tag{C-12}$$

$$\varphi = \frac{\dot{Q}}{\Delta H_{O_2} \dot{m}_{O_2}} = \frac{4934 \text{ kW}}{13{,}100 \text{ kJ/kg} \times 0.77 \text{ kg/s}} \cong 0.5 \tag{C-13}$$

Figure C-14 shows the temperature comparison for the HGL and the cable surface temperature (measured inside the ERFBS) for the base case and for the case with double doors. The plots show that the results for both cases are very similar, indicating that the door size does not significantly affect the results. Nevertheless, it is consistent with experimental data that the scenario with the equivalence ratio closest to unity produces the highest enclosure temperature.

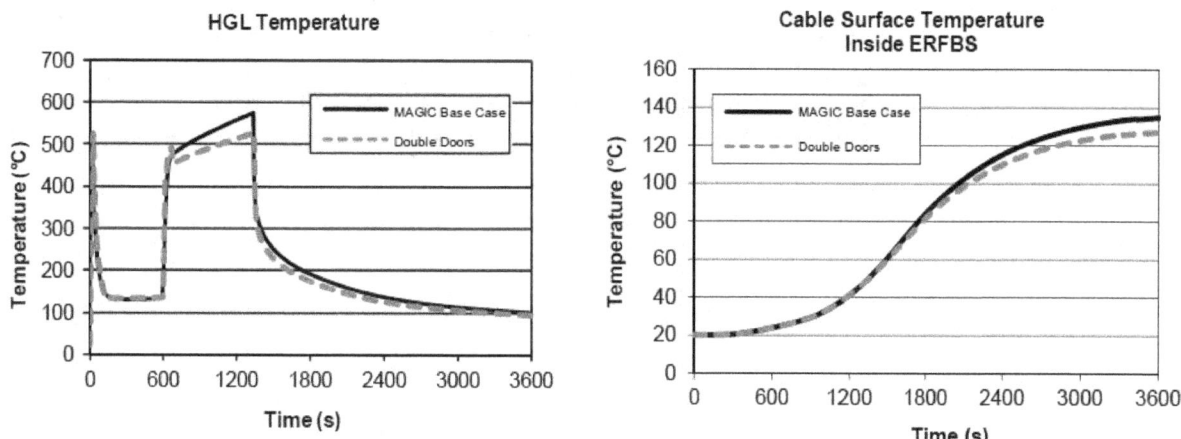

Figure C-14. Temperature predicted by MAGIC for increased door size.

Sensitivity of HRR Profile

As discussed in Section C.5.1, both MAGIC and FDS predict that the HRR decreases rapidly soon after the fire starts, due to an insufficient oxygen supply within the compartment. The models indicate that the HRR decreases from about 4900 kW to about 350 kW in approximately 2 min (Figure C-10). When the oxygen is insufficient to maintain the fire, MAGIC continues to vaporize the unburned liquid fuel, but these fuel gases do not combust (Gay, 2005). FDS also continues to vaporize the unburned fuel, and it continues to transport the fuel gas until the door is opened after 10 min, at which time this excess fuel mixes with incoming air and burns. This rapid burning of built-up excess fuel gas in an under-ventilated compartment is known as a "backdraft," and it is apparent from the HRR plot in Figure C-10. However, much of the heat from this rapid burning of fuel is immediately exhausted from the compartment and does not significantly affect the temperature of the ERFBS.

During the period of underventilation, both models specify that the liquid fuel contained in the dike continues to evaporate at the specified rate. In reality, the reduced temperature and oxygen within the compartment will probably reduce the evaporation rate of the fuel, the extent of which is difficult to predict. To account for this possibility, the fuel evaporates at a constant rate can be changed so that the evaporation rate during the time period between 2 min and 10 min is reduced to support a 350 kW fire only. The excess fuel that does not evaporate in this time period is added to the end of the specified burning period so that the total fuel mass is conserved. As shown in the plot to the right in Figure C-15, a revised HRR curve (labeled as Extended HRR) was specified as input for MAGIC. The HGL temperature for this case reaches 640 °C (1184 °F), compared to 580 °C (1076 °F) for the base case (Figure C-14), which is still significantly lower than the ASTM E119 temperature curve (Figure C-11). However, the predicted cable surface temperature is 200 °C (392 °F), falling just below the failure criterion of 205 °C (400 °F). This five-degree margin suggests that further validation may be needed to ensure that the thermal properties of the ERFBS are accurate. As shown in the previous sensitivity cases, changes in the thermal properties of the ERFBS led to an increase of more than 10 °C in ERFBS exposure temperature. Therefore, further validation of the thermal properties of the ERFBS is needed in this case to reduce the impact of parameter uncertainty on the surface temperature calculation.

The sensitivity case shows that (1) based on comparison to the ASTM E119 temperature curve, the ERFBS system is not expected to fail under the predicted exposure temperatures, and (2) based on the predicted cable surface temperature, further validation of the thermal properties of the ERFBS is warranted as the surface temperature of the cable is close to the damage criteria.

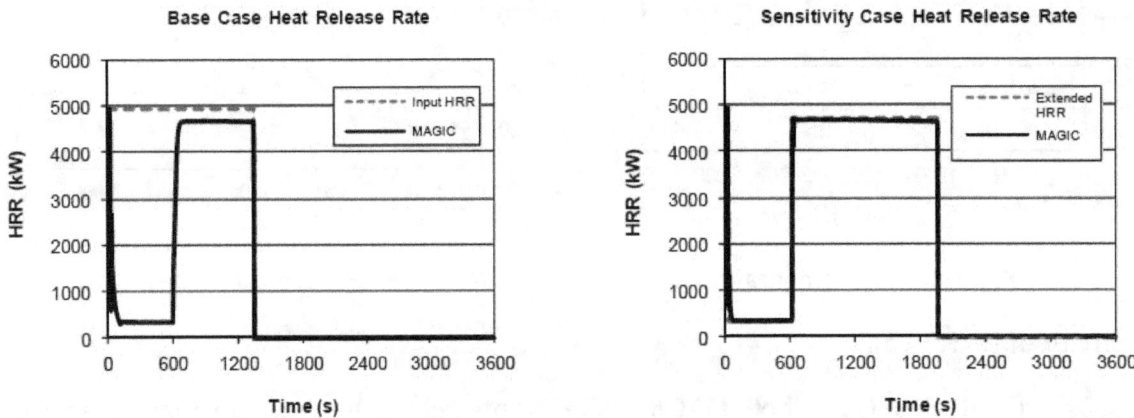

Figure C-15. HRR for base case and HRR sensitivity case.

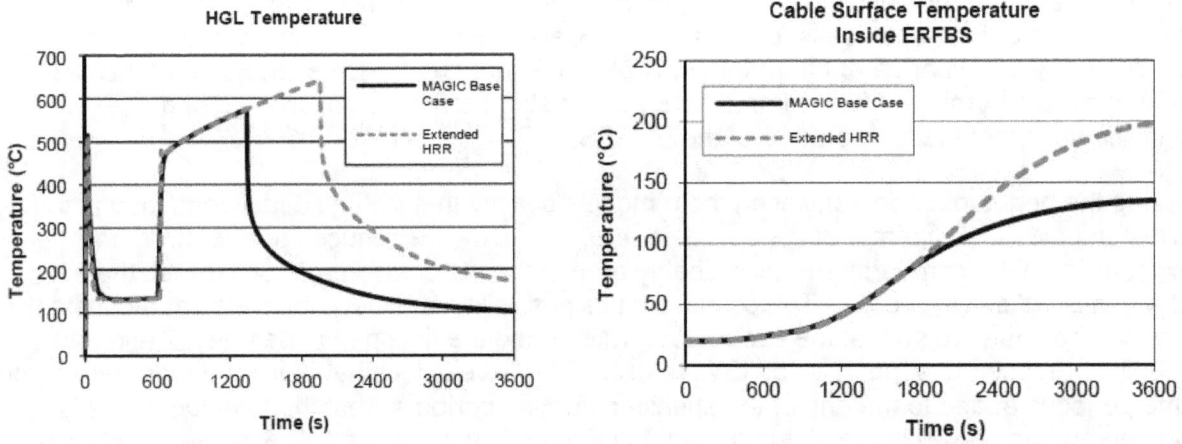

Figure C-16. Temperature for base case and HRR sensitivity case.

C.6 Conclusion

This analysis has considered the potential for a relatively large lubricating oil spill fire in a relatively small enclosure to damage a cable tray protected by an ERFBS. Algebraic calculations, the zone model MAGIC, and the CFD model FDS were all used to evaluate the fire conditions within the enclosure. MAGIC and FDS were used to calculate the thermal response of the cables to these calculated fire conditions.

Based on the specified lubricating oil spill area and burning characteristics, a fire of approximately 5 MW is expected. However, after the rapid consumption of the limited quantity of air in the room, the mechanical ventilation to the enclosure could only support a HRR of less than 1 MW before the door to the enclosure opens after 10 min. This analysis suggests that to avoid rapid fire escalation, doors to such rooms should not be opened until firefighters are prepared to suppress the fire, and, even then, the potential for rapid fire escalation should be considered.

Two different strategies were applied to assess the integrity of the ERFBS. Because the thermal and chemical properties of the insulating material are only partially known, it is necessary to implement an alternative technical approach of comparing the predicted HGL temperatures from the models with the standard temperature curve under which the ERFBS received an hour rating. Since the predicted HGL temperatures do not lie completely within the standard curve, a simple integrated heat flux calculation was performed to demonstrate that the ERFBS received approximately 10 times the thermal exposure in the standard fire endurance test than is predicted by the two models.

A second strategy for assessing the integrity of the ERFBS was to directly calculate the heat penetration through the insulating blankets using the thermal material properties of the cables and the ERFBS. Both models predicted cable temperatures below the reported critical values.

Based on the two approaches to determine its performance, the ERFBS is expected to prevent the cables from reaching temperatures that would limit their functionality in the event of a fire involving burning spilled lubricating oil. This conclusion is based on certain expected burning behavior of the lubricating oil during the under-ventilated stages. A sensitivity study on the burning behavior of the lubricating oil concluded that the results could change if the burning rate decreases during the under-ventilated stage. The results are also shown to be sensitive to the thermal properties of the ERFBS material. Further research or testing of the ERFBS thermal properties may be necessary to confirm the initial conclusion.

C.7 References

1. Gay, L., C. Epiard, and B. Gautier, *MAGIC Software Version 4.1.1: Mathematical Model*, EdF HI82/04/024/B, Electricité de France, France, November 2005.
2. Gay, L., and J. Frezabeu *Qualification File of Fire Code MAGIC version 4.1.1*, EdF HI-82/04/022/A, Electricité de France, France, December 2004.
3. Hugget, C., "Estimation of Rate of Heat Release by Means of Oxygen Consumption Measurements," *Fire and Materials*, 4:61-65, 1980.
4. Mulholland, G. W., and Croarkin, C., "Specific Extinction Coefficient of Flame Generated Smoke," *Fire and Materials*, 24:227–230, 2000.

5. NIST NCSTAR 1-5F, *Federal Building and Fire Safety Investigation of the World Trade Center Disaster: Computer Simulation of the Fires in the World Trade Center Towers*, 2005.
6. NIST SP 1018-5, *Fire Dynamics Simulator (Version 5), Technical Reference Guide, Volume 3, Experimental Validation*, 2010.
7. NUREG-1805, *Fire Dynamics Tools (FDT^s) Quantitative Fire Hazard Analysis Methods for the U.S. Nuclear Regulatory Commission Fire Protection Inspection Program*, 2004.
8. NUREG-1824 (EPRI 1011999), *Verification and Validation of Selected Fire Models for Nuclear Power Plant Applications*, 2007.
9. NUREG/CR-6850 (EPRI 1011989), *EPRI/NRC-RES Fire PRA Methodology for Nuclear Power Facilities*, 2005.
10. NUREG/CR-6931, *Cable Response to Live Fire (CAROLFIRE), Volume 3, Thermally-Induced Electrical Failure (THIEF) Model*, 2007.
11. *SFPE Handbook of Fire Protection Engineering*, 4th edition, 2008.

C.8 Attachments (on CD)

1. FDS input file: Pump_Room.fds

2. MAGIC input file:

    a. Pump_Room.cas
    b. Pump_Room_tighter_wrapping.cas
    c. Pump_Room_thinner_wrapping.cas
    d. Pump_Room_2Doors.cas
    e. Pump_Room_extendedHRR.cas

APPENDIX D
MOTOR CONTROL CENTER FIRE IN A SWITCHGEAR ROOM

D.1 Modeling Objective

The calculations described in this appendix estimate the likelihood that a motor control cabinet (MCC) fire will damage various cables and an adjacent cabinet within a switchgear room (SWGR).

D.2 Description of the Fire Scenario

General Description: The SWGR is located in the reactor building for a Boiling Water Reactor (BWR). The compartment contains multiple MCCs and some other switchgear cabinets.

Geometry: The layout of the compartment is shown in Figure D-1. Figure D-2 shows the equipment typically contained in the compartment, and Figure D-3 shows the large elevation change between the high and low ceiling areas.

Materials: Property values for the relevant materials are listed in Table 3-1. The SWGR boundaries are made of concrete that is at least 0.6 m (2 ft) thick. The cabinet housing is 1.5 mm (0.06 in) thick steel.

Cables: The cable trays are filled with cross-linked polyethylene (XPE or XLPE) insulated cables with a neoprene jacket. These are considered thermoset (TS) materials. These cables have a diameter of approximately 1.5 cm (0.6 in), a jacket thickness of approximately 2 mm (0.79 in), 3 conductors, and a mass per unit length of 0.4 kg/m. Tray locations are shown in the compartment drawing. These particular cables have been shown to fail when the temperature just underneath the jacket reaches approximately 400 °C (750 °F) (NUREG/CR-6931, Vol. 2, Table 5.10[20]). A second criterion for damage is exposure to a heat flux that exceeds 11 kW/m^2 (NUREG-1805, Appendix A, Section A.5.4). Damage criteria for the adjacent cabinet are the same as for the cable trays because the cables within the cabinet are subjected to similar thermal exposure conditions as the steel cabinet housing.

Fire: A fire originates within a MCC cabinet. The cabinet is closed and contains more than one bundle of qualified cable. The fire grows following a "t-squared" curve to a maximum value of 702 kW in 12 min and remains steady for 8 more minutes, consistent with NUREG/CR-6850 (EPRI 1011989), Appendix G. After 20 min, the heat release rate (HRR) decays linearly to zero in 19 min. A peak fire intensity of 702 kW represents the 98th percentile of the probability distribution for HRR in cabinets of this general description.

The top of the cabinet contains a louvered air vent, 0.6 m (2 ft) long and 0.3 m (1 ft) wide. The cabinet is 2.4 m (8 ft) tall. The fire burns within the interior of the cabinet, and the smoke and

---

[20] The cable failure temperature is based on experiments conducted with Cable #14, an XLPE/CSPE, 3 conductor control cable.

flames exhaust from this vent, which has an area of 0.18 m² (2 ft²) and an effective diameter[21] of 0.48 m (1.6 ft).

The heat of combustion and product yields for XLPE/neoprene cable are obtained from Table 3-4.16 of the *SFPE Handbook, 4th edition*, and are listed in Table D-1. When estimating the composition of the fire's exhaust products, the jacket and insulation material of the cable are taken as an equal-parts mixture of polyethylene (C_2H_4) and neoprene (C_4H_5Cl), with an effective chemical formula of $C_3H_{4.5}Cl_{0.5}$.

Table D-1. Products of combustion for the MCC fire.

Parameter	Value	Source
Effective Fuel Formula	$C_3H_{4.5}Cl_{0.5}$	Combination of polyethylene and neoprene
Peak HRR	702 kW	NUREG/CR-6850 (EPRI 1011989), App. G
Time to reach peak HRR	720 s	NUREG/CR-6850 (EPRI 1011989), App. G
Heat of Combustion	10,300 kJ/kg	*SFPE Handbook*, 4th Ed., Table 3-4.16
CO_2 Yield	0.63 kg/kg	*SFPE Handbook*, 4th Ed., Table 3-4.16
Soot Yield	0.175 kg/kg	*SFPE Handbook*, 4th Ed., Table 3-4.16
CO Yield	0.082 kg/kg	*SFPE Handbook*, 4th Ed., Table 3-4.16
Radiative Fraction	0.53	*SFPE Handbook*, 4th Ed., Table 3-4.16

Ventilation: The compartment is normally supplied with three air changes per hour (ACH). The supply and return vents are indicated on the drawing. The two doors are normally closed. Normal heating, ventilation, and air conditioning (HVAC) operations continue during the fire, and the doors remain closed. The volume of the compartment is 882 m³ (31,150 ft³); thus, three air changes per hour is equivalent to a volume flow rate of 0.735 m³/s.

---

[21] The effective diameter is calculated as $D = \sqrt{4A/\pi}$, where A is the area of the vent opening.

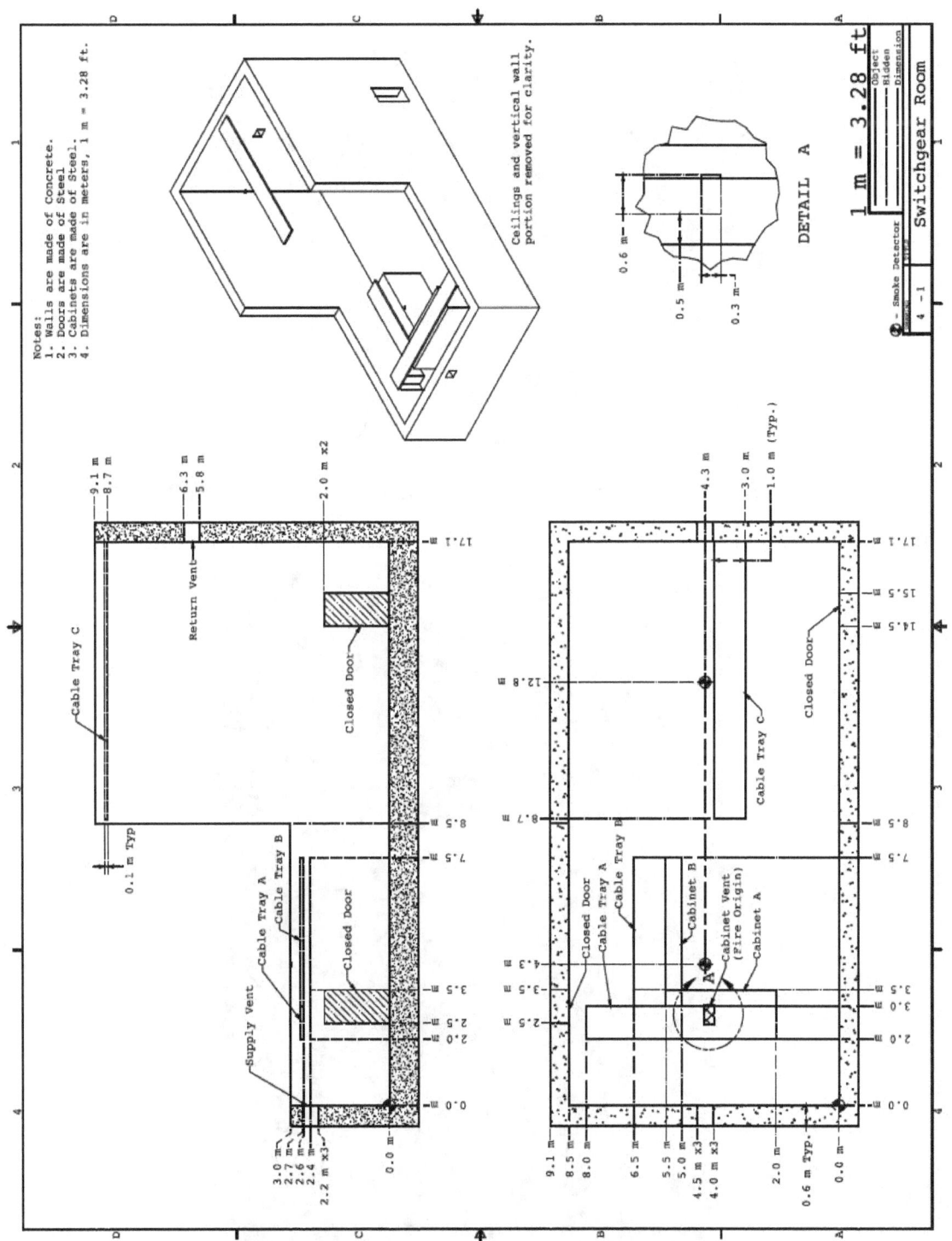

Figure D-1. Geometry of the MCC/SWGR in a BWR.

Figure D-2. Typical electrical cabinet in the lower part of the SWGR.

Figure D-3. View of the high ceiling space.

D.3 Selection and Evaluation of Fire Models

This section describes the overall modeling strategy, the selection of models, and a discussion of the validation exercises justifying the use of these models for this scenario. The discussion also separately addresses the prediction of the temperature and heat flux to the cabinet and cable targets.

D.3.1 Temperature Criterion

The primary temperature criterion of interest for this scenario is the cable temperature in Tray A, which is located 0.2 m (8 in) directly above the specified cabinet fire. The Heskestad flame height correlation included in the Fire Dynamics Tools (FDT[s]) and the Fire-Induced Vulnerability Evaluation (FIVE-Rev1) is used to show if the cables in Tray A would be engulfed in the cabinet fire flames, and would therefore be expected to fail.

The hot gas layer (HGL) temperature correlations included in the FDT[s] and FIVE-Rev1 are based on compartment fire experiments with a relatively uniform ceiling height. They are also based on fires located near floor level without flame impingement on the ceiling. For these reasons, the McCaffrey, Quintiere, and Harkleroad (MQH) and Foote, Pagni, and Alvares (FPA) temperature correlations are not appropriate for this scenario. Because the zone models Consolidated Fire Growth and Smoke Transport Model (CFAST) and MAGIC are designed primarily for compartments with relatively flat ceilings, the Fire Dynamics Simulator (FDS) will be used as a check on these simpler models. The differing ceiling heights are not an issue for FDS; the compartment geometry is input as is, with no need for further simplification. In a case like this, it may be convenient to use two rectangular meshes instead of one. The two meshes conform well to the actual geometry and enable the calculation to be run in parallel on two processors instead of one if desired.

D.3.2 Heat Flux Criterion

For this application, the point source radiation model included in the FDT[s] and FIVE-Rev1 can be used to estimate the radiation heat flux to the adjacent cabinet. However, this heat flux calculation does not consider the influence of flame extension beneath the ceiling, so it is useful primarily for screening purposes. The zone and/or the computational fluid dynamics (CFD) model are needed to calculate the heat flux from the HGL in the lower ceiling space if the HGL temperature calculation of any model indicates that the temperature of the layer is sufficiently high to cause damage.

D.3.3 Validation

The principal source of validation data justifying the use of the above-listed fire models for this scenario is the U.S. Nuclear Regulatory Commission/Electric Power Research Institute (NRC/EPRI) verification and validation (V&V) study documented in NUREG-1824 (EPRI 1011999). The National Institute of Standards and Technology (NIST) has expanded the NRC/EPRI V&V to include the latest versions of CFAST (6.1.1) (Peacock, 2008) and FDS (5.5.3) (McGrattan, 2010). Also, CFAST and FDS utilize the cable failure algorithm, THIEF, which was developed and validated in NUREG/CR-6931 (Vol. 3).

Table D-2 lists various important model parameters and the ranges for which the validation study is applicable. Two of the parameters, the Fire Froude Number and the Fire Height to Ceiling Height ratio, fall outside the listed validation ranges. The Fire Froude Number is essentially a measure of the fire's heat output relative to its base area. In this example, the fire is specified as originating at the top of the cabinet, with the vent opening serving as its base. $\dot{Q}^*$ for this scenario will be higher than would be calculated if the fire was burning completely outside of the cabinet. Thus, the high value of $\dot{Q}^*$ is the result of specifying conditions that will produce a more severe fire conditions than would be expected if the fire were to burn partially within the cabinet.

With the fire located on top of a 2.4 m (7.9 ft) high cabinet under a 3.0 m (9.8 ft) high ceiling, the flame impinges on the ceiling and consequently falls outside of the NUREG-1824 (EPRI 1011999) validation range. The extension of the flame beneath the ceiling would have an influence on the heat flux to adjacent cabinets, which needs to be addressed. The plume correlations used within the spreadsheets and the zone models are not valid when the fire impinges on the ceiling. The MQH and FPA compartment temperature correlations do not explicitly address fire elevation, but are based on experiments in which the fire source was located near floor level; they have not been validated for scenarios like this one, which have significant flame impingement on the ceiling. Zone models use plume air entrainment correlations in estimating the temperature and depth of the HGL, so they have limited validity for scenarios with significant flame impingement. For this reason, FDS is used to check the results of the simpler models.

The FDS Validation Guide (NIST SP 1018) lists two sets of experiments that involve flame impingement on the ceiling. The first set, performed in support of the NIST investigation of the collapse of the World Trade Center, involved 3 MW heptane spray fires under a 3.8 m (12.5 ft) ceiling. The value of $\dot{Q}^*$ was 2.7, and the ratio of flame to ceiling height was 1.2. A second set of experiments was performed by the Swedish fire test laboratory SP. The HRR from the propane burner was 450 kW, and the burner was elevated 0.65 m off the floor. The burner area was only 0.3 m by 0.3 m (1 ft by 1 ft), and $\dot{Q}^*$ was approximately 6. The flame to ceiling height ratio was 1.2, but, because the fire was in the corner, the degree of flame impingement was enhanced significantly. Predicted HGL temperatures for these experiments were within 10 % of the reported measurements. Predicted target temperatures were within 20 %.

The second important issue in regard to model validation is the two-tiered ceiling. Although none of the experiments used in the NRC/EPRI validation study have a similar ceiling configuration, Benchmark Exercise #2 of this study provides validation data to evaluate the models' ability to estimate the plume and HGL temperature/depth of smoke and hot gases filling a fairly large, open hall with an angled roof. Predicted temperatures were within approximately 10 % of the reported measurements.

Table D-2. Normalized parameter calculations for the MCC fire scenario.

Quantity	Normalized Parameter Calculation		NUREG-1824 Validation Range	In Range?
Fire Froude Number	$$\dot{Q}^* = \frac{\dot{Q}}{\rho_\infty c_p T_\infty D^{2.5}\sqrt{g}}$$ $$= \frac{702 \text{ kW}}{(1.2 \text{ kg/m}^3)(1.0 \text{ kJ/kg/K})(293 \text{ K})(0.5^{2.5} \text{ m}^{2.5})\sqrt{9.8 \text{ m/s}^2}} \cong 3.6$$		0.4 – 2.4	No
Flame Length, $H_f + L_f$, relative to the Ceiling Height, H_c	$$\frac{H_f + L_f}{H_c} = \frac{2.4 \text{ m} + 2.5 \text{ m}}{3.0 \text{ m}} \cong 1.6$$ $$L_f = D\left(3.7\,\dot{Q}^{*2/5} - 1.02\right) = 0.48 \text{ m }(3.7 \times 3.6^{0.4} - 1.02) \cong 2.5 \text{ m}$$		0.2 – 1.0	No
Ceiling Jet Radial Distance, r_{cj}, relative to the Ceiling Height, H_c	N/A – There are no targets like sprinklers or smoke detectors under consideration in this example.		1.2 – 1.7	N/A
Equivalence Ratio, φ, as an indicator of the Ventilation Rate	$$\varphi = \frac{\dot{Q}}{\Delta H_{O_2}\,\dot{m}_{O_2}} = \frac{702 \text{ kW}}{13,100 \text{ kJ/kg} \times 0.2 \text{ kg/s}} \cong 0.3$$ $$\dot{m}_{O_2} = 0.23\,\rho_\infty \dot{V} = 0.23 \times 1.2 \text{ kg/m}^3 \times 0.735 \text{ m}^3/\text{s} \cong 0.2 \text{ kg/s}$$		0.04 – 0.6	Yes
Compartment Aspect Ratio (Lower | Upper)	$\frac{L}{H_c} = \frac{8.5 \text{ m}}{3.0 \text{ m}} \cong 2.8$; $\frac{W}{H_c} = \frac{8.5 \text{ m}}{3.0 \text{ m}} \cong 2.8$	$\frac{L}{H_c} = \frac{8.6 \text{ m}}{9.1 \text{ m}} \cong 0.9$; $\frac{W}{H_c} = \frac{8.5 \text{ m}}{9.1 \text{ m}} \cong 0.9$	0.6 – 5.7	Yes
Target Distance, r, relative to the Fire Diameter, D	$$\frac{r}{D} = \frac{1.1 \text{ m}}{0.5 \text{ m}} \cong 2.2$$		2.2 – 5.7	Yes

Notes:

(1) The effective diameter of the fire is determined from the formula, $D = \sqrt{4A/\pi}$, where A is the area of the vent on the cabinet.

(2) The Flame Length, $H_f + L_f$, is the sum of the height of the fire from the floor and the fire's calculated flame length.

(3) The ceiling height, H_c, is the lower of the two ceiling heights.

D.4 Estimation of Fire-Generated Conditions

This section describes how each of the models is used in the analysis, including specific approximations unique to the particular model.

D.4.1 Algebraic Models

Fire: The FDT^s use a steady-state HRR in both the flame height and radiation heat flux calculation. A constant HRR of 702 kW is used for both. A fire diameter of 0.48 m (1.6 ft) is calculated from the effective vent area atop the cabinet of 0.18 m^2 (2 ft^2). Table D-2 indicates that the Heskestad flame height correlation yields a calculated flame height of 2.5 m (8.2 ft). Consequently, the cables located directly above the cabinet would be engulfed in flame and would therefore be expected to fail. This flame height calculation also shows that there would be significant flame extension beneath the ceiling, which is located just 0.6 m (2 ft) above the base of the fire.

The point source radiation model predicts the peak heat flux to the side of the adjacent cabinet, which is approximately 1.1 m (3.6 ft) from the center of the vent on top of the burning cabinet:

$$\dot{q}'' = \frac{\chi_r \, \dot{Q}}{4\pi \, r^2} = \frac{0.53 \times 702 \text{ kW}}{4\pi \times (1.1 \text{ m})^2} \cong 24.5 \text{ kW/m}^2 \tag{D-1}$$

This estimate does not include any contributions to the heat flux by the HGL or by the flame extension beneath the ceiling. However, it does indicate that the heat flux to the adjacent cabinet could exceed the critical heat flux by a relatively large margin. Consequently, this scenario would warrant more detailed analysis with either a zone model or a CFD model.

D.4.2 Zone Models

Geometry: Zone fire models subdivide the space of interest into one or more compartments connected by vents. With CFAST, the single, large compartment is modeled as two adjacent compartments, connected by a vertical vent. Figure D-4 and Figure D-5 show the geometry and input parameters of the CFAST calculation.

Fire: Following guidance in NUREG/CR-6850 (EPRI 1011989), the fire is placed near the top of the cabinet. It is positioned directly below Cable Tray A to maximize exposure. The current version of CFAST only uses the plume centerline temperature. The specified fire area, HRR, and species yields are input directly into the model. Figure D-6 shows the fire inputs for the CFAST calculation.

Cables: CFAST uses the Thermally-Induced Electrical Failure (THIEF) methodology developed as part of the Cable Response to Live Fire (CAROLFIRE) program (NUREG/CR-6931, Vol. 3). The thermal conductivity and specific heat are fixed constants. Cable density is calculated from the specified mass per unit length, m', and the cross-sectional area of the cable, A:

$$\rho = \frac{m'}{A} = \frac{0.4 \text{ kg/m}}{\pi \times (0.015/2)^2 \text{ m}^2} \cong 2264 \text{ kg/m}^3 \tag{D-2}$$

Electrical functionality is lost when the temperature just inside of the 2 mm (0.08 in) jacket reaches 400 °C (752 °F). Figure D-7 shows the target inputs for the CFAST calculation.

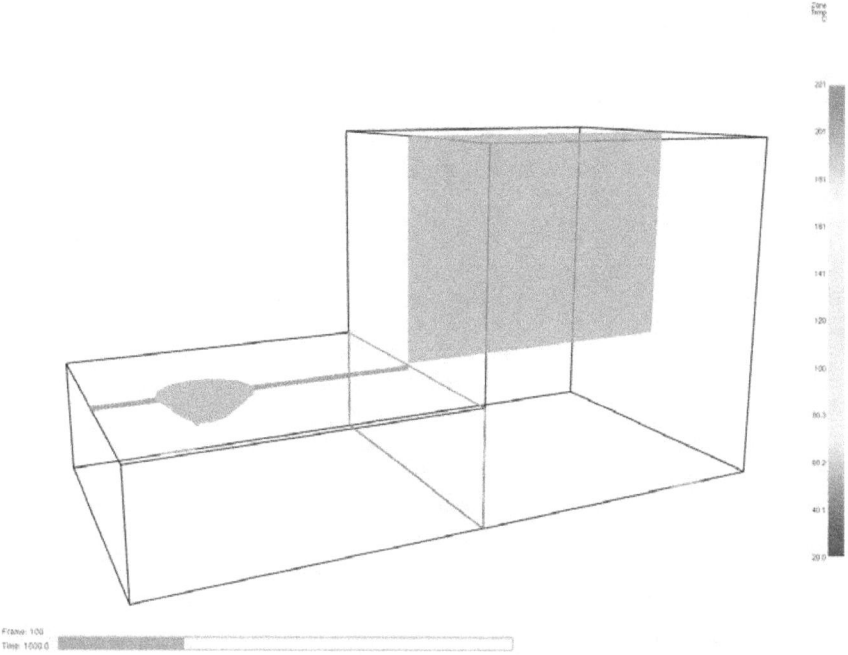

Figure D-4. Smokeview rendering of the geometry of two-height ceiling SWGR, as modeled in CFAST.

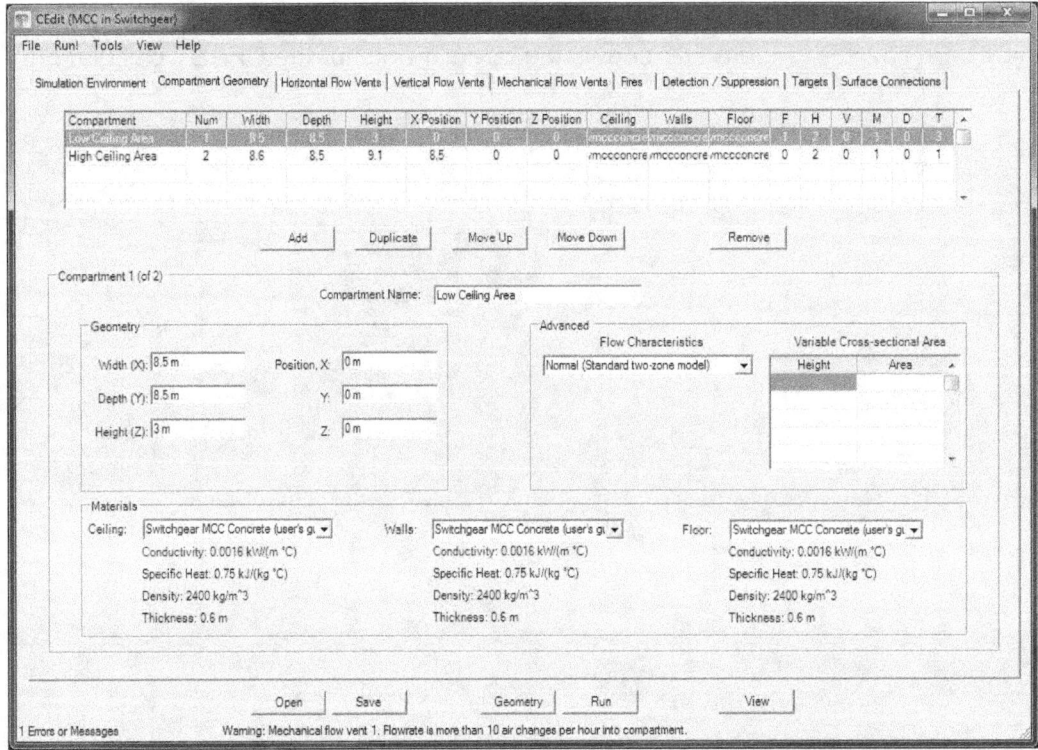

Figure D-5. CFAST inputs for compartment geometry for SWGR.

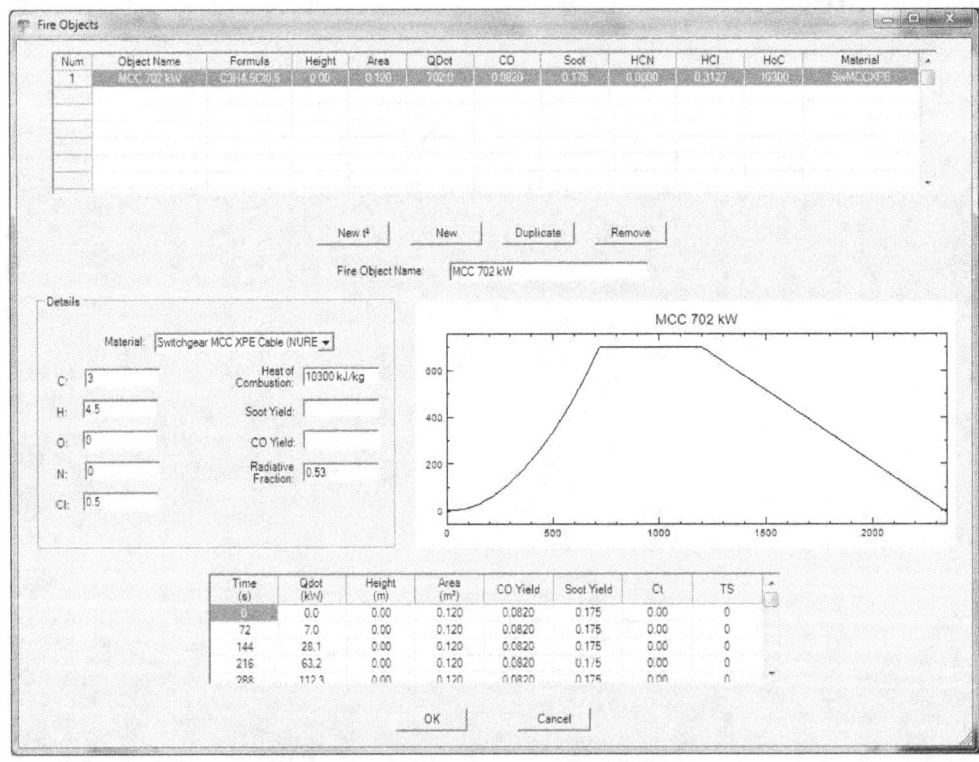

Figure D-6. CFAST fire inputs for two-height ceiling SWGR scenario.

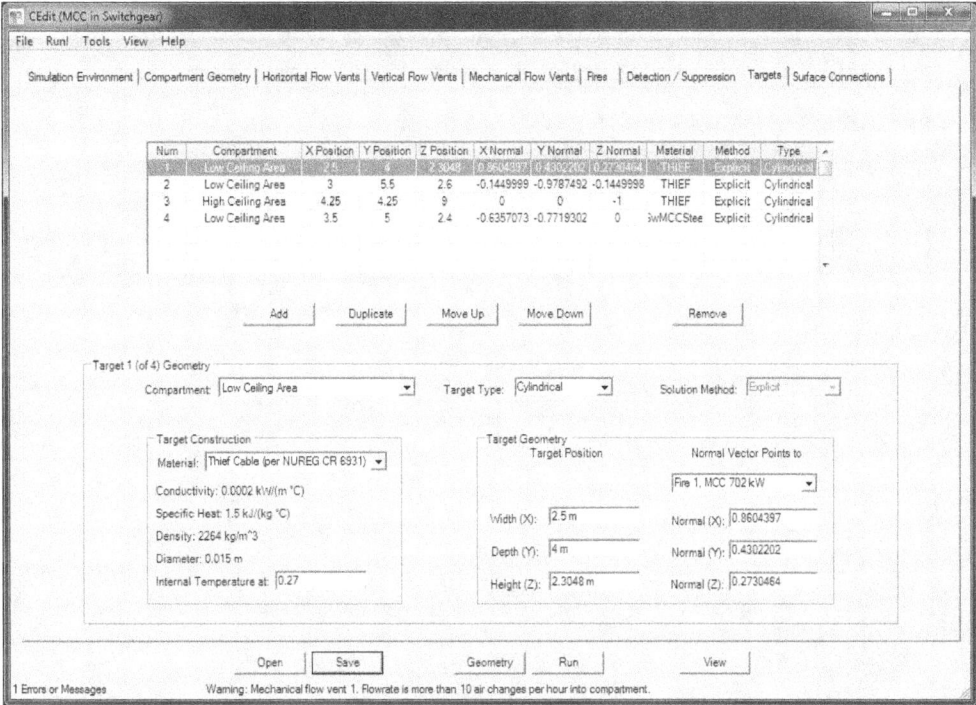

Figure D-7. CFAST target inputs for two-height ceiling SWGR scenario.

Ventilation: The two individual compartments used to model the entire room are connected by a single large vent. Although the size of this vent relative to the compartment size is not typical of a zone model application, the simple two-compartment geometry of the space and the more dominant mechanical ventilation flow from one side of the SWGR to the other should minimize any uncertainty in the calculation resulting from the large connecting vent. Mechanical ventilation is included at the specified height and with the specified volume flow applied to the single supply (in the low-ceiling space) and return (in the high-ceiling space). Additionally, since zone fire models use compartments that are completely sealed unless otherwise specified, a typical leakage vent, 13 mm (0.5 in) in height, is included at the bottom of each closed doorway to reflect the fact that the doorways are not totally airtight. Figure D-8 shows the CFAST inputs for these natural vents.

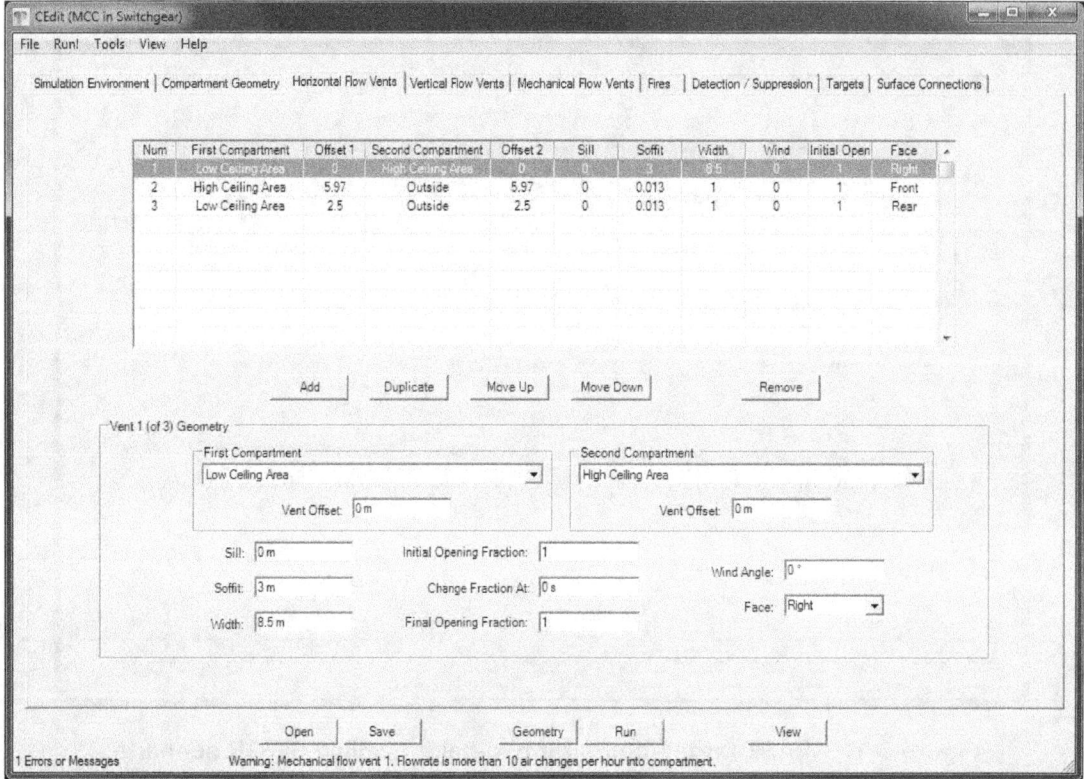

Figure D-8. CFAST inputs for vents connections in two-height ceiling SWGR scenario.

D.4.3 CFD Model

Geometry: The entire compartment is included in the computational domain. Multiple meshes are used, one for the low-ceiling space and three for the high-ceiling space. The *FDS User's Guide* (NIST SP 1019) contains detailed instructions for running the simulation on multiple computers. The concrete walls are essentially the boundaries of these two meshes. The electrical cabinets and cables are included in the simulation as simple rectangular solids, and their dimensions have been approximated to the nearest 10 cm (4 in). There is no attempt to model the details of either the cable trays or cabinets because the grid resolution is not fine enough. This is appropriate because the cables and cabinets are merely targets for which it is sufficient to know their bulk thermal properties. An FDS/Smokeview rendering of the scenario is shown in Figure D-9.

The numerical mesh consists of uniform grid cells, 10 cm (4 in) on a side. Even with this relatively fine grid, there is considerable uncertainty in the exact nature of the fire relative to the cabinet and the cables just above. This uncertainty mainly has to do with specifying that the fire originates at the top of the cabinet rather than from within.

Fire: The fire burns over an area of 0.6 m (2 ft) by 0.3 m (1 ft) on top of the cabinet, with a maximum HRR per unit area of 3,900 kW/m^2, yielding a peak HRR of 702 kW.

Cables: The cables are modeled as 1.5 cm (0.6 in) cylinders with uniform thermal properties. Following the THIEF methodology in NUREG/CR-6931, Vol. 3, electrical functionality is lost when the temperature just inside of the 2 mm (0.08 in) jacket reaches 400 °C (752 °F). Note that no attempt is made in the simulation to predict ignition and spread of the fire over the cables, which is why the in-depth heat penetration calculation is focused on a single cable. For this scenario, at least one cable per tray is expected to be relatively free of its neighbors and would heat up more rapidly than those buried deeper within the pile. This is conservative and simplifies the heat transfer calculation.

Ventilation: Three ACH are achieved with a volume flow of 0.735 m^3/s applied to the single supply and return vents. No other penetrations are included in the model.

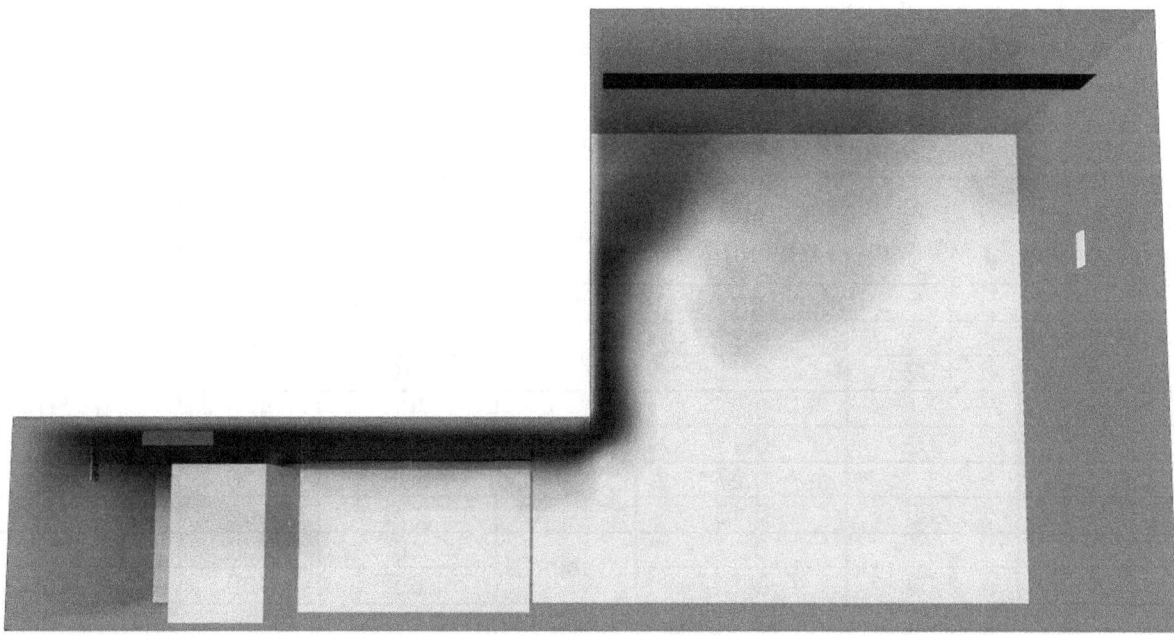

Figure D-9. FDS/Smokeview representation of the MCC/SWGR scenario.

D.5 Evaluation of Results

The purpose of the calculations described above is to predict if and when various components within the compartment become damaged due to a fire in the MCC. XLPE cables are expected to be damaged when their internal temperature exceeds 400 °C (750 °F) or the exposing heat flux exceeds 11 kW/m^2. Damage criteria for the adjacent cabinet are equivalent to those for the cables because the cables within the cabinet come in contact with the heated metal housing, and are therefore exposed to similar thermal conditions. The targets of interest are three cable trays, labeled A, B, and C, and a single electrical cabinet adjacent to the burning MCC (Figure D-1).

Table D-1 summarizes the model predictions of the target temperature and heat flux. Chapter 4 describes how to calculate the probability of exceeding the critical value. The purpose of the model summary is to readily identify the most likely targets to be damaged, and the model inputs underlying that assessment.

Table D-3. Summary of the model predictions of the MCC fire scenario.

Model	Bias Factor, δ	Standard Deviation, $\tilde{\sigma}_M$	Target	Predicted Value	Critical Value	Probability of Exceeding
Surface Temperature (°C), Initial Value = 20 °C						
CFAST	1	0.27	Cabinet	390	400	0.460
FDS	1.02	0.13		170	400	0.000*
CFAST	1	0.27	Cable A	705	400	0.950
FDS	1.02	0.13		620	400	0.997
CFAST	1	0.27	Cable B	305	400	0.112
FDS	1.02	0.13		280	400	0.000
CFAST	1	0.27	Cable C	40	400	0.000
FDS	1.02	0.13		65	400	0.000
Heat Flux (kW/m^2)						
CFAST	0.81	0.47	Cabinet	24.3	11	0.911
FDS	0.85	0.22		6.0	11	0.006*
CFAST	0.81	0.47	Cable A	104	11	0.974
FDS	0.85	0.22		75.0	11	1.000
CFAST	0.81	0.47	Cable B	15.8	11	0.823
FDS	0.85	0.22		23.0	11	0.997
CFAST	0.81	0.47	Cable C	0.2	11	0.000
FDS	0.85	0.22		2.5	11	0.000

* These results require closer scrutiny. See discussion below.

D.5.1 Damage to Cabinet

The predicted heat flux to and temperatures of the cabinet adjacent to the MCC are shown in Figure D-10. The cabinet is located approximately 1.1 m (3.6 ft) from the center of the flaming vent. The point source radiation calculation included in CFAST and the FDT^s predicts a peak heat flux of 24.5 kW/m^2 to the nearest point on the cabinet. FDS predicts the peak heat flux (and resulting surface temperature) to be significantly lower because the fire is partially obscured by the overhead cable tray and the burning MCC. FDS also accounts for the orientation of the adjacent cabinet top and side relative to the fire's location. However, the heat flux to the top of the MCC near the adjacent cabinet is substantially greater than the critical value, and a small change in the position of the fire could result in a much higher heat flux to the target. Given the sensitivity of the predicted heat flux and surface temperature to a minor change in the fire dynamics, the FDS prediction for the cabinet ought to be discounted in this case.

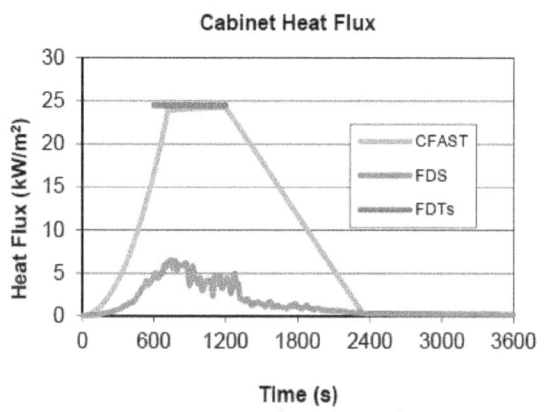 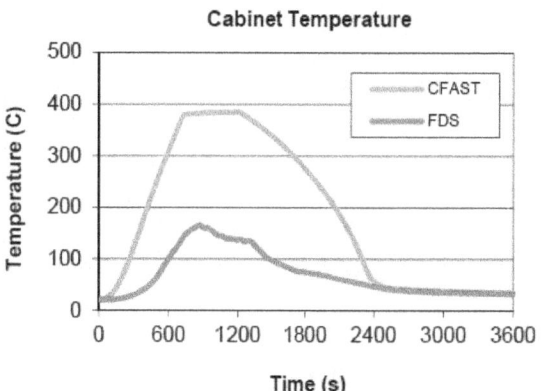

Figure D-10. Heat flux and temperature predictions for the adjacent cabinet.

D.5.2 Cable Damage Based on Temperature Alone

The predicted cable temperatures for the three trays are shown in Figure D-11. CFAST and FDS estimate cable temperatures using the THIEF methodology (NUREG/CR-6931, Vol. 3). Both models predict that the cables in Tray A are likely to fail.

Neither model predicts that the cables in Tray B will reach the failure temperature of 400 °C (750 °F), but the CFAST prediction of 300 °C (572 °F) suggests that there is a 9% probability that the cable temperature could be as high as the critical value. Note that these predictions are sensitive to the exact location of the target cable within the tray, its view of the fire, and the HGL temperature. In this case, the cables in Tray B are heated primarily by convection and radiation from the HGL. Given that the HRR is the most important parameter controlling the temperature of the HGL, how much would the HRR have to increase to increase the CFAST prediction from 300 °C (572 °F) to 400 °C (752 °F)? Table 4-3 indicates that the rise in the HGL temperature is proportional to the HRR to the 2/3 power. Following the methodology in Section 4.4.1, in order to increase the predicted HGL temperature by 100 °C (212 °F), the peak HRR, $\dot{Q}$, must increase by approximately:

$$\Delta\dot{Q} = \frac{3}{2}\dot{Q}\frac{\Delta T}{T - T_0} = \frac{3}{2}\,702\,\text{kW} \times \frac{100\,°\text{C}}{300\,°\text{C} - 20\,°\text{C}} \cong 376\,\text{kW} \tag{D-3}$$

Both FDS and CFAST predicted cable temperatures for Tray C indicate that the cables are unlikely to fail.

D.5.3 Cable Damage Based on Incident Heat Flux

The predictions of heat flux to the cables in the three trays are shown in Figure D-11. The critical value is 11 kW/m^2. Flame height correlations predict that the fire will impinge on Tray A, and both CFAST and FDS indicate that the heat flux to these cables would be well in excess of the critical value.

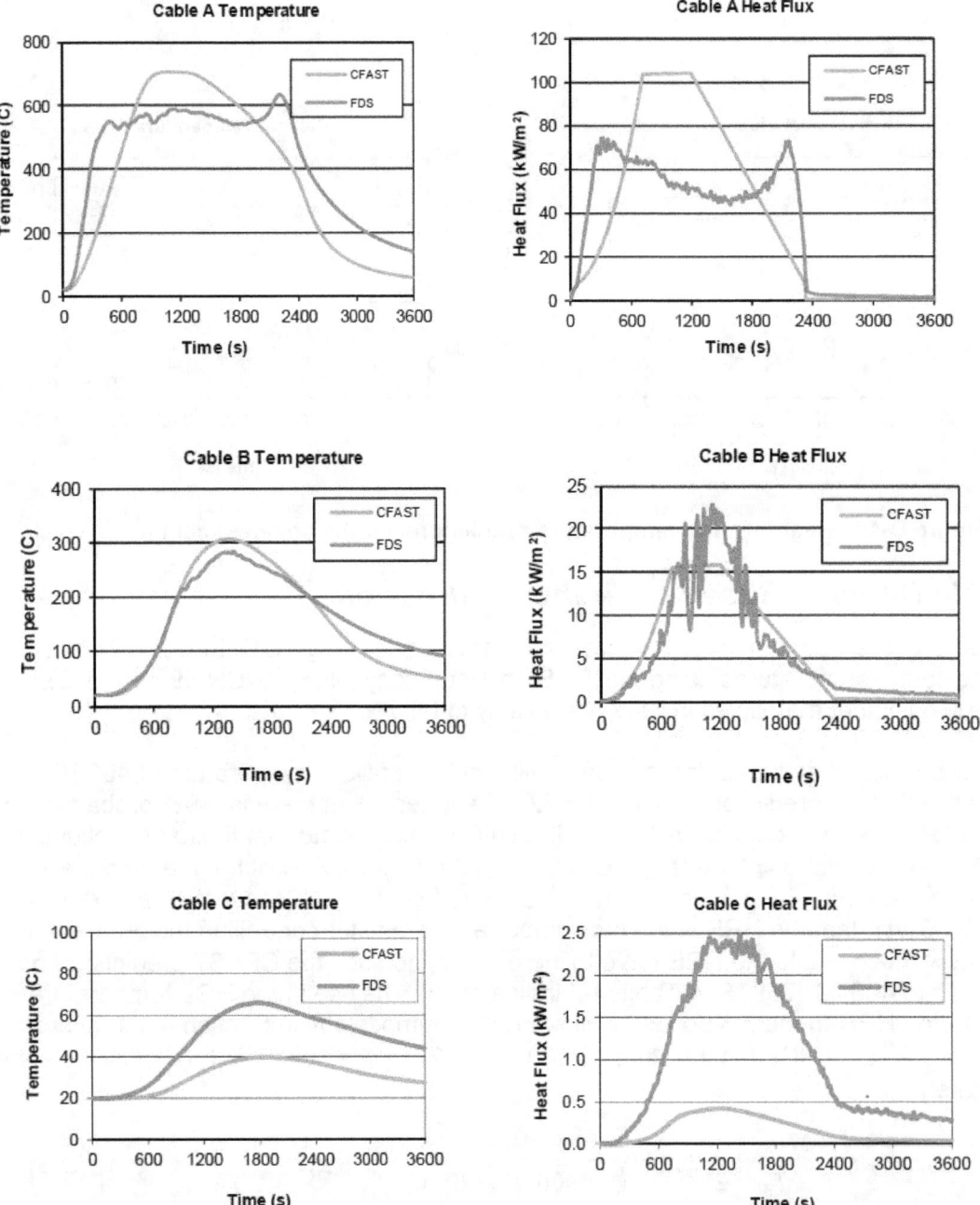

Figure D-11. Summary of the cable temperature and heat flux predictions for the MCC/SWGR.

For the cables in Tray B, both FDS and CFAST indicate a relatively high probability of failure, based partly on the fact that both models have been shown to under-predict heat flux in the NRC/EPRI fire model validation study (NUREG-1824).

For the cables in Tray C, none of the models predicts a heat flux that approaches the critical value. Note that the significant difference between FDS and CFAST is based on the fact that FDS outputs the maximum heat flux and cable temperature over the entire tray of cables, whereas CFAST predicts these values at a single target location.

D.6 Conclusion

The purpose of the calculations in this example is to predict if and when various components within a compartment will become damaged due to a fire in the MCC. The fire model analyses performed for this scenario indicate that the fire would damage the cables in Tray A because all the models (FDT^s, CFAST, FDS) predict that the flames would directly impinge on the cables themselves.

- CFAST and FDS predict that the cables in Tray B are likely to be damaged based on the heat flux criterion. However, neither model predicts that the interior cable temperatures are likely to be high enough to cause failure.

- Neither FDS nor CFAST predicts that the cables in Tray C would be damaged.

- A point source heat flux analysis indicates that the adjacent cabinet housing would be exposed to a heat flux that would cause damage. Even though FDS does not predict damage, its predictions of heat flux to surfaces very near the adjacent cabinet are sufficiently high to cast doubt on the conclusion that the cabinet would not be damaged. Small changes in the positions of various obstructions could easily change the predicted heat flux by an order of magnitude. Even though the point source method tends to over-predict the heat flux to targets close to the fire, there is too much uncertainty in the geometric configuration to accept the validity of the more detailed calculation.

D.7 References

1. NIST SP 1018-5, *Fire Dynamics Simulator (Version 5), Technical Reference Guide, Volume 3, Experimental Validation*, 2010.
2. NIST SP 1019, *Fire Dynamics Simulator (Version 5) User's Guide*, 2007.
3. NIST SP 1086, *CFAST – Consolidated Model of Fire Growth and Smoke Transport (Version 6), Software Development and Model Evaluation Guide*, 2008.
4. NUREG-1805, *Fire Dynamics Tools (FDT^s) Quantitative Fire Hazard Analysis Methods for the U.S. Nuclear Regulatory Commission Fire Protection Inspection Program*, 2004.
5. NUREG-1824 (EPRI 1011999), *Verification and Validation of Selected Fire Models for Nuclear Power Plant Applications*, 2007.
6. NUREG/CR-6850 (EPRI 1011989), *EPRI/NRC-RES Fire PRA Methodology for Nuclear Power Facilities*, 2005.
7. NUREG/CR-6931, *Cable Response to Live Fire (CAROLFIRE), Volume 2: Cable Fire Response Data for Fire Model Improvement*, 2007.
8. NUREG/CR-6931, *Cable Response to Live Fire (CAROLFIRE), Volume 3: Thermally-Induced Electrical Failure (THIEF) Model*, 2007.

9. *SFPE Handbook of Fire Protection Engineering*, 4th edition, 2008.

D.8 Attachments (on CD)

1. FDS input file: Switchgear_Room_MCC.fds

2. CFAST input files:

    a. MCC in Switchgear.in

APPENDIX E
TRANSIENT FIRE IN CABLE SPREADING ROOM

E.1 Modeling Objective

The calculations in this appendix estimate the impact that a fire in a trash receptacle inside a Cable Spreading Room (CSR) would have on safe-shutdown cables. These calculations are part of a larger fire analysis described in Chapter 11 of NUREG/CR-6850 (EPRI 1011989), Volume 2, "Detailed Fire Modeling (Task 11)." The CSR contains a large quantity of redundant instrumentation and control cables needed for plant operation. Transient combustibles have been identified as a possible source of fire that may impact the cables. The purpose of the calculation is to analyze this condition and determine whether the cable targets will fail, and, if so, at what time failure occurs. The time to smoke detector activation is also estimated. The calculation will provide information for a decision on the hazard and risk for this scenario.

E.2 Description of the Fire Scenario

General Description: The CSR contains a large quantity of redundant instrumentation and control cables needed for plant operation. The cables are installed in either ladder-back trays or conduits.

Geometry: Figure E-1 illustrates the geometry of the CSR; Figure E-2 shows a photograph of the CSR. In addition to cables, the CSR contains a fully enclosed computer compartment, ductwork, and large structural beams. There is no high- or medium-voltage equipment (switchgears or transformers) in the compartment. As shown in Figure E-3, the top 2.2 m (7.2 ft) of the compartment is filled with cable trays containing cables, or ductwork, or large structural beams.

Construction: The walls, floor, and ceiling of the CSR are constructed of normal-weight concrete. The ductwork is made of 2 mm (0.08 in) thick steel.

Materials: Thermal properties of various materials in the compartment have been taken from Table 2-3 of NUREG-1805 and are listed in Table 3-1. The important cables for this calculation are located in the third and sixth trays above the fire source, which are filled with PE-insulated, PVC-jacketed control cables important to safe shutdown. These cables have a diameter of approximately 1.5 cm (0.6 in), a jacket thickness of approximately 1.5 mm (0.06 in), and 7 conductors. The cable mass per unit length is 0.4 kg/m. There are approximately two rows of cables per tray. These cables are damaged when the internal cable temperature reaches 205 °C (400 °F) or the exposure heat flux reaches 6 kW/m^2 (NUREG-1805, Appendix A). Cable insulation thermal properties are as follows: the density is 1380 kg/m^3, the thermal conductivity is 0.192 W/m/K, and the specific heat is 1.289 kJ/kg-K (NUREG/CR-6850, Volume 2, Appendix R). The copper in the cables has the following properties: the density is 8954 kg/m^3, the thermal conductivity is 386 W/m/K, and the specific heat is 0.3831 kJ/kg-K (*SFPE Handbook, 4th Edition*). The copper mass fraction of the cable is 0.67.

TRANSIENT FIRE IN CABLE SPREADING ROOM

The bottom cable tray is protected on the lower surface by a solid metal barrier. The bottom and side surfaces of all cable trays in the CSR are of solid metal construction; top surfaces are open. The tray steel thickness is 3 mm (0.12 in).

Detection System: Smoke detectors are located on the ceiling at locations shown in Figure E-1. The detectors are UL-listed with a nominal sensitivity of 4.9 %/m.

Suppression System: A total flooding CO_2 fire suppression system is initiated automatically by cross-zoned smoke detection in the compartment, or can be operated manually. In order to maintain a proper concentration of suppression agent, CO_2 discharge causes fire dampers to close and mechanical ventilation fans to stop. Activation of the CO_2 system is not modeled in this example.

Ventilation: The CSR has two doors on the east wall that are normally closed. Each door is 2 m (6.6 ft) wide by 2 m (6.6 ft) tall, with a 1 cm (0.4 in) gap along the floor. Standard procedure calls for an operator to investigate the fire within 600 s (10 min) of an alarm condition. For this reason, the door that is farthest from the fire is opened for this investigation.

There are two supply and two return vents for mechanical ventilation, each with an area of 0.25 m² (2.7 ft²). The total air supply rate is 1.4 m³/s (3,000 cfm). All vents are located 2.4 m (8 ft) above the floor. Upon smoke detector activation, the mechanical ventilation fans stop and the dampers close.

Fire: A trash fire ignites within a cylindrical steel waste bin, 0.8 m (2.6 ft) high and 0.6 m (2.0 ft) in diameter, containing 5 kg (11 lb) of trash. The heat release rate (HRR) of the transient fire is estimated using NUREG/CR-6850 (EPRI 1011989). The fire grows following a "t-squared" curve to a maximum value of 317 kW (the 98th percentile value from Table G-1 in NUREG/CR-6850 (EPRI 1011989)) in 480 s (consistent with NUREG/CR-6850 (EPRI 1011989) Supplement 1 for a transient fire growth rate contained within a trash can). The fire burns at its maximum HRR value until the trash is consumed. To determine the duration of the fire, the total energy of the fuel, Q, is calculated as the product of the fuel mass and the heat of combustion: (5 kg)×(30,400 kJ/kg) = 152,000 kJ. The heat of combustion is described below. The time to consume the fuel is found by integrating the heat release rate over the time of the burn:

$$Q \equiv 152,000 \text{ kJ} = \int_0^{480} \dot{Q}_p \left(\frac{t}{480}\right)^2 dt + \int_{480}^{t_f} \dot{Q}_p \, dt = 317 \text{ kW} \left(\frac{480 \text{ s}}{3} + \left(t_f - 480 \text{ s}\right)\right) \qquad \text{(E-1)}$$

Here, $\dot{Q}_p$ is the peak HRR of 317 kW, which is reached following a t-squared growth curve in 480 s. Solving for t_f yields a total burning time of 799 s, which is rounded up to 800 s.

While the exact composition of the trash is unknown, typical waste at the plant includes wood scraps and polyethylene protective suits. For this scenario, the trash is specified as being comprised of equal parts of each. The radiative fraction[22] and product yields, like the heat of combustion, are taken to be averages of values for red oak and polyethylene (*SFPE Handbook*, 4th ed., Table 3-4.16) and shown in Table E-1. For the purpose of modeling, the fuel molecule is specified as $C_4H_7O_{2.5}$.

---

[22] The fraction of the fire's total energy emitted as thermal radiation.

Table E-1. Products of combustion for CSR fire.

Parameter	Value	Source
Effective Fuel Formula	$C_4H_7O_{2.5}$	Specified
Peak HRR	317 kW	NUREG/CR-6850 (EPRI 1011989), App. G
Time to reach peak HRR	480 s	NUREG/CR-6850 (EPRI 1011989), App. G
Heat of Combustion	30,400 kJ/kg	*SFPE Handbook*, 4th ed., Table 3-4.16
CO_2 Yield	2.0 kg/kg	*SFPE Handbook*, 4th ed., Table 3-4.16
Soot Yield	0.038 kg/kg	*SFPE Handbook*, 4th ed., Table 3-4.16
CO Yield	0.014 kg/kg	*SFPE Handbook*, 4th ed., Table 3-4.16
Radiative Fraction	0.40	*SFPE Handbook*, 4th ed., Table 3-4.16

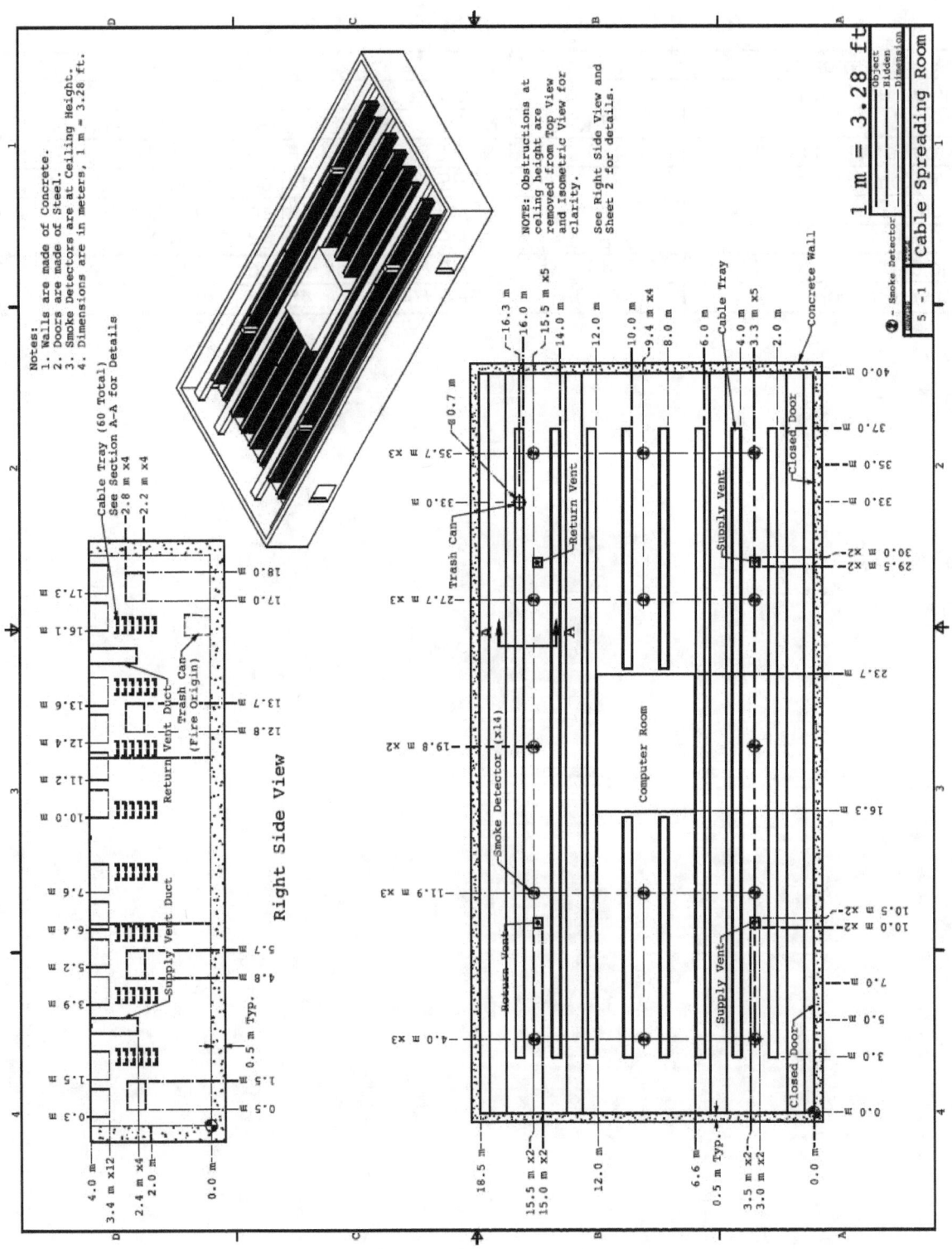

Figure E-1. Geometry of the CSR.

Figure E-2. Photograph of the CSR used for analysis. Note that the cables are located in the trays in the overhead.

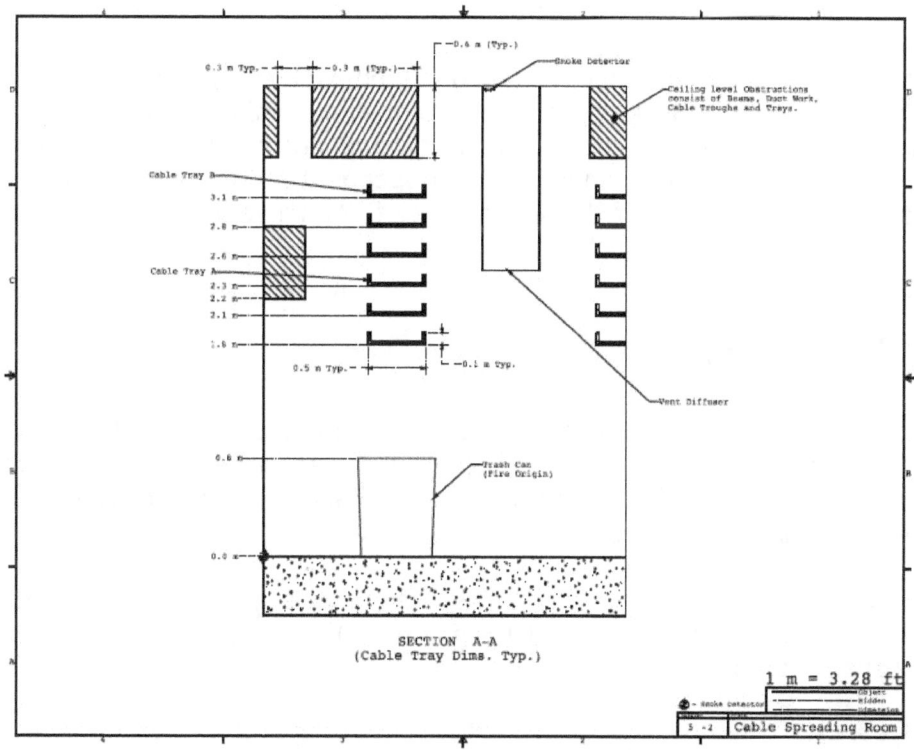

Figure E-3. Geometric detail of the CSR.

E.3 Selection and Evaluation of Fire Models

This section discusses the overall modeling strategy. In particular, it describes the process of model selection, including a discussion of the validity of these models for the given fire scenario. This scenario is a typical application of both zone models and computational fluid dynamics (CFD) models. With a single fire contained within a rectangular compartment, the application is straightforward. The general strategy is to use a plume correlation to estimate temperature at the height of the important cables. If this result indicates that the cables are likely to fail, a more detailed estimate of cable temperature will be made with zone and CFD models.

Cable failure is typically judged on both elevated temperature and heat flux criteria. These cables are damaged when the internal cable temperature reaches 205 °C (400 °F) or the exposure heat flux reaches 6 kW/m^2 (NUREG-1805, Appendix A).

E.3.1 Temperature Criterion

The algebraic models are used in this scenario to estimate the flame height and plume temperatures from the transient trash can fire to determine whether the overhead cable trays would be damaged by the trash fire. The Foote, Pagni, and Alvares (FPA) correlation is used to estimate the average hot gas layer (HGL) temperature that would result from the trash fire alone; this analysis does not consider either the potential ignition of cable trays or their potential contribution to the fire HRR.

The Consolidated Fire Growth and Smoke Transport Model (CFAST) and the Fire Dynamics Simulator (FDS) are used to estimate damage to the overhead cable trays and development of elevated temperature and heat flux from the fire and any ignited overhead cables to estimate damage to adjacent cables. CFAST imposes time-dependent conditions on the cable trays, but uses a point source for radiation from the fire sources. FDS will spread the fire in a more realistic way, which should provide more accurate estimates of temperature and heat flux to adjacent cables.

E.3.2 Heat Flux Criterion

CFAST and FDS are used to estimate heat flux to the overhead cable. The heat flux calculation in CFAST includes the radiation from the fire, upper and lower gas layers, and bounding surfaces, as well as the convection from nearby gases. Radiation from the fire sources uses a point source for the radiation calculation.

FDS models the fire in much the same way as the zone models, with the fire at the top of the trash receptacle. However, because it is a CFD model, FDS can estimate local conditions at the specific location of the target cables.

E.3.3 Validation

NUREG-1824 (EPRI 1011999) contains experimental validation results for CFAST and FDS that are appropriate for this scenario. These experiments include ventilation effects on, heat fluxes to, and temperatures of various targets, particularly cables. Fire sizes in these experiments bound those used in this scenario. For CFAST, the Software Development and Model Evaluation Guide, NIST SP 1086, includes updated validation results for the newest version of the model used for this calculation. This includes all of the validation comparisons from NUREG-1824 (EPRI 1011999), plus additional comparisons for experiments not included in the U.S. Nuclear Regulatory Commission (NRC) guide. Plume temperature calculations have been validated for a broad range of fire sizes and distances above the fire source in NIST SP 1086. Also, CFAST and FDS use the cable failure algorithm, THIEF, developed and validated in NUREG/CR-6931 (Vol. 3).

Table 2-5 of Volume 1 of NUREG-1824 (EPRI 1011999) lists various important model parameters and the ranges for which the validation study is applicable. Table E-2 below lists the values of these parameters for this fire scenario, along with their ranges of applicability.

Of these parameters, only the compartment aspect ratio is outside the range of tests included in NUREG-1824 (EPRI 1011999). In this scenario, the compartment width to height ratio is well within limits, but the length to height ratio is higher than those included in NUREG-1824 (EPRI 1011999). For a zone model, this "longer than typical" compartment is mainly a concern early in the fire development, before a reasonably uniform layer has formed. Thus, predictions of events that occur early in the fire (such as smoke detection) may be expected to have a higher uncertainty if they are located farther from the fire source than those that occur later in the fire (such as ignition of cables above the initiating fire source) once the fire is more fully developed. For this scenario, smoke detectors are included throughout the compartment, but the primary ones of concern are those which would naturally respond faster (i.e., those nearest the fire).

Table E-2. Normalized parameters and their ranges of applicability to NUREG-1824.

Quantity	Normalized Parameter Calculation	NUREG-1824 Validation Range	In Range?
Fire Froude Number	$$\dot{Q}^* = \frac{\dot{Q}}{\rho_\infty c_p T_\infty D^{2.5}\sqrt{g}}$$ $$= \frac{317 \text{ kW}}{(1.2 \text{ kg/m}^3)(1.0 \text{ kJ/kg/K})(293 \text{ K})(0.6^{2.5} \text{ m}^{2.5})\sqrt{9.8 \text{ m/s}^2}} \cong 1.0$$	0.4 – 2.4	Yes
Flame Length, L_f, relative to the Ceiling Height, H_c	$$\frac{H_f + L_f}{H_c} = \frac{0.8 \text{ m} + 1.6 \text{ m}}{4.0 \text{ m}} = 0.6$$ $$L_f = D\left(3.7\,\dot{Q}^{*2/5} - 1.02\right) = 0.6 \text{ m }(3.7 \times 1.0^{0.4} - 1.02) \cong 1.6 \text{ m}$$	0.2 – 1.0	Yes
Ceiling Jet Radial Distance, r_{cj}, relative to the Ceiling Height, H_c	N/A – Ceiling jet targets are not included in simulation.	1.2 – 1.7	N/A
Equivalence Ratio, φ, as an indicator of the Ventilation Rate	$$\varphi = \frac{\dot{Q}}{\Delta H_{O_2}\,\dot{m}_{O_2}} = \frac{317 \text{ kW}}{13{,}100 \text{ kJ/kg} \times 0.4 \text{ kg/s}} \cong 0.06$$ $$\dot{m}_{O_2} = 0.23\,\rho_\infty\dot{V} = 0.23 \times 1.2 \text{ kg/m}^3 \times 1.4 \text{ m}^3/\text{s} \cong 0.4 \text{ kg/s}$$	0.04 – 0.6	Yes
Compartment Aspect Ratio	$$\frac{L}{H_c} = \frac{40 \text{ m}}{4.0 \text{ m}} = 10 \qquad \frac{W}{H_c} = \frac{18.5 \text{ m}}{4.0 \text{ m}} \cong 4.6$$	0.6 – 5.7	No
Target Distance, r, relative to the Fire Diameter, D	$$\frac{r}{D} = \frac{2.3 \text{ m}}{0.6 \text{ m}} \cong 3.8$$	2.2 – 5.7	Yes

Notes:

(1)   The "Fire Height," $H_f + L_f$, is the sum of the height of the fire from the floor and the fire's flame length.

E.4 Estimation of Fire-Generated Conditions

This section provides details specific to each model.

E.4.1 Algebraic Models

General: The general approach to using the algebraic models for this scenario is to first calculate the flame height of the transient trash fire to determine whether the flame reaches one or more of the overhead cable trays, then to calculate the plume temperatures from the trash fire to determine which of the overhead cable trays would be damaged by the trash fire alone. The general scenario is depicted schematically in Figure E-4.

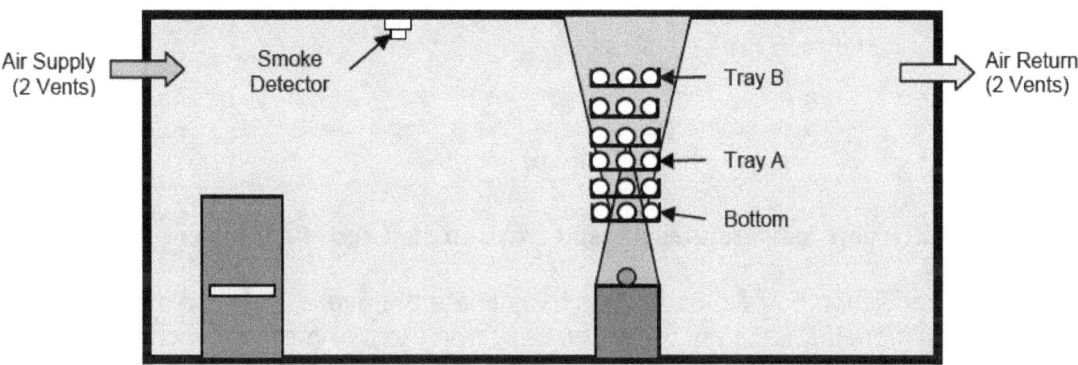

Figure E-4. Schematic diagram of transient trash fire in CSR (not to scale).

The first step to estimate the fire-generated conditions using algebraic models is to determine whether the cables in the cable trays located directly above the trash fire are likely to be damaged and potentially ignited by the trash fire. As shown in Figure E-3, the top of the trash receptacle is located at an elevation of 0.8 m (2.6 ft), and the lowest overhead cable tray is located at an elevation of 1.8 m (5.9 ft), which is 1.0 m (3.3 ft) above the top of the trash receptacle. As shown in Table E-2, the flame length of the cabinet fire is calculated to be 1.6 m (5.2 ft) at the peak heat release rate of 317 kW, so this empirical correlation for flame length can be used to confirm that at least some of the overhead cables would likely be damaged in this scenario. However, due to the metal barrier on the bottom of the lowest cable tray, algebraic models cannot be used to predict whether the cables in the lowest cable tray would ignite.

The next step is to calculate the fire plume temperatures that develop from the trash fire to determine whether cable trays A or B located above the trash fire would be damaged by the trash fire alone. The Heskestad plume temperature correlation included in the Fire Dynamics Tools (FDT^s) and the Fire-Induced Vulnerability Evaluation (FIVE-Rev1) was used to calculate the plume centerline temperature above the trash fire. The results, shown in Figure E-5, show that the plume temperature of all the cable trays would exceed the cable damage threshold temperature of 205 °C (400 °F). However, the Heskestad plume temperature correlation is based on an unobstructed plume. The obstruction caused by the position of the cable trays within the fire plume would alter the actual fire plume entrainment and temperatures. Nonetheless, these results demonstrate that the potential for damage and ignition of the cable trays located above the transient trash fire warrants more detailed analysis.

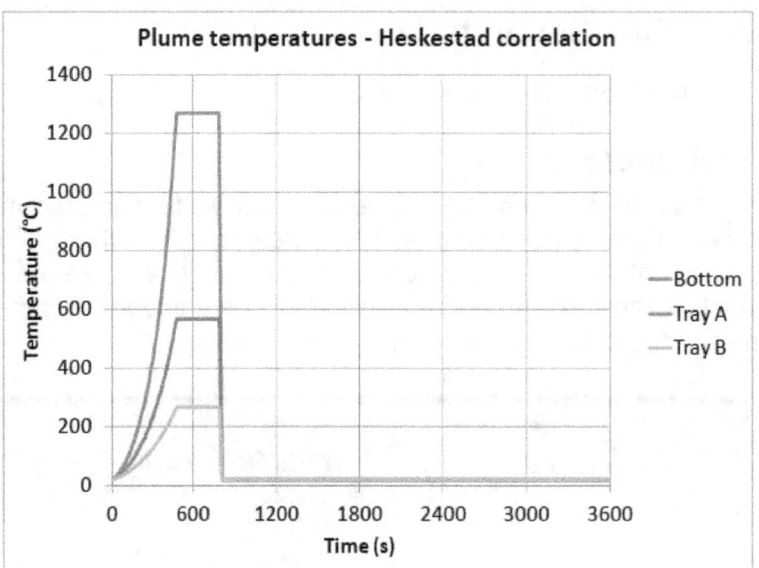

Figure E-5. Plume temperatures at cable trays located above a transient trash fire.

The FPA forced ventilation correlation is used to estimate the average HGL temperature of the CSR resulting only from the transient trash fire, based on the parameters described in the following subsections.

Geometry: The FPA correlation requires room dimensions to be specified in terms of length, width, and height. For this example, the selected compartment is a rectangular parallelepiped, as shown in Figure E-1, so its length, width, and height are specified directly.

Fire: As applied to this scenario, the FPA correlation is used with the time-dependent HRR specified for the transient trash fire. This HRR history is shown in Figure E-6.

Materials: The walls, ceiling, and floor are all specified as concrete, with the thermal properties specified in Table 3-1.

Ventilation: The ventilation rate in the smoke purge mode is 1.4 m³/s (3,000 cfm). This value is used as a direct input parameter in the FPA correlation.

Table E-3. Summary of input parameters for FPA analysis of CSR scenario.

Parameter	Value	Source
Room height (H)	4.0 m	Figure E-1
Room length (L)	40.0 m	Figure E-1
Room width (W)	18.5 m	Figure E-1
Room boundary material	Concrete	Figure E-1 See Table 3-1 for properties.
Mech. Ventilation rate ($\dot{V}$)	1.4 m³/s	From scenario description
Fire elevation (H_f)	0.8 m	From scenario description
Ambient temperature (T_a)	20°C	Specified
Fire parameters	See Table E-1	

Temperature: The FPA HGL temperature correlation for mechanically ventilated spaces is included in both the FDT[s] and FIVE-Rev1. The FPA results for the trash fire alone are shown in Figure E-6, based on the input parameters specified in Table E-3 and the HRR history shown in Figure E-6. These results show that, for the specified parameters, the average HGL temperature reaches a maximum of approximately 49 °C (120 °F) at 800 seconds, based on the peak trash fire HRR of 317 kW. These results show that cables in adjacent cable trays would not be damaged by the trash can fire alone. However, further analysis is required to determine the potential impact of overhead cable ignition on the potential for damage to cables in the adjacent cable trays. CFAST and FDS are used to perform this more detailed analysis.

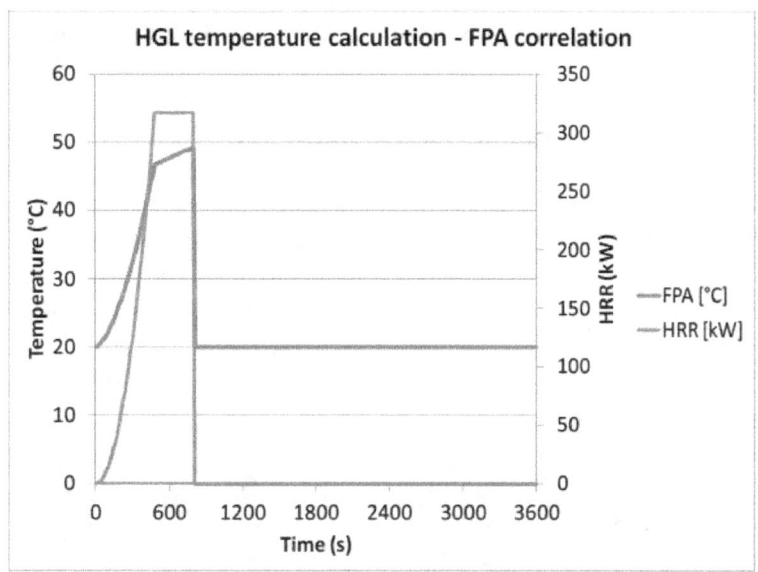

Figure E-6. Average HGL temperature (red line) from FPA correlation and heat release rate (blue line) for the CSR trash fire scenario.

E.4.2 Zone Model

General: In CFAST, the CSR is modeled as a single compartment with obstructions accounted for by modifying the cross-sectional area of the compartment as a function of height, as described in the geometry section below. Figure E-7 shows the scenario, as modeled by CFAST.

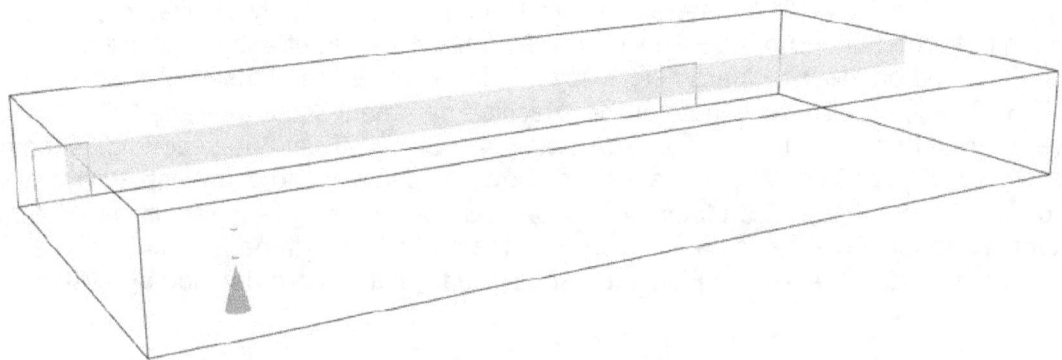

Figure E-7. CFAST/Smokeview rendering of the CSR scenario.

Geometry: Since zone models are concerned with volumes and not physical length and width, the volume of the computer compartment, as well as the numerous cable trays, ductwork, and beams, was modeled in CFAST with a cross-sectional area that varies with height. Table E-4 shows the cross-sectional area as a function of the height calculated from the compartment geometry shown in Figure E-1. Figure E-8 shows the CFAST inputs for compartment geometry.

Table E-4. Cross-sectional area as a function of height used for CFAST calculation.

	Height (m)	Area (m^2)
Floor Level	0	700.04
Bottom of Cable Trays	1.8	635.74
Bottom of Obstructions	2.2	483.74
HVAC Ductwork	2.4	514.89
Top of Obstructions	2.8	634.74
Top of Cable Trays	3.2	699.04
Ceiling Level Obstructions	3.6	291.46
Ceiling Level	4	291.46

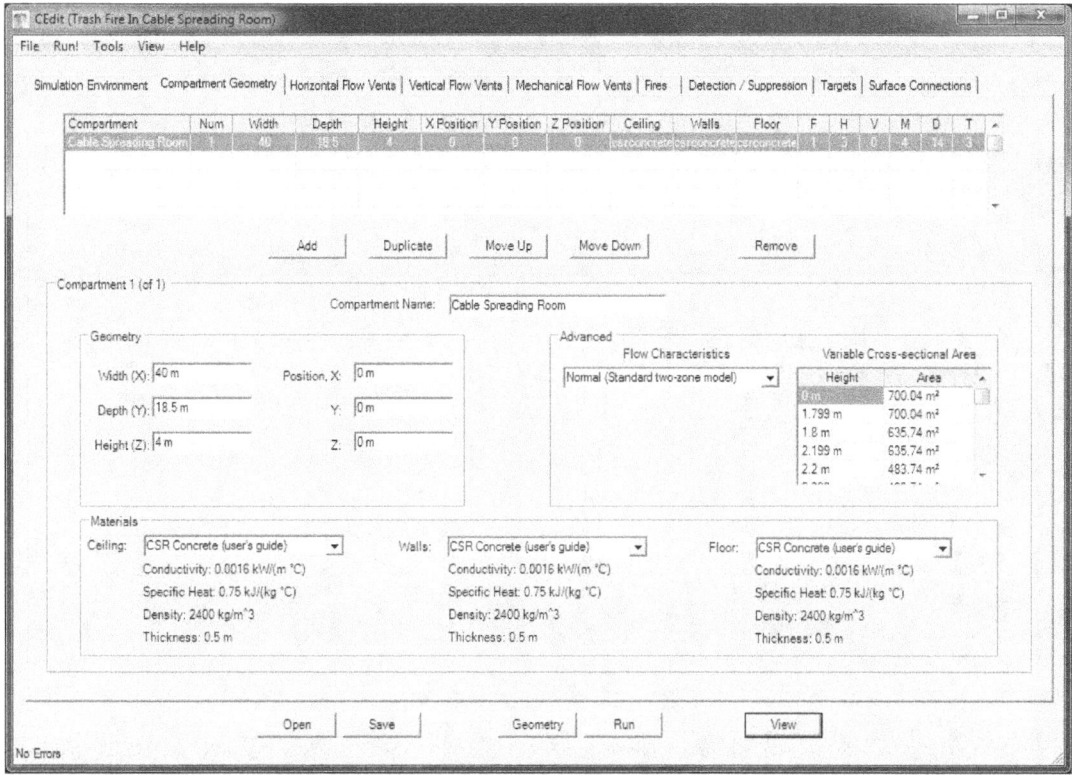

Figure E-8. CFAST inputs for compartment geometry for the CSR scenario.

Fire: The specified fire is input directly. The combustion chemistry in CFAST is described, at a minimum, by the production rates of CO and soot. Figure E-9 shows the CFAST inputs for the fire location in the scenario taken directly from Table E-1.

Cables: In CFAST, target temperatures are calculated using a one-dimensional cylindrical heat transfer calculation based on the material properties and cable diameter, as specified in the scenario description. Following the Thermally-Induced Electrical Failure (THIEF) methodology in NUREG/CR-6931, Vol. 3, electrical functionality is lost when the temperature just inside the jacket of a thermoplastic cable reaches 205 °C (400 °F). Specific heat and thermal conductivity for the cables are taken to be 1.5 kJ/kg/K and 0.2 W/m/K, as specified in the THIEF model. Cable density is determined from the mass per unit length and cross-sectional area of the cable from the scenario description, $\rho = 0.4 / \pi \left(0.015/2\right)^2 = 2264 \, \mathrm{kg/m^3}$. To account for the shielding of the cables on the lower surface of the cable tray, the CFAST input for the normal vector from the cable surface is directed upwards. This effectively shields the cables from the fire below while exposing them to the surrounding gas temperature for convection and to the hot upper gas layer for radiation. Figure E-10 shows the CFAST inputs for the three cable targets above the fire.

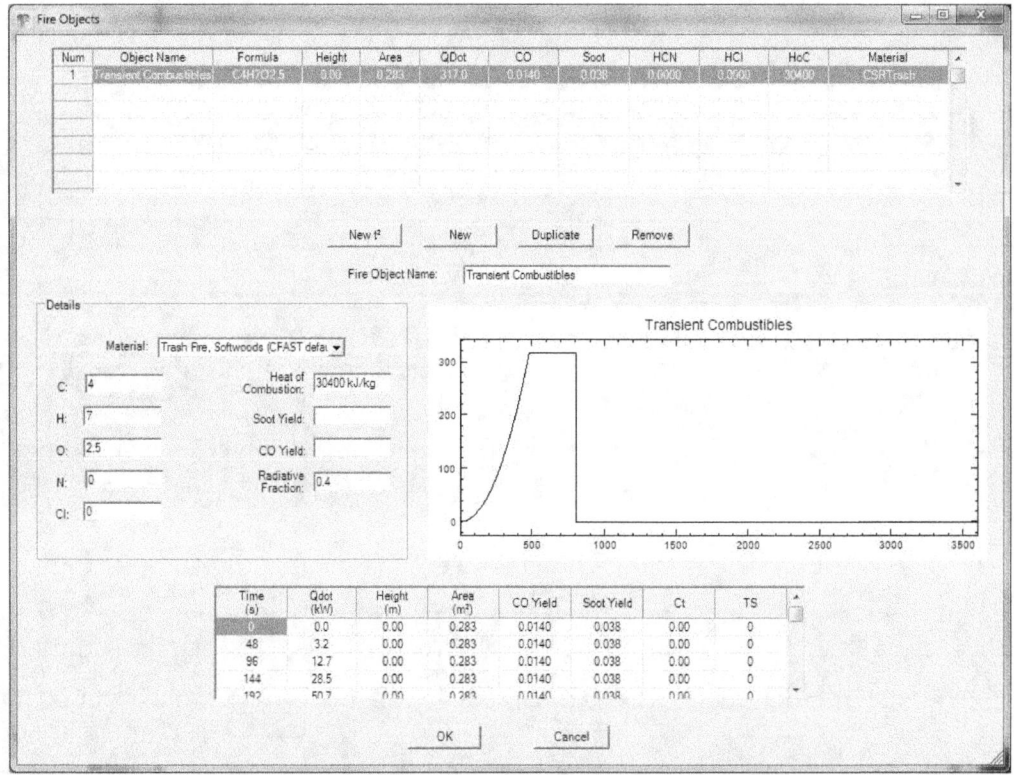

Figure E-9. CFAST inputs for the fire in the CSR scenario.

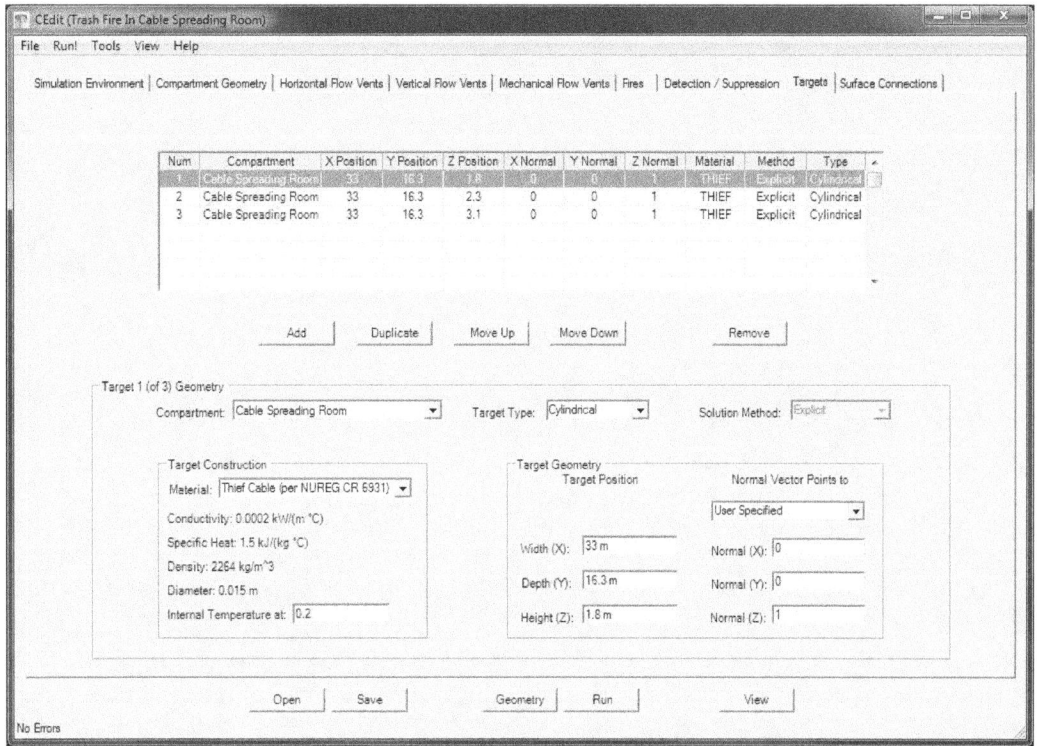

Figure E-10. CFAST inputs for cable targets above the fire source in the CSR scenario.

Ventilation: The supply and return airflow rates are input directly into CFAST. Upon smoke detector activation, mechanical ventilation fans stop and dampers are closed. Thus, before a stop time for the fans can be specified, the time to smoke detector activation is needed. This requires that CFAST be run with the fans on for the entire time to find the first smoke detector activation. The model is then re-run using the smoke detector activation time as the fan stop time. Figure E-11 shows the CFAST inputs for the natural ventilation.

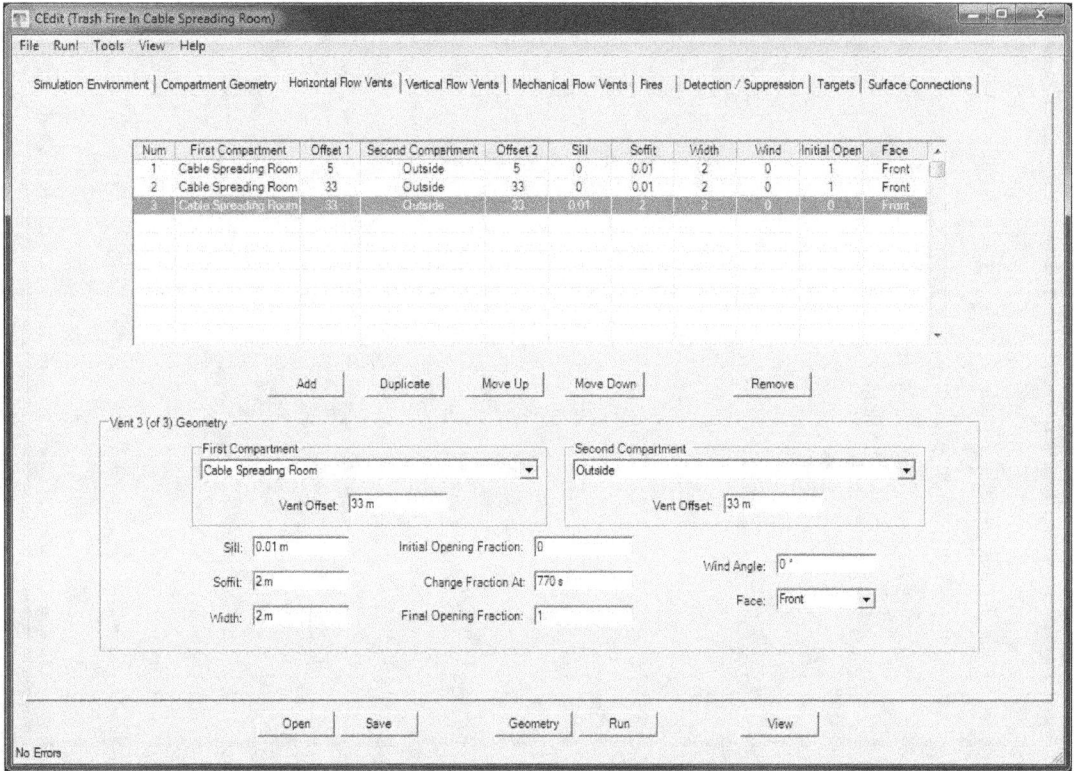

Figure E-11. CFAST inputs for natural ventilation for the CSR scenario. Note the opening of the doorway at 170 s, the time estimated for smoke detectors to activate.

Fire Detection: Although there are multiple smoke detectors in the space, the closest detector is the only one that needs to be modeled to determine time to detection, based on the fact that the nearest detector will be exposed to the greatest concentration of smoke products and the highest gas temperatures, thus leading to the earliest response. There are no geometric or ventilation features that would prevent this from being the case in the example considered. In CFAST, there is no direct way of calculating smoke density for smoke detector activation. Instead, the smoke detector is modeled as a sprinkler with a very low activation temperature and response time index (RTI). An activation temperature of 30 $^{\circ}$C (86 $^{\circ}$F) and an RTI of 5 (m/s)$^{1/2}$ were used for this scenario. Figure E-12 shows the CFAST inputs for the detectors in the scenario.

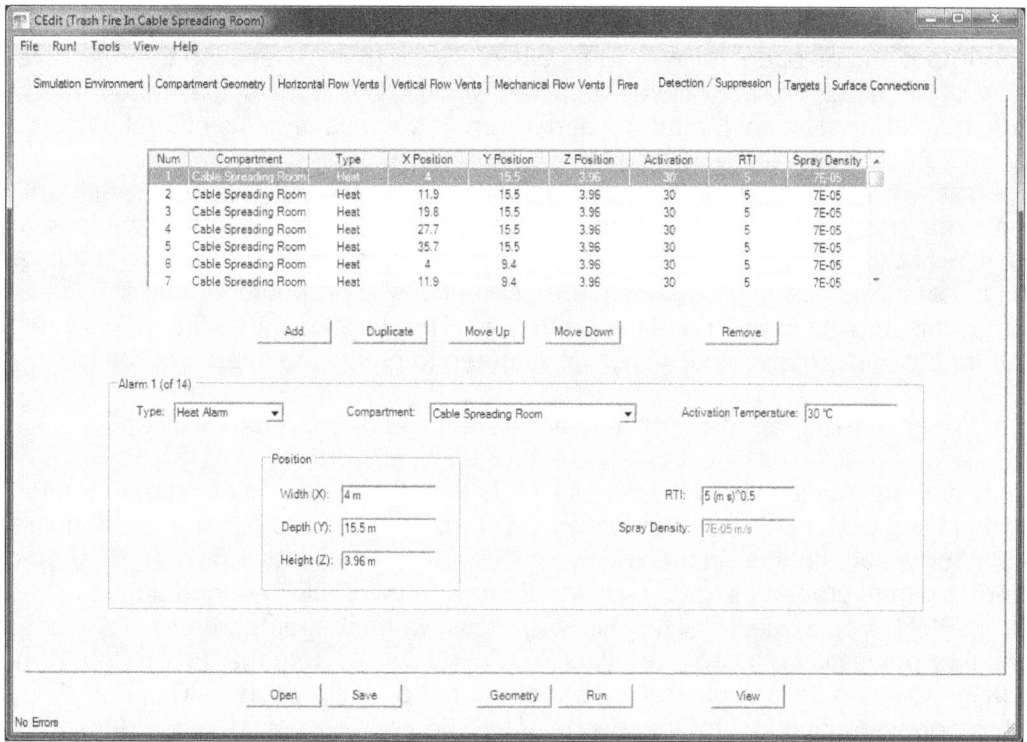

Figure E-12. CFAST inputs for smoke detectors in the CSR scenario.

E.4.3 CFD Model

General: This scenario is notable because it includes a considerable amount of "clutter," that is, the space has a relatively large number of obstructions. Figure E-13 illustrates the FDS simulation with all the blockages. Because the cable trays are regularly spaced in both the horizontal and vertical directions, it is easy in FDS to simply replicate a single tray as many times as necessary. Another interesting feature of the scenario is the automatic shutdown of the ventilation system at the time of any smoke detector activation. FDS models this by associating the creation or removal of obstructions or the activation/deactivation of a vent with actions taken by any number of fire protection devices.

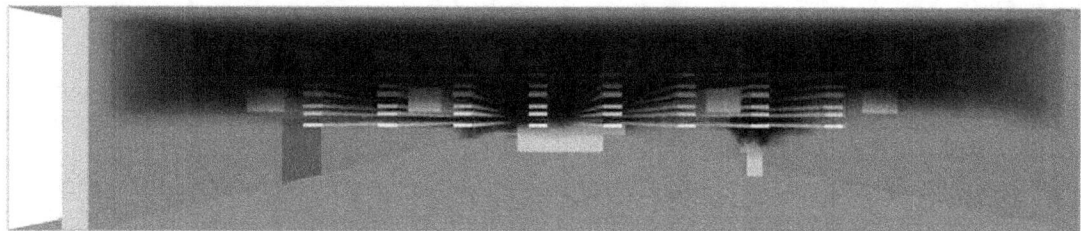

Figure E-13. FDS/Smokeview rendering of the CSR scenario.

Geometry: The interior of the compartment is modeled, and all obstructions have been included. To get increased resolution in the area of interest, multiple meshes are used. The finest mesh has a 10 cm (4 in) resolution and spans a volume surrounding the trash can, which is 6 m (20 ft) long, 3 m (10 ft) wide, and 4 m (13 ft) high. Coarse meshes cover the remainder of the compartment and adjacent hallway with cells of 20 cm (8 in). Because the objective of the calculation is to estimate time to failure for cables within stacked trays, it is important to have at least a 10 cm (4 in) resolution, the typical dimension of the rails of conventional cable trays.

Fire: The trash can is modeled with a square, rather than round, cross section with an equivalent area to the round cross section and a height equal to the height of the trash can. The local flow features around the fire are not capable of affecting the outcome of this example calculation; therefore, the transformed square geometry is appropriate for this application. The specified HRR is applied to the top of the trash can. The duration of the fire is 635 s, as was computed for the zone model input. There is no need to model the interior of the can.

Materials: The thermal properties of the walls are applied directly, as specified.

Cables: The primary objective of the calculation is to estimate the potential damage to the cables within the trays. FDS is limited to only 1-D heat transfer into either a rectangular or cylindrical obstruction. In this simulation, the cables are modeled as 1.5 cm (0.6 in) cylinders with uniform thermal properties, given above. Following the THIEF methodology in NUREG/CR-6931, Vol. 3, electrical functionality is lost when the temperature just inside the jacket of a thermoplastic (TP) cable reaches 205 °C (400 °F). Note that no attempt is made in the simulation to estimate ignition and spread of the fire over the cables. The THIEF methodology does not account for the effects of bundled cables, which may reduce the overall heat-up of a single cable.

For the bottommost cable tray, the bottom surface of the tray is protected by a solid metal surface. To model this in FDS, which models the trays as rectangular obstructions, individual properties for the top, sides, and bottom are specified. The top has the same properties as the cables; the sides and bottom are specified as having the same metal properties.

Detection: FDS has a smoke detection algorithm that predicts the smoke obscuration within the detection chamber based on the smoke concentration and air velocity in the grid cell in which the detector is located. The detector itself is not modeled, as it is merely a point within the computational domain. The two parameters needed for the model are the obscuration at alarm, which is given by the manufacturer, and an empirically determined length scale from which a smoke entry time lag is estimated from the outside air velocity. The *SFPE Handbook, 4th Edition,* provides a nominal value of 1.8 m (5.9 ft) for this length scale. The obscuration at alarm is 4.9 %/m, a typical sensitivity for smoke detectors.

Ventilation: The supply and return airflow rates are input directly into FDS. The ducts are represented by rectangular obstructions with thin plates just below (one grid cell) the vent itself to represent the diffusing effect of the grill. The resolution of the grid is not fine enough to capture this effect directly. FDS is capable of stopping the ventilation system upon the activation of any smoke detector.

E.5 Evaluation of Results

The purpose of the calculations described above is to estimate smoke detector activation times and potential cable damage from a trash can fire in the CSR.

The compartment itself is relatively large, and the relatively small fire (317 kW) does not substantially heat it up. Figure E-14 shows the HRR and estimated HGL temperature for the CSR scenario for CFAST and FDS. Differences between the two models likely result from FDS's ability to locally account for all the blockages in the room. HGL temperature in CFAST is a spatially average value intended to represent the bulk conditions throughout the compartment, while the FDS values are calculated based on a single vertical profile of temperature at a fixed location within the room (in this scenario placed several meters away from the fire location to eliminate local effects of the fire plume on the temperature profile).

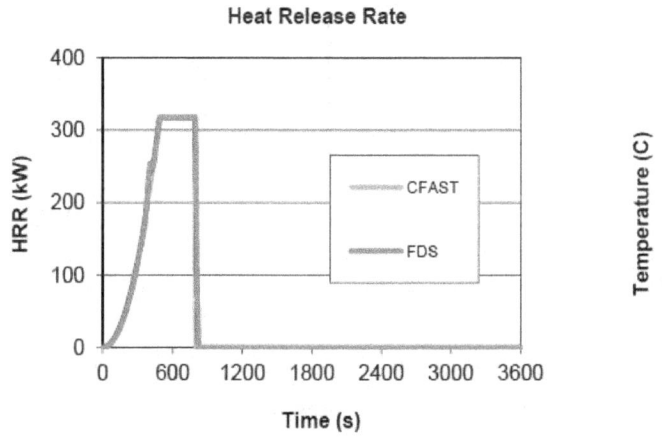
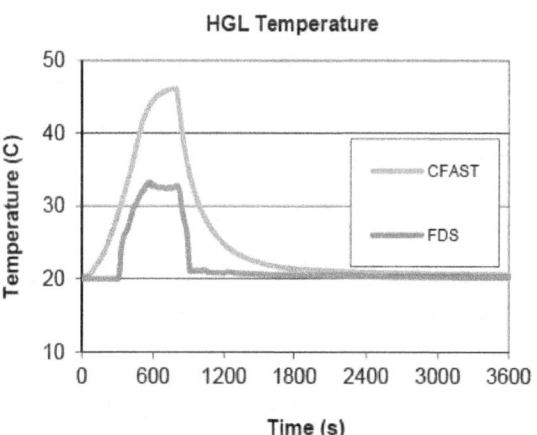

Figure E-14. HRR and estimated HGL temperature for CSR scenario.

Because of the uncertainty in all models' smoke detector activation predictions and the uncertainty associated with the possible ignition of cables in the trays just above the fire, it is difficult to predict whether or not the CO_2 suppression system would be activated in time to prevent possible cable ignition. No validation results are available that address time to detector activation. Thus, the analysis does not consider activation of the suppression system.

Table E-5 summarizes the results of all the models for the chosen damage criteria. For each predicted value, a calculation is performed to determine the probability of exceeding the critical value. The procedure for calculating this probability is given in Chapter 4, and it accounts for the model bias and scatter. The purpose of this table is to highlight the criterion that is most likely to be exceeded so that further analysis can be focused on this criterion and the model or models that predict it. Each criterion is discussed in greater detail in the following subsections.

Table E-5. Summary of the model predictions for the CSR scenario.

Model	Bias Factor, δ	Standard Deviation, σ_M	Location	Predicted Value	Critical Value	Probability of Exceeding
Temperature (°C), Initial Value = 20 °C						
CFAST	1	0.27	Bottom Cable	298	205	0.893
FDS	1.02	0.13		54	205	0.000
CFAST	1	0.27	Cable A	202	205	0.472
FDS	1.02	0.13		36	205	0.000
CFAST	1	0.27	Cable B	126	205	0.003
FDS	1.02	0.13		61	205	0.000
Heat Flux (kW/m²)						
CFAST	0.81	0.47	Bottom Cable	4.2	6	0.367
FDS	0.85	0.22		1.0	6	0.001
CFAST	0.81	0.47	Cable A	3.0	6	0.091
FDS	0.85	0.22		0.3	6	0.000
CFAST	0.81	0.47	Cable B	2.0	6	0.000
FDS	0.85	0.22		0.8	6	0.001

E.5.1 Smoke Detection

Table E-6 shows the models' predictions for smoke detector activation. The models provide similar estimates of the detector activation time. CFAST models smoke detector actuation as a heat detector with a relatively low thermal inertia and activation temperature. However, there is no consensus in the fire literature for the appropriate RTI value and activation temperature. Given the presence of beam pockets and obstructions, even a CFD model like FDS, which uses actual smoke concentration rather than temperature in its detector algorithm, is subject to significant uncertainty. For this reason, no credit is given for smoke detection in the final design, and there is no need to consider either its uncertainty or any further impact on the design.

Table E-6. Smoke detector activation times, CSR.

Model	Time (s)
CFAST	170 s
FDS	160 s

E.5.2 Temperature Criterion

Figure E-15 shows the estimated impact of the fire on the cable trays above the fire. The bottom cable is located at least 1 m (3.3 ft) above the base of the waste bin fire. With an estimated flame height of 1.7 m (5.7 ft), ignition may occur from flame impingement. The algebraic tools and CFAST both predict a temperature well above the cable failure temperature. The plume temperature estimated by the algebraic tool is naturally higher than the CFAST calculation, since CFAST is able to account for the heat transfer into the cable and thus provide an estimate of the actual cable temperature. However, both of these calculations are based on the cable being directly exposed to the high gas temperature of the fire plume directly above the fire and do not account for the protection afforded by the solid metal lower surface of the cable tray. FDS calculations take this into account and show a much lower temperature of the bottommost cable.

For the upper cables, CFAST's predicted temperature and heat flux are higher than FDS's because CFAST does not account for the fact that the cable trays of interest are shielded by trays below or that the burning spreads outward from the ignition point.

E.5.3 Heat Flux Criterion

Heat flux estimates for both CFAST and FDS in Figure E-15 show values well below the critical value of 6 kW/m^2. Still, the CFAST-estimated heat flux to the bottom cable tray of 4.2 kW/m^2 (Table E-5), combined with the estimated flame height (Figure E-15), does show the importance of the solid lower surface of the cable tray.

It should also be noted that these damage criteria are intended to indicate electrical failure, but are routinely used as ignition criteria. In newer studies in NUREG/CR-7010, cable ignition was not observed at fluxes below 25 kW/m^2, and most often only with direct flame impingement. Handbook values for minimum ignition flux for power and communication cables are reported in the range of 15 kW/m^2 to 35 kW/m^2 (*SFPE Handbook*, 4th ed., Table 3-4.2).

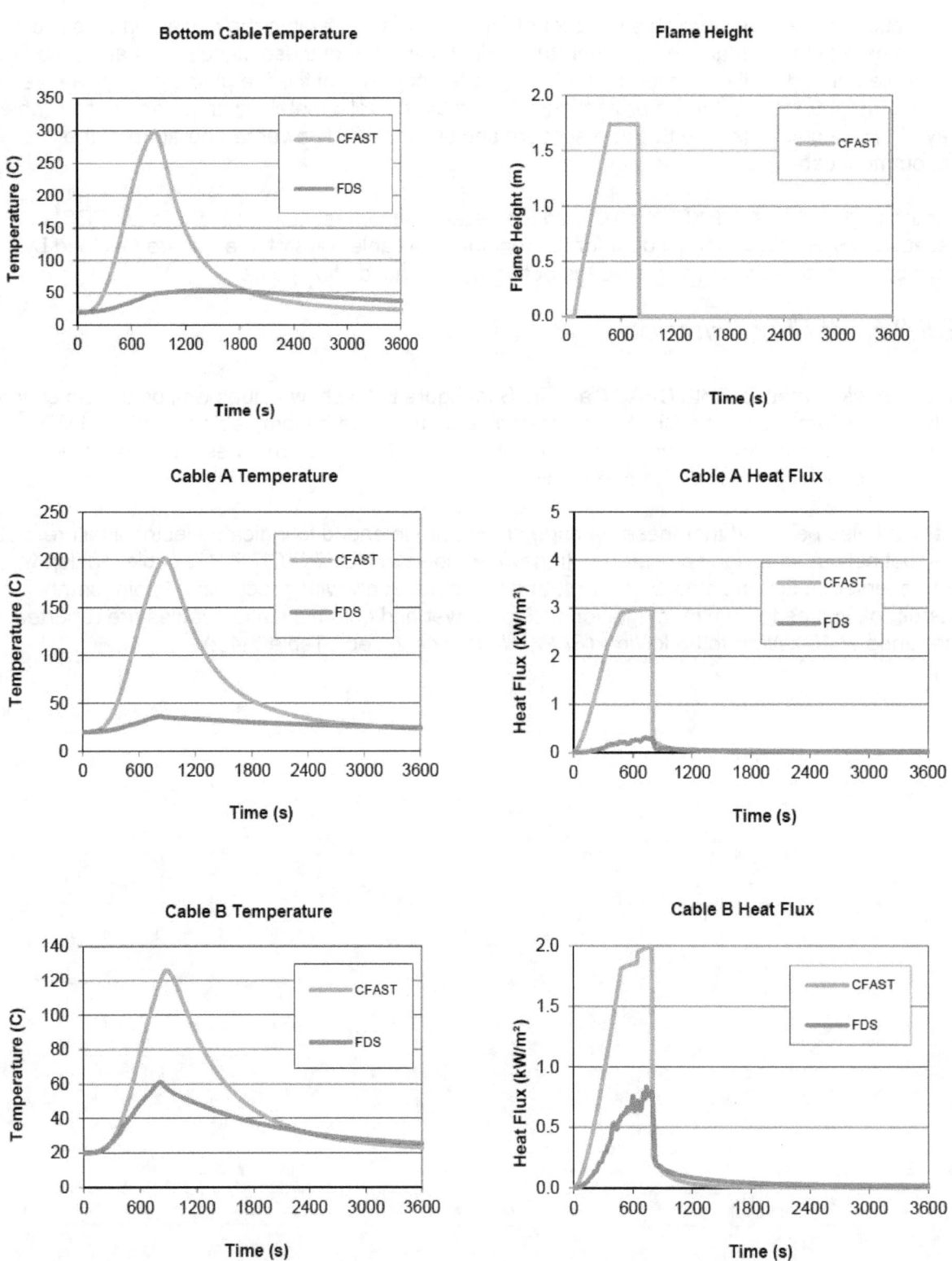

Figure E-15. Estimated cable conditions for the CSR.

E.5.4 Parameter Uncertainty Propagation

The analysis above has shown that a 98th percentile trash fire is unlikely to damage cables in certain trays overhead because of the presence of a protective barrier beneath the lowest tray. However, for some PRA applications, it may be necessary to calculate the probability of cable damage for *any* fire, not just the 98th percentile fire, with or without a protective barrier. To illustrate this concept, let's consider a simple plume temperature correlation.

Figure E-16 displays the distribution[23] of peak heat release rates for transient combustibles (NUREG/CR-6850, Appendix E). The analysis described above made use of the 98th percentile fire from this distribution, whose peak is 317 kW.

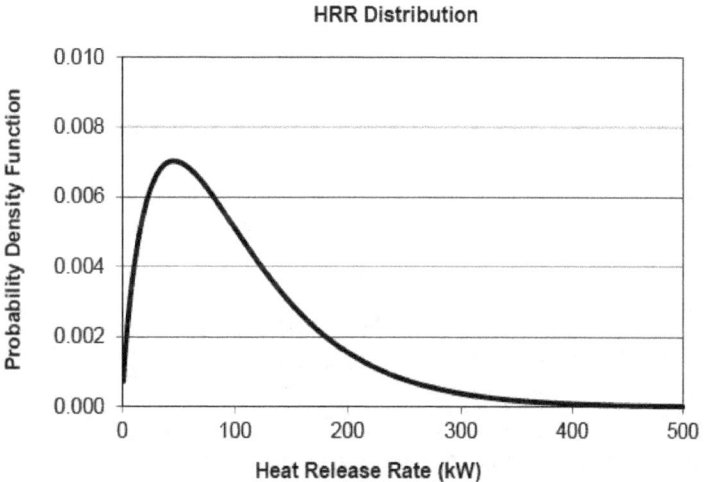

Figure E-16. Distribution of HRR for a trash fire.

Following the methodology described in Chapter 4 and Appendix E of NUREG/CR-6850, Heskestad's plume temperature correlation[24] is applied to the entire range of HRR, resulting in distributions of plume temperatures at the locations of Tray A and Tray B, as shown in Figure E-17. Tray A is 1.5 m above the top of the trash bin. The probability that the plume temperature at this height from a randomly chosen fire will exceed the cable failure temperature (205 °C; 400 °F) is equal to the area beneath the left hand curve in Figure E-17 for temperatures greater than 205 °C, or approximately 0.42. Similarly, the probability that the plume temperature at Cable Tray B (2.3 m above the bin) will exceed 205 °C is approximately 0.08. Consistent with the guidance in NUREG/CR-6850, this resulting probability can be used as the "severity factor" for the quantification of corresponding fire ignition frequencies.

Note that this analysis neglects the effect of the protective barrier or cable mass. It just estimates the likelihood that gas temperatures in the vicinity of the critical components could reach levels that might cause damage.

---

[23] NUREG/CR-6850 specifies gamma distributions for the various types of combustbles found within an NPP. Microsoft Excel® provides a built-in function (GAMMA.DIST) that calculates the probability density function given the parameters α and β. In this case, these parameters are 1.8 and 57.4, respectively.
[24] A value of 0.35 was used for the radiative fraction in determining the convective heat release rate.

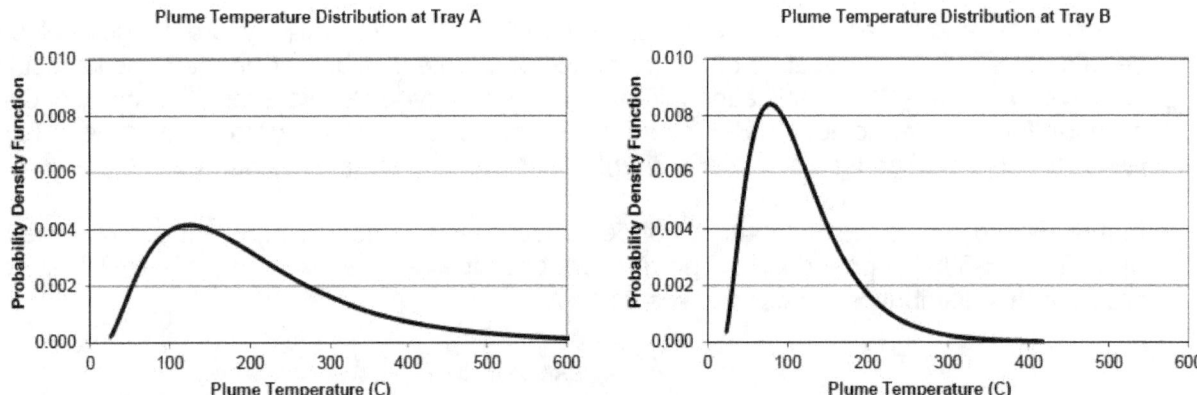

Figure E-17. Distribution of plume temperatures at Trays 3 and 6, respectively.

E.6 Conclusion

This analysis considers the potential impact of a transient trash fire on a stack of horizontal cable trays located directly above the transient fire source; it also considers the potential for the cables in these trays to ignite.

Algebraic equations from the FDT^s and FIVE-Rev1, including the Heskestad flame height correlation and the Heskestad fire plume temperature correlation, were used to evaluate the potential for the transient trash fire to ignite or damage cables in the cable trays located directly above this fire source. These calculations showed that the unobstructed flame height would reach multiple cable trays at the bottom of the stack; they also showed that the unobstructed plume temperatures would exceed the cable damage temperature in all trays located above the fire source. However, these calculations do not account for the impact of the cable trays on the actual flame height and plume temperatures, so more detailed analysis with CFAST and FDS is needed.

The FPA correlation included in the FDT^s and FIVE-Rev1 was used to evaluate whether the transient trash fire alone would cause damage to cable trays not located directly above this fire source. This calculation showed that the HGL temperature from the transient trash fire alone is well below the damage temperatures for the cables.

The more detailed analysis with CFAST and FDS shows that a 317 kW waste bin fire beneath a vertical array of horizontal cable trays is unlikely to damage cables in the trays located three (Tray A) and six (Tray B) levels above the fire. Both CFAST and FDS estimate peak temperatures and heat fluxes below the failure criteria for cables in the third tray from the bottom. Based on FDS calculations, temperatures and heat fluxes for the protected lowest cable tray are well below critical values. Estimates from CFAST for unprotected cables demonstrate the importance of the protection afforded by the solid metal lower surface of the cable trays.

E.7 References

1. NIST SP 1086, *Consolidated Model of Fire Growth and Smoke Transport, CFAST (Version 6), Software Development and Model Evaluation Guide*, 2008.
2. NUREG-1805, *Fire Dynamics Tools (FDT^s) Quantitative Fire Hazard Analysis Methods for the U.S. Nuclear Regulatory Commission Fire Protection Inspection Program*, 2004.
3. NUREG-1824 (EPRI 1011999), *Verification and Validation of Selected Fire Models for Nuclear Power Plant Applications*, 2007.
4. NUREG/CR-6850 (EPRI 1011989), *EPRI/NRC-RES Fire PRA Methodology for Nuclear Power Facilities*, 2005.
5. NUREG/CR-6931, *Cable Response to Live Fire (CAROLFIRE) Volume 3: Thermally-Induced Electrical Failure (THIEF) Model*, 2008.
6. *SFPE Handbook of Fire Protection Engineering*, 4th edition, 2008.

E.8 Attachments (on CD)

1. FDS input files: Cable_Spreading_Room.fds

2. CFAST input files:

    a. Trash Fire In Cable Spreading Room.in

3. Algebraic model input files:

    a. FPA_AppE.xlsx

APPENDIX F
LUBRICATING OIL FIRE IN A TURBINE BUILDING

F.1 Modeling Objective

The purpose of the calculations described in this appendix is to estimate the temperatures of unprotected structural steel members due to a large lubricating oil fire in a turbine building. For this scenario, a catastrophic failure of the turbine results in a large lubricating oil spill fire. The purpose of this example is to evaluate the turbine building structural steel response to internal lubricating oil fire in one of two potential curbed locations in the turbine building. The calculations provide information for a decision on the hazard posed to the structural steel for each potential fire location, thus serving as a basis for a plant modification.

F.2 Description of the Fire Scenario

General Description: The turbine building is a large structure that is approximately 100.3 m (329 ft) long by 99.5 m (326 ft) wide by 21 m (69 ft) tall, as shown in Figure F-1. The ambient temperature is 36 °C (97 °F), inside and outside the building. Although the building has multiple levels, for this example, only two levels the main turbine deck and the level just below it are evaluated. The lowest portion of the lower level floor is about 1.2 m (4 ft) below grade in the area near the fire, but most of the lower level floor elevation is at the 0 m elevation. The ceiling height of the lower level is 4.6 m (15 ft) above grade. The floor of the turbine deck is at the 5.6 m (18 ft) elevation, and the ceiling for this upper level is at the 19.8 m (65 ft) elevation. The turbine deck has a smaller floor area than the lower level; the turbine deck level is approximately 90 m (295 ft) long by 70 m (230 ft) wide. The building contains the turbine generators (see Figure F-1) and a heating, ventilating, and air conditioning (HVAC) room.

Each turbine generator contains about 3,000 L (792.5 gal) of lubricating oil in a single tank, as illustrated in Figure F-2. The proposed curbed is 6.1 m (20 ft) long by 4.6 m (15 ft) wide. Two potential locations have been identified on the lower level where the curbing could be installed. These locations are shown in Figure F-1. Either of the curbed areas would be designed to contain the entire volume of 3,000 L (792.5 gal) from one of the turbine generators main lubricating oil tanks.

There are 40 unprotected steel support columns in a rectangular configuration (four rows of ten each) around the lubricating oil tank. Figure F-3 shows a typical unprotected steel column. The columns are all W14×145 standard wide flange members, as shown in Detail A of Figure F-1 (American Institute of Steel Construction, 2006). The six columns identified for analysis are denoted in Figure F-1 as A, B, C, D, E, and F. A structural analysis indicates that the loss of any of these six columns could lead to partial collapse of the turbine building.

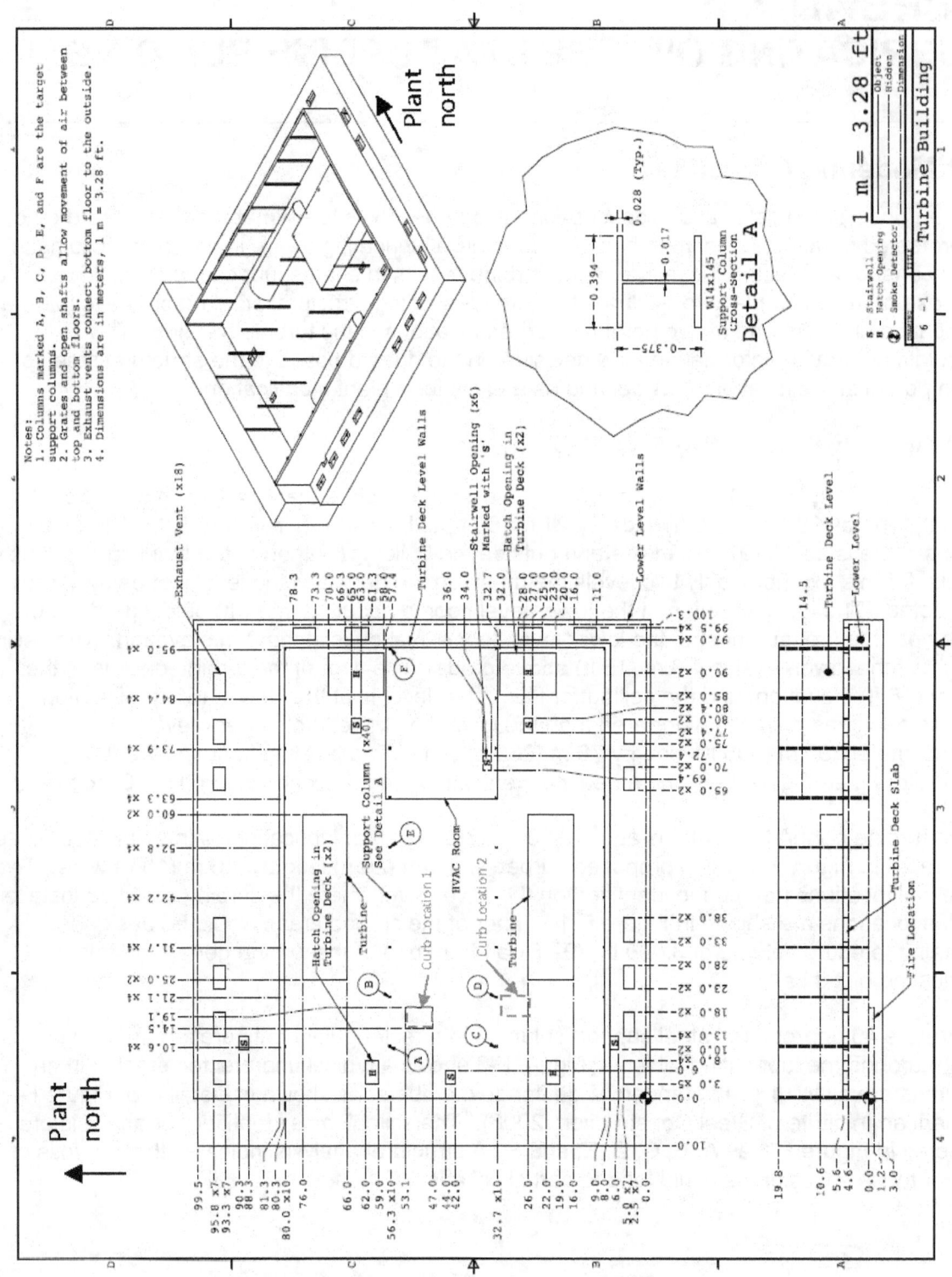

Figure F-1. Geometry of the turbine building.

Figure F-2. Main turbine lubricating oil tanks in the turbine building.

Figure F-3. Typical steel column in the turbine building.

LUBRICATING OIL FIRE IN A TURBINE BUILDING

Geometry: Plan and elevation views of the turbine building are shown in Figure F-1. The area of interest involves the two levels shown in Figure F-1, which are separated by the concrete turbine deck. There are several stairwell, hatch, and exhaust vent penetrations throughout the turbine deck. These penetrations are identified in Figure F-1 by the **H** symbol for a hatch and the **S** symbol for a stairwell.

Construction: The turbine deck is made of 1 m (3.3 ft) thick normal-weight concrete. The floor and walls of the lower level are constructed of 0.3 m (1 ft) thick normal-weight concrete. Numerous areas and landings in the turbine building are made of metal grating. The floor in the area of the lubricating oil tank is 1 m (3.3 ft) thick normal-weight concrete. The walls and ceiling of the upper level of the turbine building are made of 3 mm (0.12 in) thick corrugated steel. The structural columns, steel grating, and corrugated steel are fabricated from steel containing 0.5 percent carbon.

Materials: Thermal properties of the normal-weight concrete and steel are listed in Table 3-1. The damage criteria for the structural steel are listed in Table F-1; these criteria are based on the acceptance criteria for a standard fire resistance test (ASTM E119-10a, 2010). The maximum cross-sectional temperature threshold normally applies to unprotected structural steel because of its high thermal conductivity.

Table F-1. Structural steel failure criteria (ASTM E119-10a).

Member	Maximum Cross-Section Average Temperature °C (°F)
Beam	593 (1,099)
Column	538 (1,000)

Detection System: There is no safety-related equipment in the turbine building, thus, no detection or suppression systems are credited for this analysis. If under-deck sprinklers were installed, they would not be credited when considering the potential for worst-case structural failure and the possible need for passive structural fire protection.

Ventilation: The turbine building is an open area configuration with all forced ventilation intentionally shut down at the start of the fire for reasons unrelated to the fire. There are eighteen roof-mounted horizontal exhaust vents around the perimeter of the turbine deck level, as shown in Figure F-1. The exhaust vents each measure 5.0 m (16.4 ft) long by 2.5 m (8.2 ft) wide. There are no other internal or external openings beyond those already noted in the drawings.

Fire: A large, confined spill fire involving 3,000 L (792.5 gal) of lubricating oil from the main turbine lubricating oil tank is postulated on the lower level. The lubricating oil has been preheated prior to the spill, such that the growth rate of the fire would be relatively short compared to the total time required to burn the spill volume. The total spill area is approximately 28.1 m² (300 ft²), as shown in Figure F-1; the spill depth is calculated to be 0.11 m (0.35 ft), based on the specified spill area and fuel volume.

The lubricating oil is dominated by alkanes, that have the chemical formulae of C_nH_{2n+2}, where n ranges between 12 and 15 (centered around 14) and a soot yield of 0.059. Fuel properties for the lubricating oil, obtained from Tables 3-2 and 3-4 of NUREG-1805 and from Table 3-4.16 of the *SFPE Handbook, 4th edition*, are summarized in Table F-2. The *Handbook* identifies the fuel as a "Hydrocarbon." NUREG-1805 lists properties and states that lubricating and transformer oils are similar.

Table F-2. Data for lubricating oil fire.

Parameter	Value	Source
Effective Fuel Formula	C_nH_{2n+2}	Developed from fuel chemistry (n in range of 12-15)
Mass burning rate	0.039 kg/s.m^2	NUREG-1805 Table 3-4
Fuel volume	3,000 L	Specified
Density	760 kg/m^3	NUREG-1805 Table 3-4
Heat of Combustion	46,000 kJ/kg	NUREG-1805 Table 3-4
CO_2 Yield	2.64 kg/kg	*SFPE Handbook*, 4th ed., Table 3-4.16
Soot Yield	0.059 kg/kg	*SFPE Handbook*, 4th ed., Table 3-4.16
CO Yield	0.019 kg/kg	*SFPE Handbook*, 4th ed., Table 3-4.16
Radiative Fraction	0.34	*SFPE Handbook*, 4th ed., Table 3-4.16
Mass Extinction Coefficient	8,700 m^2/kg	Mulholland and Croarkin (2000)

The peak heat release rate (HRR), $\dot{Q}$, is computed from the fuel mass burning rate, $\dot{m}''$, the heat of combustion, ΔH, and the specified area of the spill, A:

$$\dot{Q} = \dot{m}'' \, \Delta H \, A = 0.039 \, \text{kg/m}^2/\text{s} \times 46,000 \, \text{kJ/kg} \times 28.1 \, \text{m}^2 \cong 50,400 \, \text{kW} \qquad \text{(F-1)}$$

The fire duration, Δt, is determined from the pool depth, δ, density, ρ, and burning rate, $\dot{m}''$:

$$\Delta t = \frac{\delta \, \rho}{\dot{m}''} = \frac{0.11 \, \text{m} \times 760 \, \text{kg/m}^3}{0.039 \, \text{kg/m}^2/\text{s}} \cong 2,144 \, \text{s} \quad (35.7 \, \text{min}) \qquad \text{(F-2)}$$

F.3 Selection and Evaluation of Fire Models

This section describes the process of model selection, including a discussion of their validity for this scenario. To the temperature of the steel it is necessary to calculate the heat flux from the fire to and the resulting temperature of the steel. It is also necessary to determine whether flame extension beneath the ceiling or a hot gas layer (HGL) would develop and contribute significantly to the column heating.

F.3.1 Heat Flux and Target Temperature Algorithms

The turbine building fire under consideration is not a typical compartment fire scenario because it exhibits both pre- and post- flashover conditions simultaneously. The 50 MW fire is expected to fully engulf the lower deck from floor to ceiling in the vicinity of the curbed areas and resemble a compartment fire after the onset of flashover. Away from the curbed areas, a distinct smoke layer is expected to form and the space would exhibit pre-flashover fire conditions. The empirical correlations and zone models are only capable of modeling one or the other condition, not both. However, the heat flux and target temperature algorithms used by these simpler models may be applicable. In particular, the point source and solid flame radiant heat flux models can be used to estimate the thermal exposure condition to the near-field and far-field columns, respectively.

The CFD model, FDS, does have the necessary physics to model the scenario, but it is necessary first to assess if it has been validated for such conditions.

F.3.2 Validation

Table F-3 provides a summary of the normalized parameter calculations for the turbine building fire scenario. A number of parameters fall outside of the parameter space of the NRC/EPRI V&V study (NUREG-1824 (EPRI 1011999)), including the flame length/ceiling height ratio, the compartment aspect ratios and some of the target distance to fire diameter ratios. In addition, the equivalence ratio cannot be calculated with the simple equations that have been used for various other examples.

The calculation of the equivalence ratio is challenging because natural ventilation is provided through the 18 roof vents located around the perimeter of the turbine deck level. To evaluate the potential impact of ventilation on the fire for this scenario, the quantity of oxygen available in the turbine building is compared to the amount of oxygen that would be consumed by the specified lubricating oil fire. Given a total volume of approximately 209,600 m^3, the mass of oxygen within the turbine building is estimated to be:

$$m_{O_2,\text{tot}} = \rho V Y_{O_2} = 1.1 \text{ kg/m}^3 \times 209,600 \text{ m}^3 \times 0.23 \cong 53,030 \text{ kg} \tag{F-3}$$

The mass of oxygen required to burn all the fuel is estimated to be:

$$m_{O_2,\text{req}} = \frac{\dot{Q} \, \Delta t}{\Delta H_{O_2}} = \frac{50,400 \text{ kW} \times 2,144 \text{ s}}{13,100 \text{ kJ/kg}} \cong= 8,249 \text{ kg} \tag{F-4}$$

Thus, the specified fire would consume less than 16% of the oxygen available within the turbine building. Consequently, the fire would not be expected to be ventilation-limited on a global basis, even without ventilation from the outside environment through the roof vents.

Table F-3. Normalized Parameter Calculations for the Turbine Building Fire Scenario.

Quantity	Normalized Parameter Calculation	NUREG-1824 Validation Range	In Range?
Fire Froude Number	$\dot{Q}^* = \dfrac{\dot{Q}}{\rho_\infty c_p T_\infty D^{2.5}\sqrt{g}} = \dfrac{50,400\ \text{kW}}{(1.1\ \text{kg/m}^3)(1.0\ \text{kJ/kg/K})(309\ \text{K})(6.0^{2.5}\ \text{m}^{2.5})\sqrt{9.8\ \text{m/s}^2}} \approx 0.524$	0.4 – 2.4	Yes
Flame length, L_f, relative to ceiling height, H_c	$L_f = D\left(3.7\,\dot{Q}^{*2/5} - 1.02\right) = 6.0\ \text{m}\left(3.7\times 0.52^{0.4} - 1.02\right) \approx 11\ \text{m}$; $\quad \dfrac{L_f}{H_c} = \dfrac{11.0\ \text{m}}{4.6\ \text{m}} \approx 2.4$	0.2 – 1.0	No
Ceiling jet radius relative to the ceiling height, H_c	N/A	1.2 – 1.7	N/A
Equivalence ratio based on opening area	See Section F.3.2 for discussion of this parameter.	0.04 – 0.6	Yes
Compartment aspect ratios	$\dfrac{L}{H_c} = \dfrac{100.3\ \text{m}}{4.6\ \text{m}} \approx 21.8$; $\quad \dfrac{W}{H_c} = \dfrac{99.5\ \text{m}}{4.6\ \text{m}} \approx 21.6$	0.6 – 5.7	No
Target distance to fire diameter (Columns A,B,C,D,E,F)	$\dfrac{8.5\ \text{m}}{6.0\ \text{m}} \approx 1.4$ $\quad \dfrac{7.2\ \text{m}}{6.0\ \text{m}} \approx 1.2$ $\quad \dfrac{18.8\ \text{m}}{6.0\ \text{m}} \approx 3.1$ $\quad \dfrac{18.3\ \text{m}}{6.0\ \text{m}} \approx 3.1$ $\quad \dfrac{36.5\ \text{m}}{6.0\ \text{m}} \approx 6.1$ $\quad \dfrac{78.\ \text{m}}{6.0\ \text{m}} \approx 13.1$; $\dfrac{28.0\ \text{m}}{6.0\ \text{m}} \approx 4.7$ $\quad \dfrac{26.9\ \text{m}}{6.0\ \text{m}} \approx 4.5$ $\quad \dfrac{8.8\ \text{m}}{6.0\ \text{m}} \approx 1.5$ $\quad \dfrac{3.9\ \text{m}}{6.0\ \text{m}} \approx 0.7$ $\quad \dfrac{43.3\ \text{m}}{6.0\ \text{m}} \approx 7.2$ $\quad \dfrac{80.\ \text{m}}{6.0\ \text{m}} \approx 13.5$	2.2 – 5.7	Yes/No

Notes: The effective diameter of the fire is determined from the formula, $D = \sqrt{4A/\pi}$, where A is the curbed area of the dike.

The flame length to ceiling height ratio of 2.4 calculated for this scenario indicates that the flames would be impinging on the ceiling and spreading out over a relatively large distance beneath the ceiling. This configuration is beyond the capability of the FDT^s, FIVE-Rev1, and CFAST, and could affect the plume entrainment, the HGL temperature and depth, and the radiant heat flux to a target. In fact, the fire in this scenario is so large that for columns within a few (say 3) fire diameters, the fire can be taken as fully-engulfing the columns. For columns at greater distances, the fire can be viewed as a point source of radiation.

As part of the work performed at NIST for the investigation of the World Trade Center (WTC) disaster, FDS was validated against large-scale fire experiments involving liquid fuel fires impinging on the ceiling of a large compartment with a flat ceiling. Measurements included the heat flux to and the temperatures of bare and insulated steel structural members. The NIST experiments and the FDS simulations are described in NIST NCSTAR 1-5F. The WTC validation work did not involve a compartment with aspect ratios as large as those in the scenario under consideration here, but other validation exercises have shown that the model can handle hallway flows with comparable length to height ratios. The FDS Validation Guide (NIST Special Publication 1018-5) includes the details of these validation studies, including the Bureau of Alcohol, Tobacco and Firearms (ATF) Corridor Experiments and the National Bureau of Standards (NBS) Multi-room Experiments.

F.4 Estimation of Fire-Generated Conditions

This section provides details of how the different models are applied to the scenario.

F.4.1 Algebraic Models

General: The fire is depicted schematically in Figure F-4. As discussed in Section F.3, the flame length is expected to be approximately 2.4 times the ceiling height, indicating that significant flame extension will occur beneath the turbine deck.

The concept of ceiling flame impingement is illustrated in Figure 3-8 of NUREG-1805 and is discussed in Karlsson and Quintiere (2000). This reference indicates that the radial length of flame beneath a ceiling is approximately equal to the length of flame truncated by the ceiling, as illustrated in Figure F-4. This estimate is based on data from Heskestad and Hamada (1993). With a calculated flame length of 11 m (36 ft) and a ceiling height of 4.6 m (15 ft), this suggests that the flame beneath the ceiling would have a radial extension of approximately 6.4 m (21 ft) from the centerline of the fire.

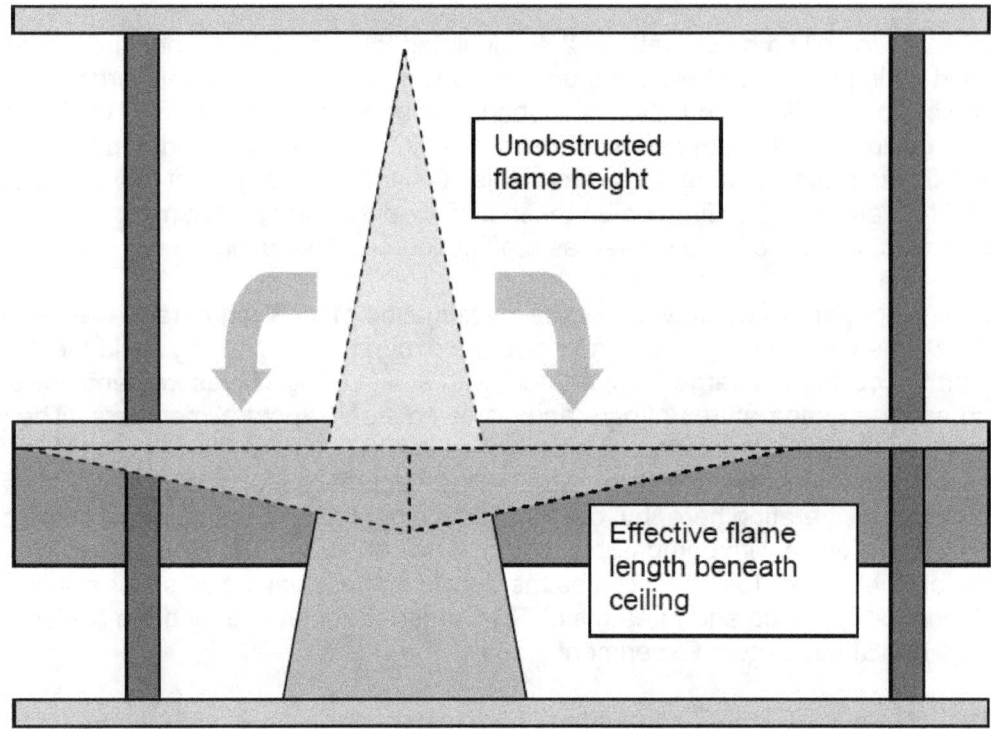

Figure F-4. Schematic diagram of the fire impinging on the ceiling.

Figure F-5 shows the approximate radial flame extension of 6.4 m (21 ft) superimposed (yellow circle) on a detail of Figure F-1 near the two proposed curb locations. As shown in Figure F-5, Column B is at the edge of the estimated flame extension for Curb Location 1, and Column D is well within the estimated flame extension for Curb Location 2, which means that the upper part of Column D would be engulfed in flames. Column A is near the edge of the estimated flame extension for Curb Location 1, and Column C is near the edge of the estimated flame extension for Curb Location 2.

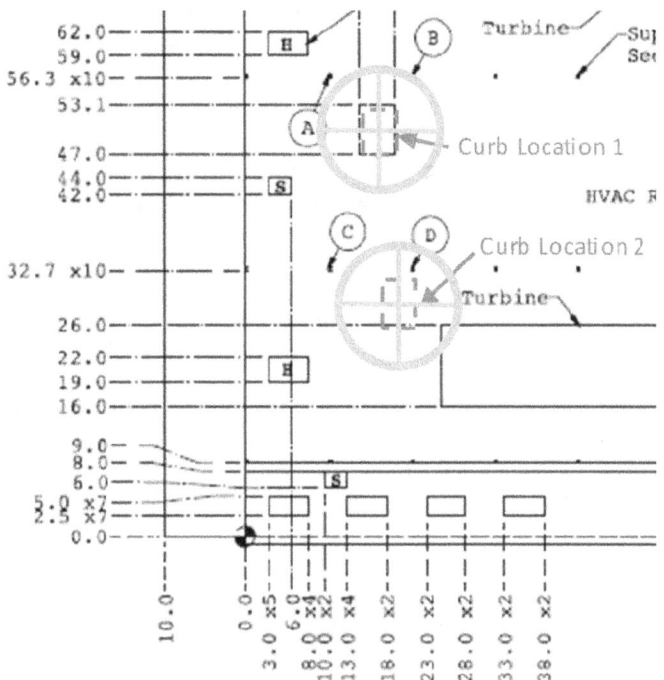

Figure F-5. Detail from Figure F-1 with estimated flame extension beneath ceiling superimposed.

The heat flux from the fire to a nearby column can be estimated using one of the FDT[s] (NUREG-1805, 2005), specifically:

- 05.1_Heat_Flux_Calculations_Wind_Free.xls (Point Source)

- 05.1_Heat_Flux_Calculations_Wind_Free.xls (Solid Flame 1)

For the point source method, the estimated peak HRR is 50,400 kW, the radiative fraction is 0.33, and the horizontal distance from the center of the lubricating oil pool to the nearest column (Column D) is approximately 4.2 m (13.8 ft):

$$\dot{q}''_r = \frac{\chi_r \, \dot{Q}}{4\pi \, r^2} = \frac{0.34 \times 50{,}400 \text{ kW}}{4\pi \times 4.2^2 \text{ m}^2} \cong 75.0 \text{ kW/m}^2 \tag{F-5}$$

The solid flame calculation in the FDT[s] is based on the flame being approximated as "a cylindrical, blackbody, homogeneous radiator with an average emissive power" (Beyler, *SFPE Handbook, 4th edition*). However, Beyler points out that the solid flame calculation "represents the average emissive power over the whole flame and is significantly less than the emissive power that can be attained locally. Thus, the solid flame calculation is inappropriate for estimating the heat flux to columns that are potentially engulfed in flames. In fact, neither calculation is appropriate for estimating the heat flux to columns that are potentially in flames. Neither of these simple radiation calculations account for the flame extension beneath the ceiling, so it is likely that they underestimate the actual heat flux at the column, particularly for the case of the solid flame heat flux calculation.

LUBRICATING OIL FIRE IN A TURBINE BUILDING

Even though the simple heat flux calculation methods are questionable for close targets, it is still worthwhile to estimate the time for the column to heat up to temperatures that could potentially cause it to fail. If so, then it makes sense to continue analysis with a more detailed model like FDS.

In order to estimate an approximate time for a column to reach the specified failure temperature of 538 °C (1,000 °F) when subjected to a given radiant heat flux, a simple energy balance is used to calculate the steel's rate of temperature:

$$\rho_s c_s V_s \frac{dT_s}{dt} = \dot{q}_r'' A_s \tag{F-6}$$

The subscript s refers to steel. The temperature of the steel is modeled as changing uniformly through the steel cross-section due to its relatively high thermal conductivity. This is sometimes called a "lumped capacitance" or "low Biot number" approximation. For a constant net heat flux, this differential equation can be readily integrated to yield the steel temperature as a function of time:

$$T_s - T_0 = \frac{\dot{q}_r'' \, t}{\rho_s c_s (V_s/A_s)} \tag{F-7}$$

To calculate the time, t_{crit}, when the steel failure temperature is reached, this equation is rearranged, with the critical steel temperature, T_{crit}, inserted for the steel temperature.

$$t_{\text{crit}} = \frac{\rho_s c_s (V_s/A_s)(T_{\text{crit}} - T_0)}{\dot{q}_r''} = \frac{c_s (W/D)(T_{\text{crit}} - T_0)}{\dot{q}_r''} \tag{F-8}$$

The term V_s/A_s is sometimes called the section factor, and is the effective thickness of the steel member; it is calculated as the cross-sectional area of a steel member divided by the heated perimeter of the member. In the U.S., it is more common to use a parameter referred to as the W/D ratio, which is simply the section factor multiplied by the steel density. For a W14x145 steel column, the W/D ratio has a value of approximately 96.2 kg/m^2 (1.64 lb/ft/in). Using this value for the W/D ratio, the time to reach the critical steel temperature for the column can be estimated using the radiant heat flux estimated in Equation F-5:

$$t_{\text{crit}} = \frac{(0.465 \text{ kJ/kg/°C})(96.2 \text{ kg/m}^2)(538 \text{ °C} - 36 \text{ °C})}{75.0 \text{ kW/m}^2} \cong 300 \text{ s} \tag{F-9}$$

Because this time is much shorter than the estimated burning duration of the lubricating oil fire (~2,100 s), this analysis suggests that it may be necessary to protect the columns nearest the proposed curb locations with fire resistant coatings to prevent the columns from reaching the specified steel failure temperature.

Because a number of conservative approximations were used for this analysis using algebraic models, FDS is also used to analyze this scenario.

F.4.2 CFD Model

Geometry: The entire turbine hall is included in the simulation (see Figure F-6). One mesh covers the lower deck, and one covers the upper turbine deck. The numerical mesh consists of uniform grid cells with a resolution of 1 m (3.3 ft). While this mesh appears to be fairly coarse, the fire is so large that the ratio of D^* (the characteristic fire diameter) to the cell size is about five.

The columns cannot be resolved on the relatively coarse grid, and are approximated as steel plates with the given thickness of the actual columns. The column obstructions are one cell thick, which allows the boundary on the surface opposite the fire to be exposed to the room conditions. Even though the column obstruction is one cell thick in the domain mesh, the thickness of the steel surface through which heat is transferred is equal to the thickness of the column web. Note that FDS only performs a one-dimensional heat transfer calculation within solid obstructions, which is why there is little to be gained by resolving the column further. The neglect of lateral heat conduction within the solid tends to produce a slight over-prediction of the average column cross section temperature, but, because the heat flux from the fire is expected to be fairly uniform over the width of the column, a more detailed thermal conduction calculation is not warranted.

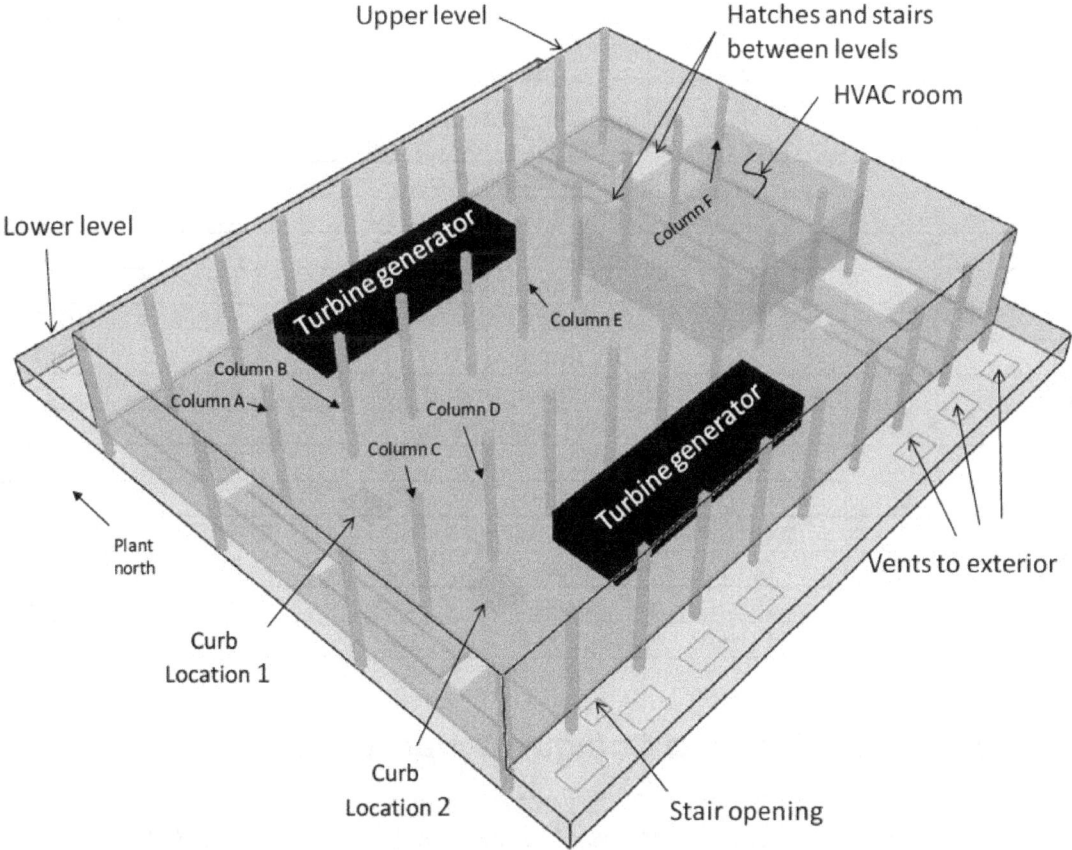

Figure F-6. FDS geometry for the turbine building fire scenario.

Materials: The material properties are applied directly as specified to the walls, floor, and ceiling.

Fire: The fire is specified in the curbed areas, as described in Section F.2. The HRR, soot yield, and molecular weight are as described in Section F.2, and are provided directly as inputs to FDS. A ten-second growth rate is used to allow the flows to develop over a finite time interval.

Ventilation: The openings to the exterior and between the upper and lower level are modeled at the locations, as shown in Figure F-6. It should be noted that the point of including the lower and upper levels of the turbine building in the simulation is to check whether there would be sufficient make-up air drawn through the various vents to sustain a steady-state 50 MW fire.

F.5 Evaluation of Results

The purpose of the calculations described above is to estimate the steel temperature of six large columns in the turbine building to determine whether any would lose the ability to carry their design load in the event of a large fire in the curbed area around a tank of lubricating oil. A structural steel column is considered to fail if the average cross section temperature of the steel exceeds 538 °C (1,000 °F), as described in Section F.2. The results of the FDS simulations are summarized in Table F-4.

Table F-4. Summary of results for the turbine building fire scenarios.

Model	Bias Factor, δ	Standard Deviation, $\tilde{\sigma}_M$	Target	Predicted Value	Critical Value	Probability of Exceeding
Surface Temperature (°C), Initial Value = 36 °C						
Curb Location 1						
FDS	1.02	0.13	Column A	270	538	0.000
FDS	1.02	0.13	Column B	260	538	0.000
FDS	1.02	0.13	Column C	170	538	0.000
FDS	1.02	0.13	Column D	150	538	0.000
FDS	1.02	0.13	Column E	90	538	0.000
FDS	1.02	0.13	Column F	50	538	0.000
Curb Location 2						
FDS	1.02	0.13	Column A	130	538	0.000
FDS	1.02	0.13	Column B	120	538	0.000
FDS	1.02	0.13	Column C	400	538	0.001
FDS	1.02	0.13	Column D	620	538	0.828
FDS	1.02	0.13	Column E	75	538	0.000
FDS	1.02	0.13	Column F	50	538	0.000

F.5.1 Column Heat Flux and Column Temperature

The predicted column temperatures for Curb Location 1 and Curb Location 2 are shown in Figure F-7 and Figure F-8. According to the FDS analyses, only Column D is threatened significantly by a fire at Curb Location 2.

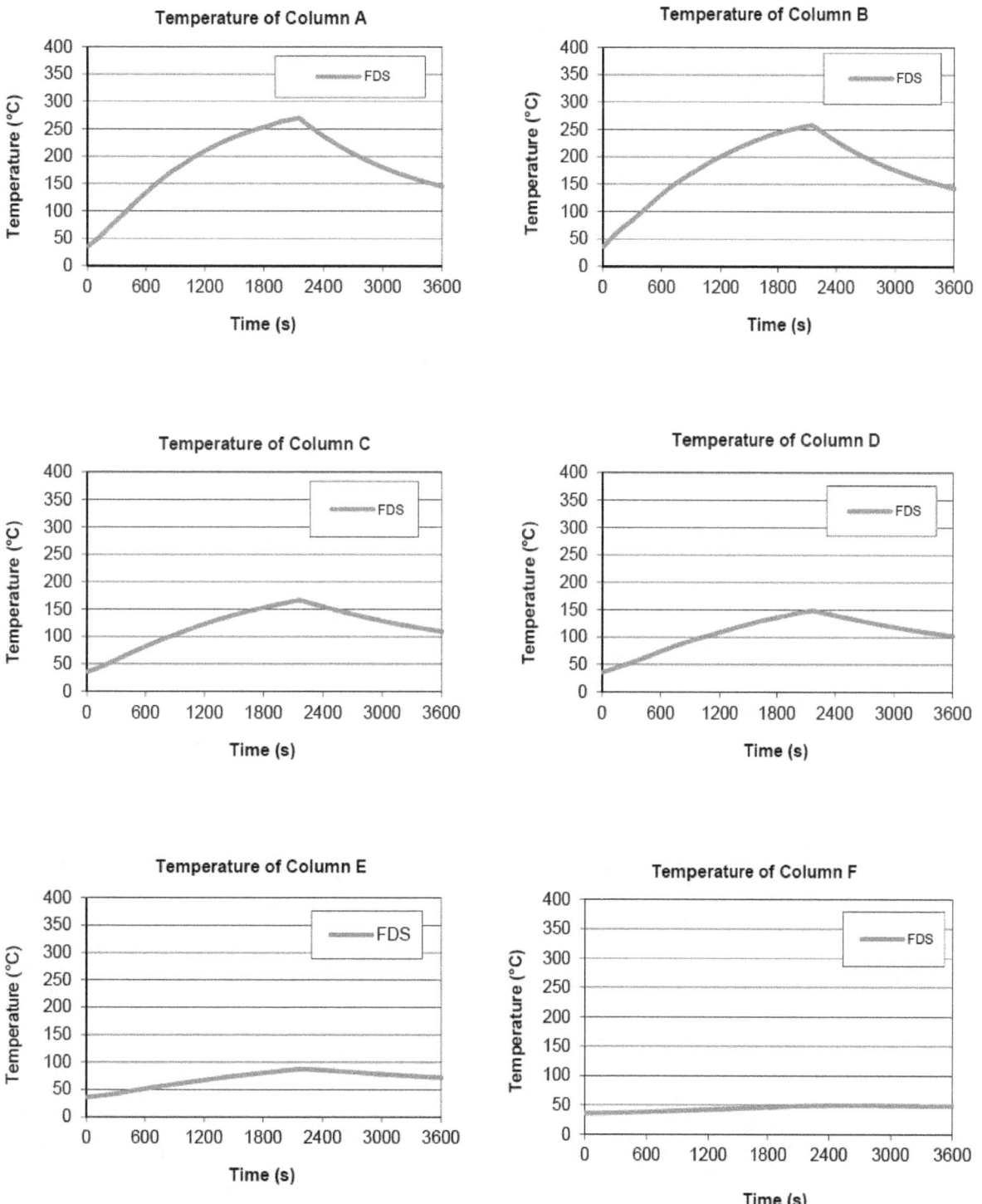

Figure F-7. Temperatures of columns for Curb Location 1.

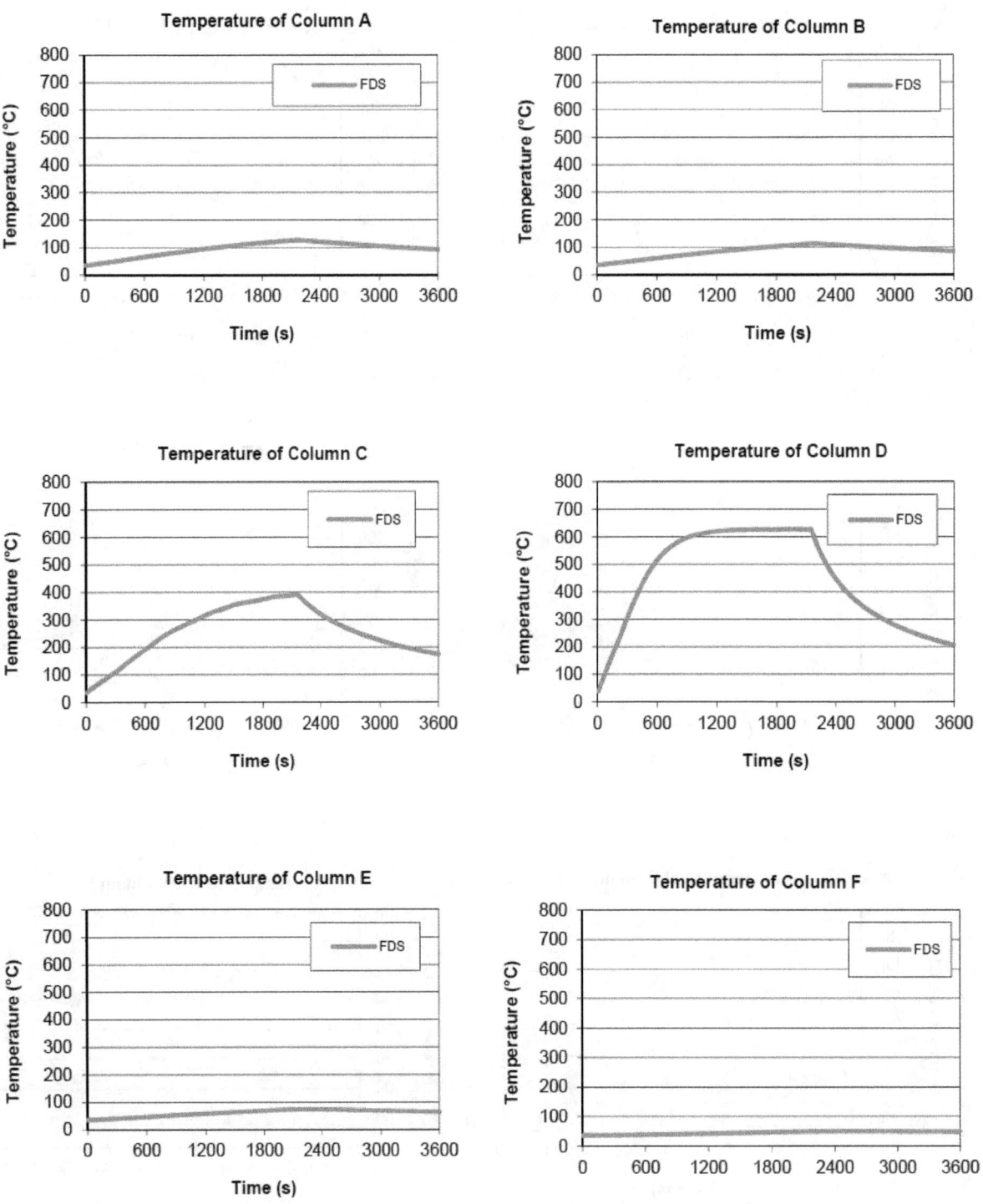

Figure F-8. Temperatures of columns for Curb Location 2.

F.6 Conclusion

This analysis has addressed the potential for a relatively large lubricating oil fire to damage exposed structural steel in a turbine building. The analysis is complicated by the significant flame impingement on the ceiling caused by an oil fire spread over a relatively large area. This type of fire behavior is beyond the validation range addressed in NUREG-1824 (EPRI 1011999).

Algebraic calculations were performed to estimate the extent of flame extension beneath the ceiling. These algebraic calculations indicate that at least one of the columns (Column D) would be engulfed in the flames extending from the fire at Curb Location 2 These calculations also indicate that other columns would be located near the outer extent of flames from Curb Locations 1 and 2. Algebraic calculations were also performed to estimate the time to reach the critical steel temperature of the nearest column. These calculations indicate that damage could occur within a time frame of approximately five minutes. These calculations indicate that a more detailed analysis is warranted. The CFD model, FDS, was used to perform this more detailed analysis because zone models do not have the necessary physical models to simulate the postulated fire.

Based on the FDS simulation of this scenario, a 50 MW lubricating oil fire in Curb Location 1 is not predicted to cause the steel columns to exceed a temperature of 538 °C (1,000 °F). This is not the case for the proposed Curb Location 2, which is located closer to Column D. Consequently, the recommendation for the design package is to install the curbed area at Curb Location 1.

F.7 References

1. American Institute of Steel Construction, *Steel Construction Manual*, 13th Edition, New York, 2006.
2. ASTM E119-10a. *Standard Test Methods for Fire Tests of Building Construction Materials, American Society for Testing and Materials*, West Conshohocken, PA, 2010.
3. Heskestad, G., and Hamada, T., "Ceiling jets of strong fire plumes," *Fire Safety Journal*, Vol. 21, No. 1, pp. 69-82, 1993.
4. Karlsson, B., and J.G. Quintiere, *Enclosure Fire Dynamics*, CRC Press LLC, Boca Raton, FL, 2000.
5. Mulholland, G.W., and C. Croarkin, "Specific Extinction Coefficient of Flame Generated Smoke," *Fire and Materials*, 24:227-230, 2000.
6. NIST NCSTAR 1-5F, *Federal Building and Fire Safety Investigation of the World Trade Center Disaster: Computer Simulation of the Fires in the World Trade Center Towers*, 2005.
7. NIST SP 1018-5, *Fire Dynamics Simulator (Version 5), Technical Reference Guide, Vol. 3, Experimental Validation*, 2010.
8. NUREG-1805, *Fire Dynamics Tools (FDTs) Quantitative Fire Hazard Analysis Methods for the U.S. Nuclear Regulatory Commission Fire Protection Inspection Program*, 2004.
9. NUREG-1824 (EPRI 1011999), *Verification and Validation of Selected Fire Models for Nuclear Power Plant Applications*, 2007.
10. *SFPE Handbook of Fire Protection Engineering*, 4th edition, 2008.

F.8 Attachments (on CD)

1. FDS input files:

      a. Lube_oil_fire_in_TB_Location_1.fds
      b. Lube_oil_fire_in_TB_Location_2.fds

2. Algebraic model files:

      a. 05.1_Heat_Flux_Calculations_Wind_Free_AppF.xls (Point Source)
      b. 05.1_Heat_Flux_Calculations_Wind_Free_AppF.xls (Solid Flame 1)

APPENDIX G
TRANSIENT FIRE IN A MULTI-COMPARTMENT CORRIDOR

G.1 Modeling Objective

The calculations described in this appendix predict room temperatures in a multi-compartment corridor after the accidental start of a transient fire in a stack of burning pallets with two trash bags on top that were left after a plant modification. The purpose of the calculation is to determine whether cables in adjacent compartments will fail, and, if so, at what time failure occurs. The time to smoke detector activation is also estimated.

G.2 Description of the Fire Scenario

General Description: After a plant modification, a fire ignites in a stack of pallets with two trash bags on top, in a corner of a corridor containing multiple compartments, with different door heights and soffits. Various important cables are routed through the corridor compartments.

Geometry: This area consists of interconnected compartments and corridors on the same level. All boundary surfaces are 0.5 m (1.6 ft) thick, as shown in Figure G-1. Figure G-1 and Figure G-2 illustrate the geometry.

Materials: The walls, ceiling, and floor are made of concrete. The thermal properties of the materials in the compartment are listed in Table 3-1. The cable trays contain cross-linked polyethylene (XPE or XLPE)-insulated cables with a Neoprene jacket. These thermoset (TS) type cables are considered damaged when the internal temperature just underneath the jacket reaches 330 °C (625 °F) (NUREG-6850 (EPRI 1011989), Table H-1). The tray locations are shown in Figure G-2.

Fire Protection Systems: There are nine smoke detectors, located as shown in Figure G-1. The detectors are Underwriters Laboratory (UL) listed, with a sensitivity of 4.9%/m (UL 217, 2006). There is no automatic fire suppression.

Ventilation: The ventilation system supplies the combined space at a rate of 1.67 m³/s (3,54 ft³/min). The supply and return vents are shown in the drawing. There are three doors leading into the space, all of which are closed during normal operation but each has a 1.3 cm (0.5 in) gap between the floor and its base.

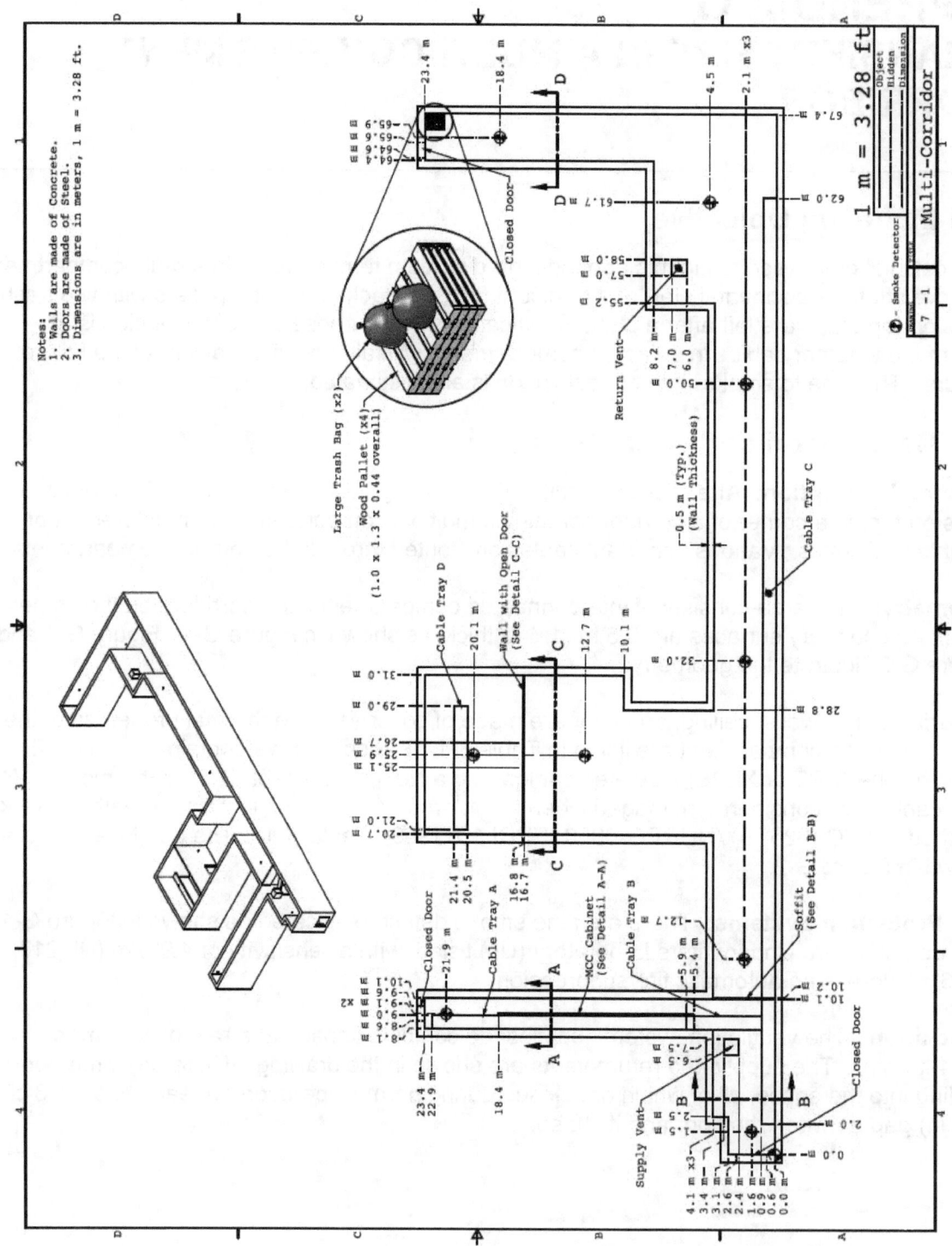

Figure G-1. Geometry of the multi-compartment corridor.

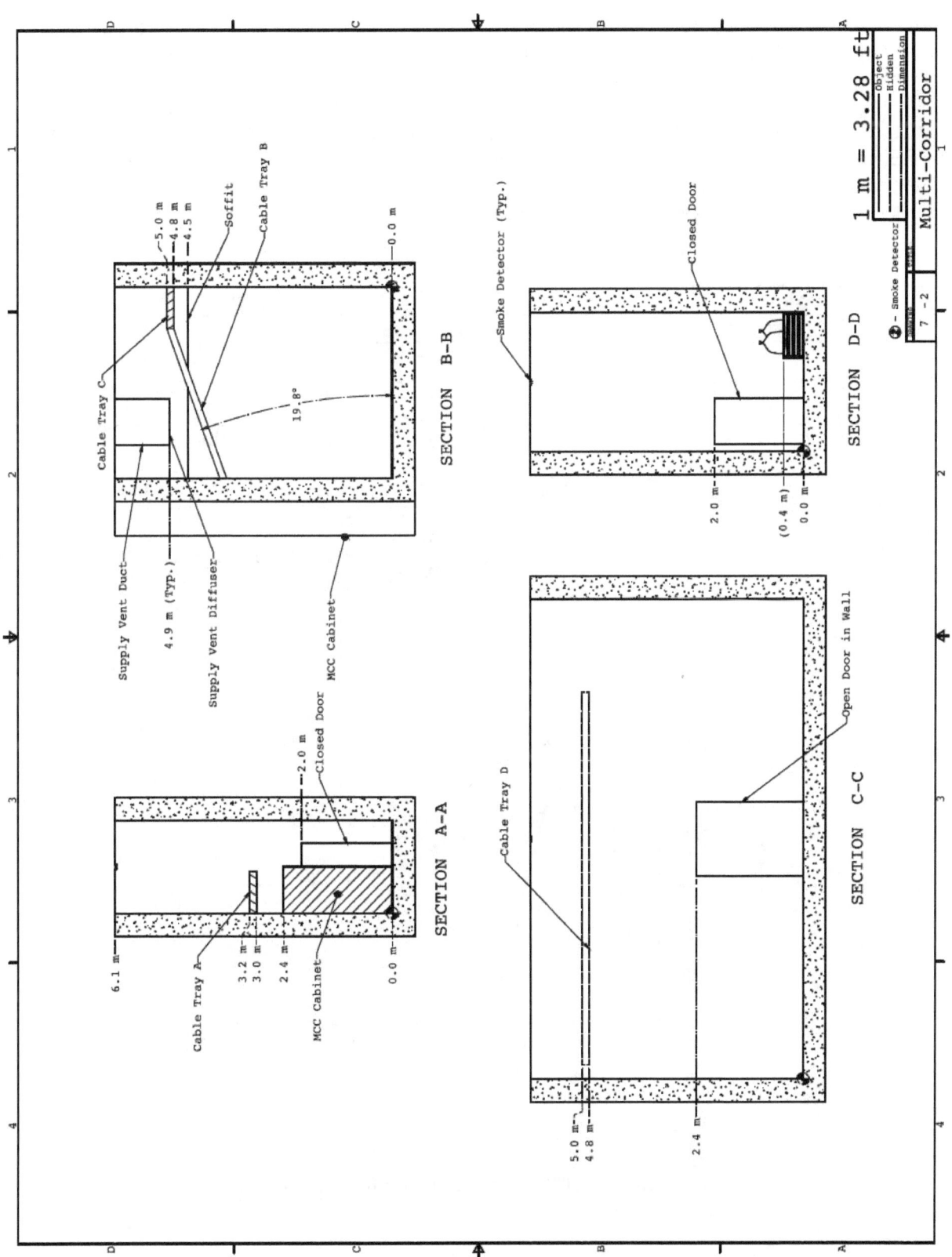

Figure G-2. Geometry details of the multi-compartment corridor.

Fire: The fire, a stack of four wood pallets with two trash bags, is located in the corner, as shown in Figure G-1 and Figure G-2. The pallet stack is 0.44 m (1.4 ft) tall. The fire HRR follows a "t-squared" curve to a maximum value of 2,500 kW in seven minutes and remains steady for eight additional minutes. The heat release rate (HRR) is estimated by combining separate estimates for the stack of wood pallets and the trash bags and using data from the *SFPE Handbook, 4th edition*. After that, the fire's HRR decays linearly to zero in eight minutes. The heat of combustion and product yields are based on values for red oak, and are listed in Table G-1. The HRR curve is shown in Figure G-3.

Table G-1. Products of combustion for a wood pallet fire.

Parameter	Value	Source
Effective Fuel Formula	$C_6H_{10}O_5$	Cellulose
Peak HRR	2,500 kW	*SFPE Handbook*, 4th Ed., Figs. 3-1.65, 3-1.100
Time to reach peak HRR	420 s	*SFPE Handbook*, 4th Ed., Figs. 3-1.64
Heat of Combustion	17,100 kJ/kg	*SFPE Handbook*, 4th Ed., Table 3-4.16
Heat of Combustion per unit mass of oxygen consumed	13,100 kJ/kg	Hugget 1980, Average value
CO_2 Yield	1.27 kg/kg	*SFPE Handbook*, 4th Ed., Table 3-4.16
Soot Yield	0.015 kg/kg	*SFPE Handbook*, 4th Ed., Table 3-4.16
CO Yield	0.004 kg/kg	*SFPE Handbook*, 4th Ed., Table 3-4.16
Radiative Fraction	0.37	*SFPE Handbook*, 4th Ed., Table 3-4.16

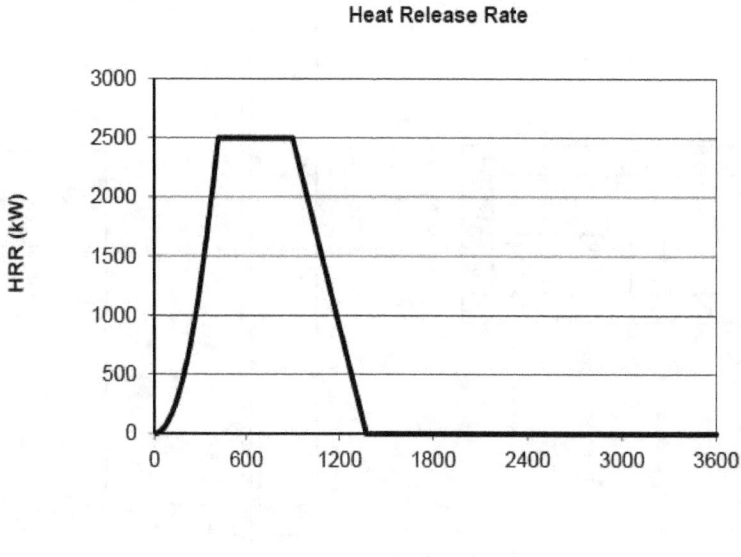

Figure G-3. HRR for the corridor fire scenario.

G.3 Selection and Evaluation of Models

This section discusses the overall modeling strategy. In particular, it describes the process of model selection, including a discussion of the validity of these models for the given fire scenario.

G.3.1 Temperature Criterion

The McCaffrey, Quintiere, and Harkleroad (MQH) correlation incorporated in the Fire-Induced Vulnerability Evaluation (FIVE-Rev1) is used to estimate the hot gas layer (HGL) temperature in the fire room. The correlation applies only to a single room; however, if the temperature in the fire room is less than the damage temperature, the conclusion can be made that the temperature will be even lower in the adjacent compartment, where the closest cable tray is located.

The time to smoke detection can be calculated using FIVE-Rev1 with the ceiling jet temperature unconfined flow correlation of Alpert and the confined flow correlation of Delichatsios.

Zone models can calculate the time-dependent HGL properties in multi-compartment scenarios, as well as the activation times of smoke detectors. Although the geometry in this scenario is somewhat complex, it can be handled by zone modeling, since it is largely a group of interconnected compartments. The zone model MAGIC is used in this scenario.

G.3.2 Validation

A source of validation data justifying the use of the fire models discussed above for this scenario is the U.S. Nuclear Regulatory Commission/Electric Power Research Institute (NRC/EPRI) verification and validation (V&V) study documented in NUREG-1824 (EPRI 1011999). The National Bureau of Standards (NBS, now the National Institute of Standards and Technology (NIST)) Multi-Room Test Series consists of 45 fire experiments in two relatively small rooms connected via a relatively long corridor. The fire source, a gas burner, was located against the rear wall of one of the small compartments, and fire tests of 100 kW, 300 kW, and 500 kW were conducted. The present scenario has a larger fire in a larger, longer compartment. The V&V study assessed the MQH HGL temperature correlation, the Alpert ceiling jet correlation, and the zone model, MAGIC. The unconfined ceiling jet correlation of Delichatsios is included in this analysis to assess the sensitivity of various input parameters.

Table G-2 lists various important model parameters and the ranges for which the NRC/EPRI validation study is applicable. The calculations in Table G-2 are for the fire room (compartment 8, as shown in Figure G-4). The ceiling jet radial distance relative to the ceiling height is not within the range of validation. As a sensitivity case, an additional detector was added at a distance that would fall within the validation range (6.8 m (22.3 ft)). In addition, the room-width geometry is not within the range of validation. A sensitivity case with the height reduced to a value within the validation range (5 m (16.4 ft)) was performed. The remaining model parameters are within the applicable validation ranges.

Table G-2. Normalized parameter calculations for the multi-compartment corridor fire scenario.

Quantity	Normalized Parameter Calculation	NUREG-1824 Validation Range	In Range?
Fire Froude Number	$$\dot{Q}^* = \frac{\dot{Q}}{\rho_\infty c_p T_\infty D^{2.5}\sqrt{g}}$$ $$= \frac{2500\ \text{kW}}{(1.2\ \text{kg/m}^3)(1.0\ \text{kJ/kg/K})(293\ \text{K})(1.29^{2.5}\ \text{m}^{2.5})\sqrt{9.8\ \text{m/s}^2}} = 1.2$$	0.4 – 2.4	Yes
Flame Length, L_f, relative to the Ceiling Height, H_c	$$\frac{H_f + L_f}{H_c} = \frac{0.44\ \text{m} + 3.8\ \text{m}}{6.1\ \text{m}} = 0.7$$ $$L_f = D\left(3.7\,\dot{Q}^{*2/5} - 1.02\right) = 1.29\ \text{m}\ (3.7 \times 1.2^{0.4} - 1.02) = 3.8\ \text{m}$$	0.2 – 1.0	Yes
Ceiling Jet Horizontal Radial Distance, r_{cj}, relative to the Ceiling Height, H_c	$$\frac{r_{cj}}{H_c - H_f} = \frac{4.46\ \text{m}}{6.1\ \text{m} - 0.44\ \text{m}} = 0.8$$	1.2 – 1.7	No
Equivalence Ratio, φ, as an indicator of the Ventilation Rate	$$\varphi = \frac{\dot{Q}}{\Delta H_{O_2}\dot{m}_{O_2}} = \frac{2500\ \text{kW}}{13,100\ \text{kJ/kg} \times 0.46\ \text{kg/s}} = 0.4$$ $$\dot{m}_{O_2} = 0.23\,\rho_\infty \dot{V} = 0.23 \times 1.2\ \text{kg/m}^3 \times 1.67\ \text{m}^3/\text{s} \cong 0.46\ \text{kg/s}$$	0.04 – 0.6	Yes
Compartment Aspect Ratios	$$\frac{L}{H_c} = \frac{15.2\ \text{m}}{6.1\ \text{m}} = 2.49 \ ; \ \frac{W}{H_c} = \frac{3.0\ \text{m}}{6.1\ \text{m}} = 0.49$$	0.6 – 5.7	No
Target Distance, r, relative to the Fire Diameter, D	N/A	2.2 – 5.7	N/A

Notes:

(1) The non-dimensional parameters are explained in Table 2-5.

(2) The effective diameter of the base of the fire, D, is calculated using $D = \sqrt{4A/\pi}$, where A is the area of the pallets.

(3) The "Fire Height," $H_f + L_f$, is the sum of the height of the fire from the floor and the fire's flame length.

G.4 Estimation of Fire-Generated Conditions

G.4.1 Algebraic Models

This scenario concerns the prediction of cable damage at a location outside the compartment of fire origin. The temperature of the HGL in the compartment of fire origin can be modeled as a potential screening tool. If the HGL temperature within the compartment of origin is not likely to cause damage to cables in that compartment, damage to cables outside the fire compartment is even more unlikely. As a part of this approach, the cable surface temperature is modeled as exactly matching the HGL temperature (i.e., heat-up of the cable is immediate).

FIVE-Rev1 was used for the MQH room temperature analysis. The inputs to the model are found in Table 3-1, Table G-1, and Table G-3. The calculation is applied to the fire room only, with the opening to the next compartment treated as an opening with an area (height × width) equal to 6.1 × 3 = 18.3 m^2 (197 ft^2). To correct the MQH temperature correlation for a fire in the corner, the results from FIVE-Rev1 are multiplied by a factor of 1.7, as suggested by Karlsson and Quintiere (Equation 6.23).

For the time to detection, the Alpert ceiling jet temperature calculation is used. The approach is to calculate the time at which the ceiling jet temperature at the heat detector is 30 °C (86 °F). The additional inputs for this correlation are the horizontal radial distance from the centerline of the fire plume to the detector, which is 4.5 m (14.8 ft), and the fire location factor of 4, because of the fire in the corner. Because the fire room is corridor-shaped, the flow is likely to be confined; therefore, the confined flow correlation by Delichatsios is also used. The additional input is the corridor half-width of 1.5 m (4.9 ft).

G.4.2 Zone Model

This is a classic application of a zone fire model with a fire in one compartment connected to a number of additional compartments with doorway-like vents. Outputs of primary interest in the simulation include temperatures in the compartments, activation of smoke detectors in the compartments, and the temperature of cable targets in the compartments. This scenario was modeled using the zone model MAGIC (Gay et. al., 2005).

Geometry: To simplify the process of modeling the multi-compartment geometry, the layout was divided into eight areas, as illustrated in Figure G-4. Note that the small indentation in compartment 1 was ignored for the MAGIC calculations. Compartments were connected by a door (compartments 5 to 6), a soffit (compartments 2 to 3), or were left open by using a full-wall opening (compartments 1 to 2, 3 to 4, 4 to 5, 3 to 7, and 7 to 8).

Table G-3 summarizes the compartment dimensions used for zone modeling. A graphical depiction of the scenario, as modeled in MAGIC, is shown in Figure G-5.

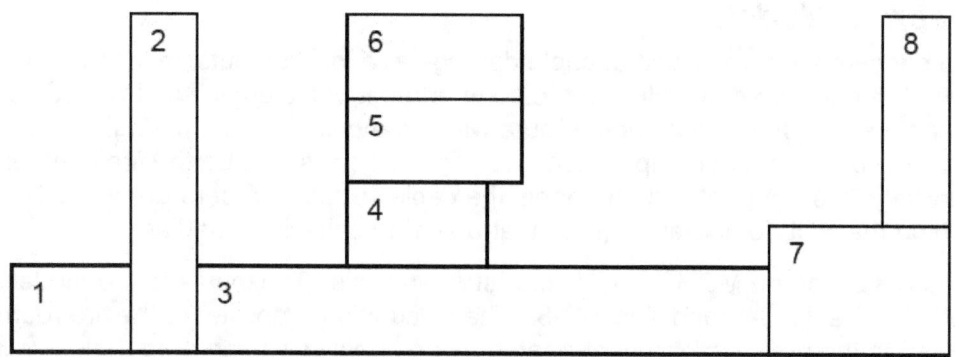

Figure G-4. Effective corridor layout for implementation in zone models (not to scale).

Table G-3. Compartment dimensions for corridor scenario.

Comp.	Length m (ft)	Width m (ft)	Area m^2 (ft^2)
1	8.1 (26.6)	4.1 (13.5)	33.2 (357)
2	2.0 (6.6)	23.4 (76.8)	46.8 (504)
3	45.1 (148)	4.1 (13.5)	184.9 (1990)
4	8.1 (26.6)	6.0 (19.7)	48.6 (523)
5	10.3 (33.8)	6.6 (21.7)	68.0 (732)
6	10.3 (33.8)	6.6 (21.7)	68.0 (732)
7	12.2 (40)	8.2 (26.9)	100.0 (1076)
8	3 (9.8)	15.2 (49.9)	45.6 (491)

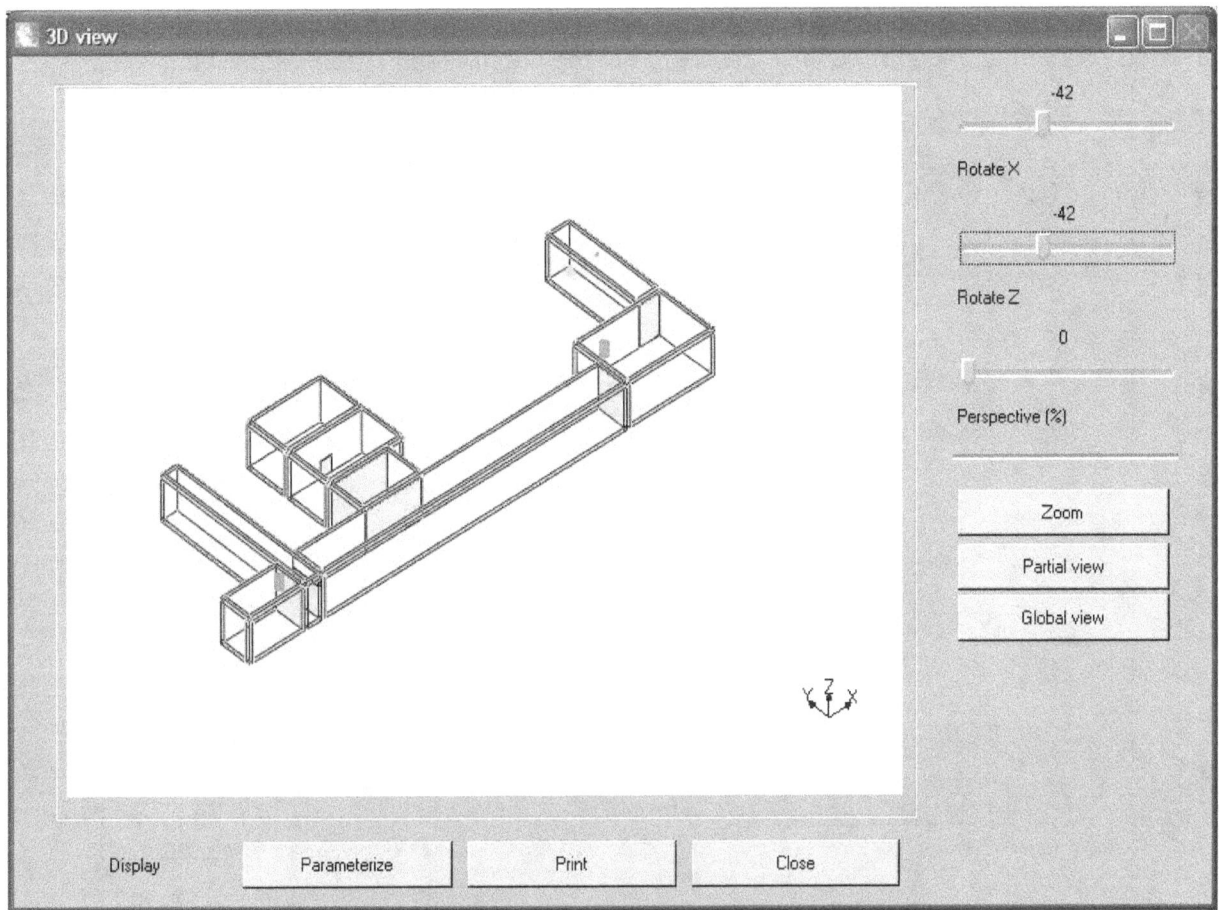

Figure G-5. MAGIC rendering of the corridor scenario.

Fire: The fire is modeled as a 1.3 m^2 (14 ft^2) source (equivalent diameter of 1.29 m (4.2 ft)) at an elevation of 0.44 m (1.4 ft) (see Figure G-2). The stoichiometric oxygen to fuel ratio, ψ, is calculated using Equation 22 from Chapter 4 in Section 3 of the *SFPE Handbook, 4th edition* and the values from Table G-1, as follows:

$$\psi = \frac{\Delta H}{\Delta H_{O_2}} = \frac{17.1 \text{ kJ/g}}{13.1 \text{ kJ/g}} \cong 1.3 \text{ g/g} \tag{G-1}$$

where ΔH is heat of combustion per unit mass of fuel consumed and ΔH_{O_2} is the heat of combustion per unit mass of oxygen consumed. One of the inputs required by MAGIC is the specific area, s (NUREG-1824, Volume 6, Section 3.2.7):

$$s = y_s k_m = 0.015 \times \frac{7,600 \text{ m}^2}{\text{g}} = 114 \text{ m}^2/\text{g} \tag{G-2}$$

where k_m is a constant and y_s is the soot yield, as listed in Table G-1. Products of combustion for a wood pallet fire.. The pyrolysis rate (g/s) is calculated for input to MAGIC by dividing the HRR values (kW) at each time step (as shown in Figure G-3) by the heat of combustion

(17.1 kJ/g). Other inputs needed for MAGIC are listed in Table G-1. Products of combustion for a wood pallet fire.. Figure G-6 is a screenshot of the source fire in the MAGIC input file.

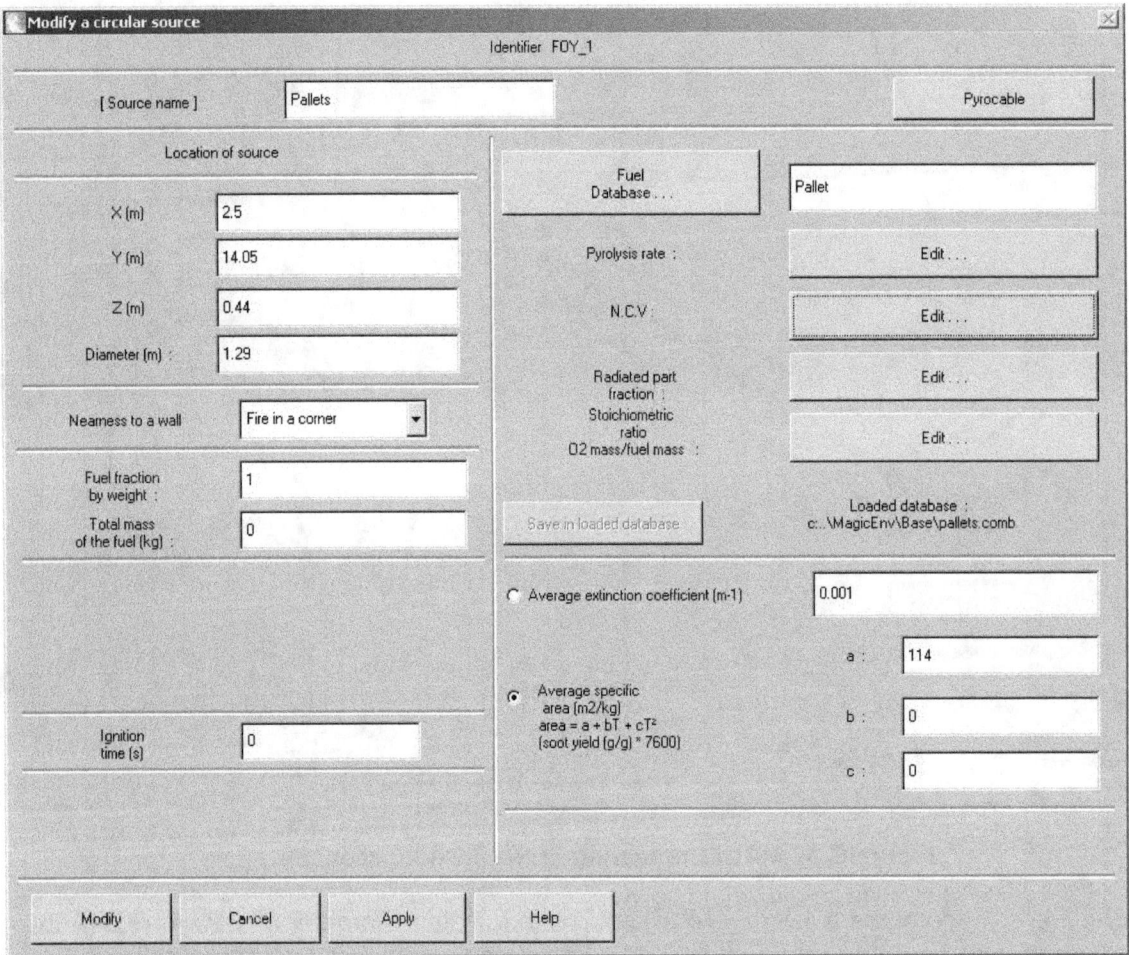

Figure G-6. MAGIC screen capture of the fire in corridor scenario.

Ventilation: The ventilation rate is given in Section G.2. The vents in Figure G-1 are square, but, because MAGIC uses round vents, an equivalent diameter of 1.13 m (3.7 ft) was used as input. In room-to-room connections with the same ceiling height, a shallow (0.1 m) soffit was added to allow smoother model execution. Finally, the only leakage from the space occurs via a 1.3 cm (0.5 in) crack under each of the three doors. The inputs for the supply air vent are shown in Figure G-7.

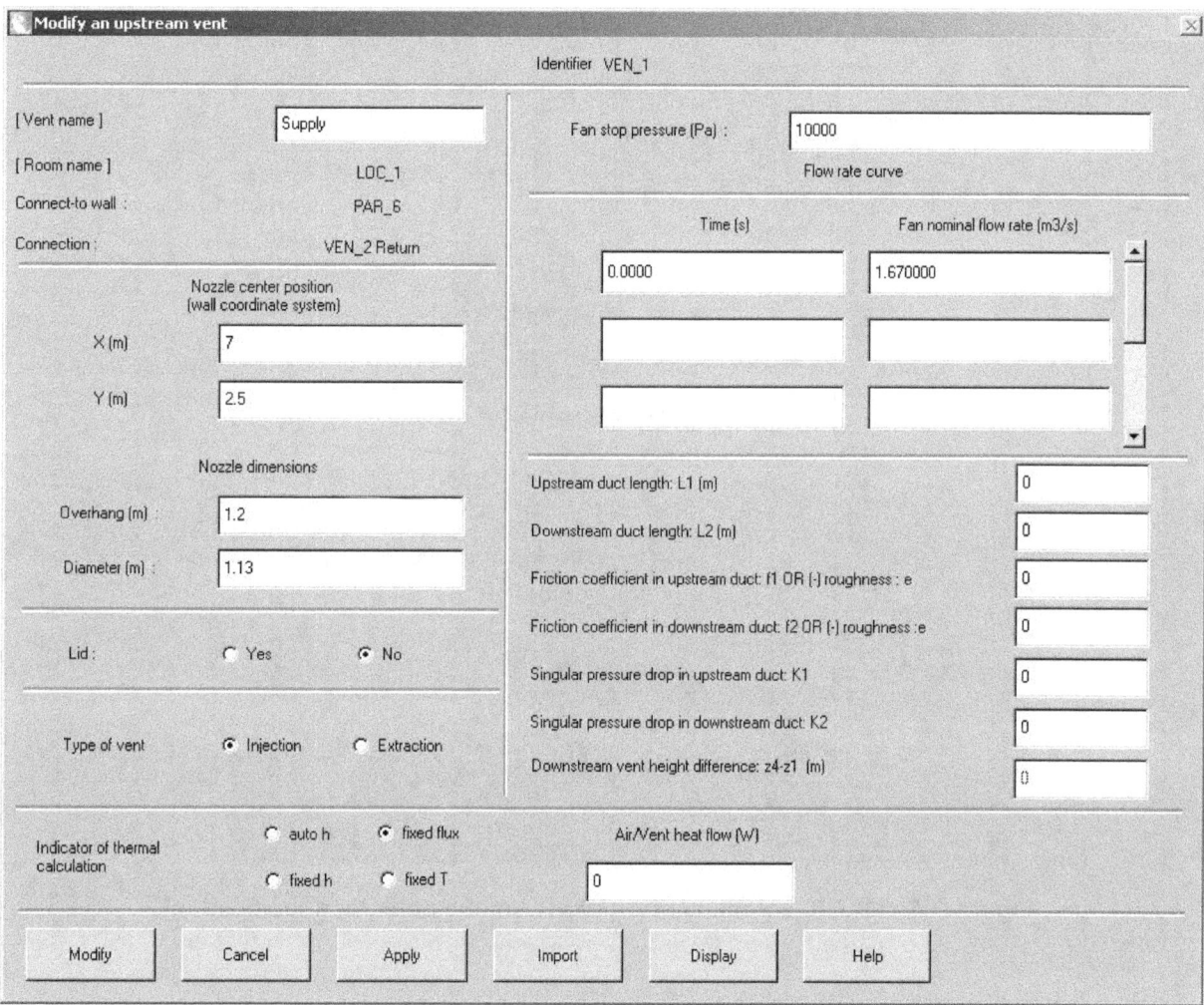

Figure G-7. MAGIC screen capture for supply vent specification.

Fire Protection Systems: In MAGIC, there is no direct way of calculating smoke density for smoke detector activation. Consistent with NUREG-1805, the recommended approach given by the developers is to model the smoke detector as a sprinkler with a very low activation temperature and response time index (RTI). An activation temperature of 30 °C (86 °F) and an RTI of 5 (m/s)$^{1/2}$ was selected. The location of the smoke detector closest to the fire was entered into the input file, as shown in Figure G-8.

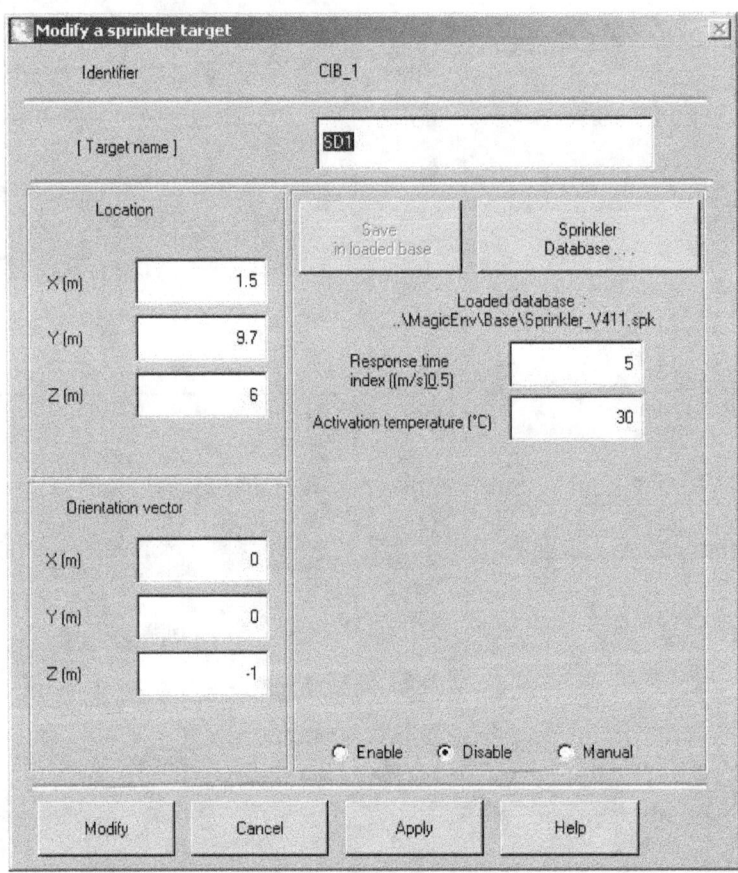

Figure G-8. MAGIC screen capture of the smoke detector specification.

G.5 Evaluation of Results

The purpose of the calculations is to determine whether a stack of burning pallets plus two trash bags in a corridor could generate gas temperatures in adjacent compartments that are capable of damaging cables and electrical equipment. Smoke detector activation is also estimated. The algebraic model FIVE-Rev1 and the zone model MAGIC were used for this scenario. In general, the results demonstrate that the fire is not capable of generating damaging conditions, even in the compartment of fire origin; as a result, there is no need for detailed modeling of the targets in remote locations. The following sections describe the results in greater detail.

Table G-4 summarizes the results of the models for the damage criteria. For each predicted value, a calculation is performed to determine the probability of exceeding the critical value. The procedure for calculating this probability is given in Chapter 4, and it accounts for the model bias and scatter. The purpose of this table is to highlight the criterion that is most likely to be exceeded so that further analysis can be focused on this criterion and on the model or models that predict it. The criterion is discussed in greater detail in the following subsections.

Table G-4. Summary of the model predictions of the corridor scenario.

Model	Bias Factor, δ	Standard Deviation, $\tilde{\sigma}_M$	Ventilation	Predicted Value	Critical Value	Probability of Exceeding
HGL Temperature (°C), Initial Value = 20 °C						
FIVE-Rev1 (MQH)	1.56	0.32	Natural	256	330	0.001
MAGIC	1.01	0.07	Mechanical	240	330	0.000

G.5.1 Temperature Criterion

The MQH correlation in FIVE-Rev1 predicts a peak temperature of 256 °C (493 °F) in the corridor where the fire is located. MAGIC predicts 240 °C (464 °F) (see Figure G-9). Both predictions are below the cable damage temperature threshold of 330 °C (625 °F). MAGIC results show that the HGL temperatures for the other corridors are substantially lower; for example, the next closest corridor (compartment 7) reaches a peak temperature of 160 °C (320 °F), as shown in Figure G-9. A comparison of Figure G-9 and Figure G-3 shows that the timing and the shape of the HGL temperature curve closely follows the HRR curve. The HRR produced by MAGIC is unmodified from the one based on the input (i.e., there is no oxygen starvation).

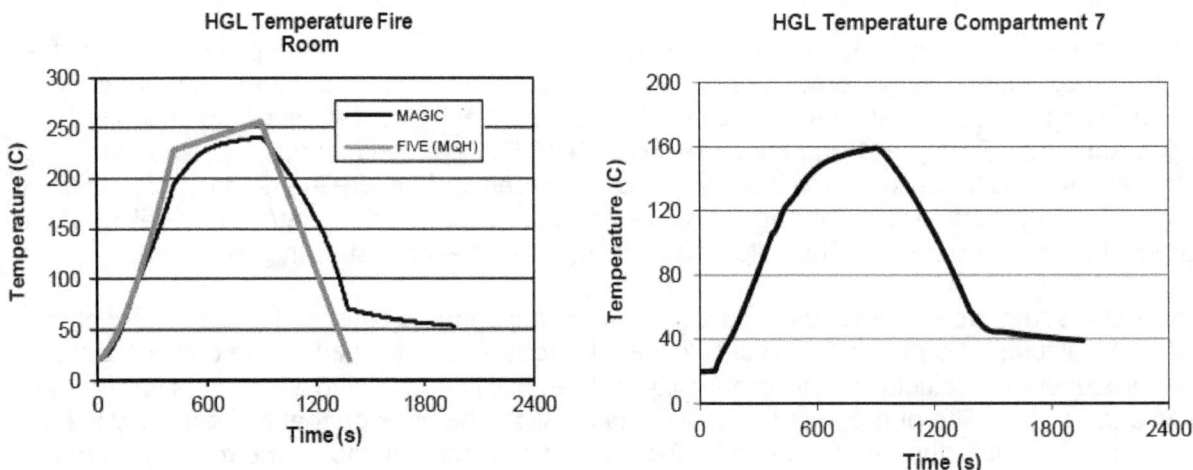

Figure G-9. HGL temperature predictions by MAGIC for the corridor scenario.

Chapter 4, Model Uncertainty, provides guidance on how to express the uncertainty of the MAGIC predictions. MAGIC predicted a temperature rise of M = 240 – 20 = 220 °C (428 °F). As shown in Table 4-1, the NRC/EPRI V&V study (NUREG-1824 (EPRI 1011999)) found that MAGIC predictions of HGL temperature rise are, on average, 1% greater than corresponding measurements, with a relative standard deviation of 7%. The adjusted model prediction is μ = 220/1.01 = 218, and the standard deviation is σ = 0.07 × (220/1.01) = 15.2 °C. Therefore, the probability that the cable temperature would exceed 330 °C (625 °F) is:

$$P(T > 330) = \frac{1}{2} \, \text{erfc}\left(\frac{330 - 20 - 218}{15.2\sqrt{2}}\right) \cong 0 \tag{G-3}$$

The same uncertainty calculation for the MQH correlation in FIVE-Rev1 results in a probability of exceeding the damage temperature of 0.001, or 0.1%. In other words, there is a near-zero probability of exceeding the damage temperature threshold for cables within the compartment of fire origin based on a surrounding HGL temperature, according to the FIVE-Rev1 and MAGIC predictions. This demonstrates that detailed analyses of the cables outside the compartment of origin are not warranted.

Sensitivity of the Room Width to Height Ratio

As shown in Table G-2, the room width to height ratio is not within the validation bounds. A sensitivity case is run in which the room height is reduced until the ratio is within the bounds. All inputs are the same, except that the height is reduced from 6.1 m (20 ft) to 5 m (16.4 ft) in the fire room. The results, plotted in Figure G-10, show that although the temperatures are slightly higher in the fire room, there is no change in the adjacent room, which contains the target. This shows that the room width to height ratio is not a significant influence on the results in this scenario and that the MAGIC results are applicable, even though the ratio falls outside of the validation bounds.

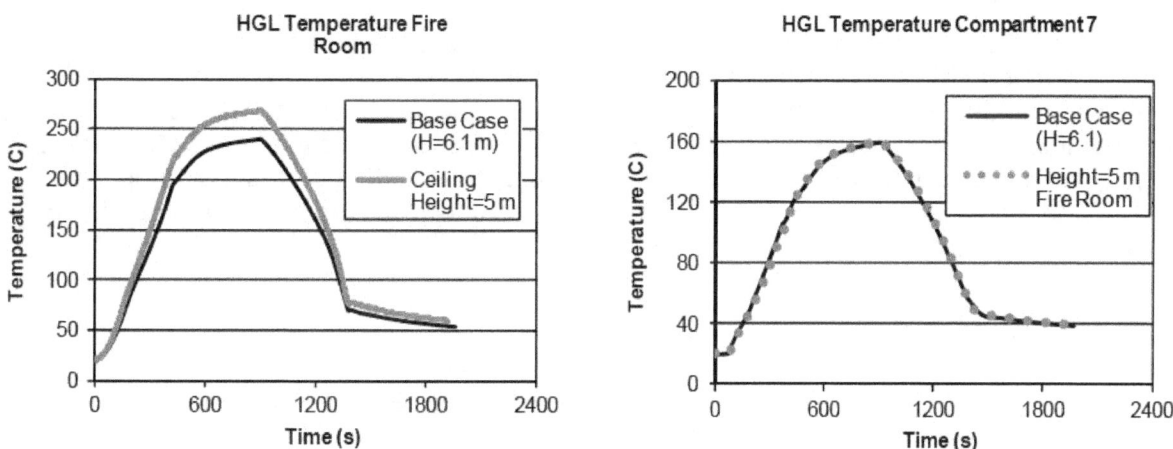

Figure G-10. HGL temperature for reduced ceiling height by MAGIC.

G.5.2 Smoke Detection

The smoke detector activation time in the corridor containing the fire is based on the time for the ceiling jet temperature to reach 30 °C (86 °F) at the detector location. The results, plotted in Figure G-11, show that the two correlations from FIVE-Rev1 produce identical results of 50 s, and MAGIC predicts 40 s.

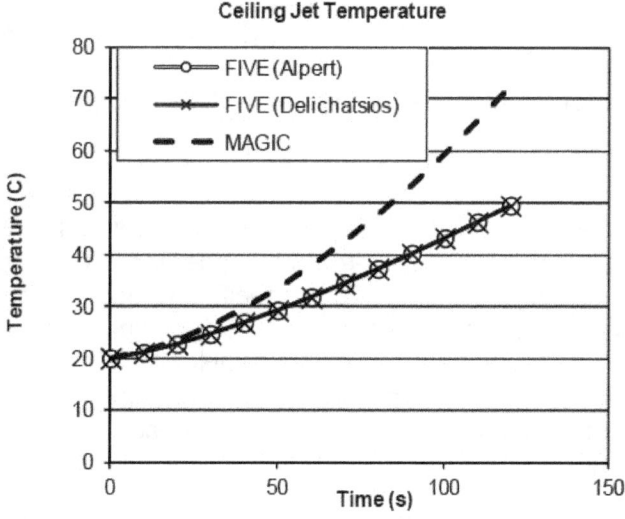

Figure G-11. Detector temperature prediction by MAGIC for fire corridor.

Sensitivity of the Ceiling Jet Radial Distance Relative to the Ceiling Height

As shown in Table G-2, the ceiling jet radial distance relative to the ceiling height is not within the validation bounds. A sensitivity case is performed in which a detector is placed at a further distance so that the ratio is within the validation bounds. The results, plotted in Figure G-12, show that the difference in time to detection varies from 40 s for the base case to 50 s for the detector located slightly further away. This small difference in time indicates that for this scenario, the results are valid, even though the ratio of radius to ceiling height falls outside of the validation bounds.

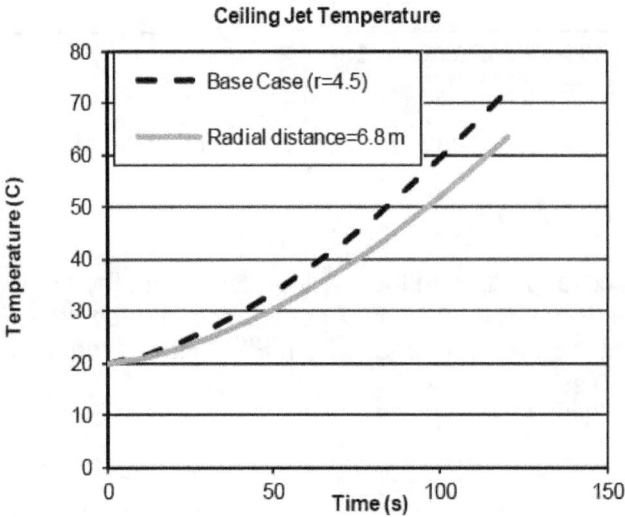

Figure G-12. Detector temperature for two radial distances predicted by MAGIC.

G.6 Conclusion

This analysis addresses a transient fire in a relatively long, narrow room, identified as a corridor, which is connected with other rooms in a multi-room arrangement. The targets include cables in rooms located beyond the corridor where the fire is specified to occur. The MQH correlation included in FIVE-Rev1 was used to estimate temperature conditions within the corridor, and it was found that the HGL temperature predicted by the MQH correlation was lower than the cable damage temperature. Because the cables are located in rooms beyond the corridor where the fire occurs and it can be reasoned that the temperatures in these rooms will be lower than in the corridor, cable damage would not be expected.

The zone model MAGIC was used to predict the HGL temperatures in all of the interconnected compartments, resulting from the stack of burning pallets and trash bags. These MAGIC calculations also demonstrate that the HGL temperature is not high enough to cause cable damage in any compartment or corridor, including the corridor of fire origin. These predictions account for model uncertainty in the temperature predictions and the sensitivity to variations in the heat release rate. Based on a simplified method for smoke detector activation, smoke detector operation is expected to occur between 40 s and 50 s after the ignition of the fire.

G.7 References

1. Gay, L., C. Epiard, and B. Gautier, *MAGIC Software Version 4.1.1: Mathematical Model*, EdF HI82/04/024/B, Electricité de France, France, November 2005.
2. Huggett, C., "Estimation of Rate of Heat Release by Means of Oxygen Consumption Measurements," *Fire and Materials*, 4:61-65, 1980.
3. Karlsson, B., and Quintiere, J. G., *Enclosure Fire Dynamics*, CRC Press, Boca Raton, 2000.
4. NUREG-1805, *Fire Dynamics Tools (FDTs) Quantitative Fire Hazard Analysis Methods for the U.S. Nuclear Regulatory Commission Fire Protection Inspection Program*, 2004.
5. NUREG-1824 (EPRI 1011999), *Verification and Validation of Selected Fire Models for Nuclear Power Plant Applications*, 2007.
6. NUREG/CR-6850 (EPRI 1011999), *EPRI/NRC-RES Fire PRA Methodology for Nuclear Power Facilities*, 2005.
7. *SFPE Handbook of Fire Protection Engineering*, 4th edition, 2008.
8. UL 217, *Single and Multiple Station Smoke Alarms*, 6th edition, Underwriters Laboratories, Inc., 2006.

G.8 Attachments (on CD)

1. Algebraic model file:

    a. Appendix G FIVE.xlsm
    b. MQH_AppG.xlsx

2. MAGIC input files:

    a. Corridor.cas
    b. Corridor_reduced_H.cas

APPENDIX H
CABLE TRAY FIRE IN ANNULUS

H.1 Modeling Objective

The calculations described in this appendix assess the potential for damage to redundant safe-shutdown cables due to a fire in an adjacent cable tray in the annulus region of a pressurized water reactor (PWR) containment building. In addition, the calculations provide information on the effectiveness of a sprinkler system installed in a non-standard configuration.

H.2 Description of the Fire Scenario

General Description: The annulus is the region between the primary containment steel structure and the secondary concrete containment (shield) building. The primary and secondary containments are cylindrical with domes on top. The annulus space contains a variety of penetrations from the reactor to the external support systems. One of these penetrations contains two trays with redundant cables that control systems in both trains of safety equipment. A fire starts in one tray and spreads vertically and horizontally. The purpose of the calculation is to determine if a fire in one train of a cable tray (e.g., Train A) can damage the redundant train of equipment (e.g., Train B).

Geometry: The layout of the annulus is shown in Figure H-1. The exterior wall is made of concrete, while the interior wall and cable trays are made of steel. The cable tray locations are shown in Figure H-2. The trays are 0.6 m (2 ft) wide.

Materials: Property values for the relevant materials are listed in Table 3-1. The annulus wall thicknesses are indicated in the drawing. The cable tray steel is approximately 2 mm (0.079 in) thick.

Cables: The cable trays are filled with PE-insulated, PVC-jacketed control cables. These cables have a diameter of approximately 1.5 cm (0.6 in), a jacket thickness of approximately 1.5 mm (0.06 in), and 7 conductors. There are approximately 120 cables in each tray. The mass of each cable is 0.4 kg/m. The mass fraction of copper is 0.67. These cables fail when the internal temperature just underneath the jacket reaches approximately 205 °C (400 °F) or when the exposure heat flux exceeds 6 kW/m^2 (NUREG-1805, Appendix A).

Fire: The heat of combustion and product yields for PE/PVC cables are taken from Table 3-4.16 of the *SFPE Handbook, 4th edition* and listed in Table H-1. Note that the non-metallic components of the cables are a mixture of PE (C_2H_4) and PVC (C_2H_3Cl). Because the mixture consists of approximately the same mass of each, the cable materials are modeled using an effective chemical formula of $C_2H_{3.5}Cl_{0.5}$.

The fire ignites at the base of the lower cable train in the vicinity of the bend at the inner wall. From the results of the CHRISTIFIRE project (NUREG/CR-7010, Vol. 1), the heat release rate (HRR) per unit area of this thermoplastic (TP) cable is estimated to be 250 kW/m^2. The fire spreads vertically at a rate of 258 mm/s (10 in/s) and horizontally at a rate of 0.9 mm/s

(0.035 in/s) (NUREG/CR-6850 (EPRI 1011989), Appendix R). The peak HRR is 945 kW once all of the cables in the first tray are burning (see Figure H-3).

Table H-1. Products of combustion for a PE/PVC cable fire.

Parameter	Value	Source
Heat of Combustion	20,900 kJ/kg	*SFPE Handbook*, 4th ed., Table 3-4.16
CO_2 Yield	1.29 kg/kg	*SFPE Handbook*, 4th ed., Table 3-4.16
Soot Yield	0.136 kg/kg	*SFPE Handbook*, 4th ed., Table 3-4.16
CO Yield	0.147 kg/kg	*SFPE Handbook*, 4th ed., Table 3-4.16
Radiative Fraction	0.49	*SFPE Handbook*, 4th ed., Table 3-4.16

Local burnout of the fire occurs when the cable plastic is consumed. The time to burnout is calculated as follows. First, determine the combustible mass per unit area of tray, m_c'':

$$m_c'' = \frac{n\,Y_p\,(1-v)m'}{W} = \frac{120 \times 0.33 \times (1-0) \times 0.4\,\text{kg/m}}{0.6\,\text{m}} \cong 26.4\,\text{kg/m}^2 \tag{H-1}$$

where n is the number of cables per tray, Y_p is the mass fraction of combustible (i.e., non-metallic or plastic) material in the cable, v is the residue yield (which is conservatively set to zero in this calculation), m' is the total mass per unit length of a single cable, and W is the tray width. Next, calculate the burnout time, Δt:

$$\Delta t = \frac{m_c''\,\Delta H}{5\,\dot{q}_{\text{avg}}''/6} = \frac{26.4\,\text{kg/m}^2 \times 20,900\,\text{kJ/kg}}{5/6 \times 250\,\text{kW/m}^2} \cong 2648\,\text{s} \tag{H-2}$$

where ΔH is the heat of combustion and $\dot{q}_{\text{avg}}''$ is the average HRR per unit area of tray. The HRR per unit area ramps linearly to its average value over a time period of $\Delta t/6$, remains steady for a time period of $2\,\Delta t/3$, and then decreases linearly to zero over a time period of $\Delta t/6$. The linear ramp-up and ramp-down are typical ways of approximating the time history of an item's HRR.

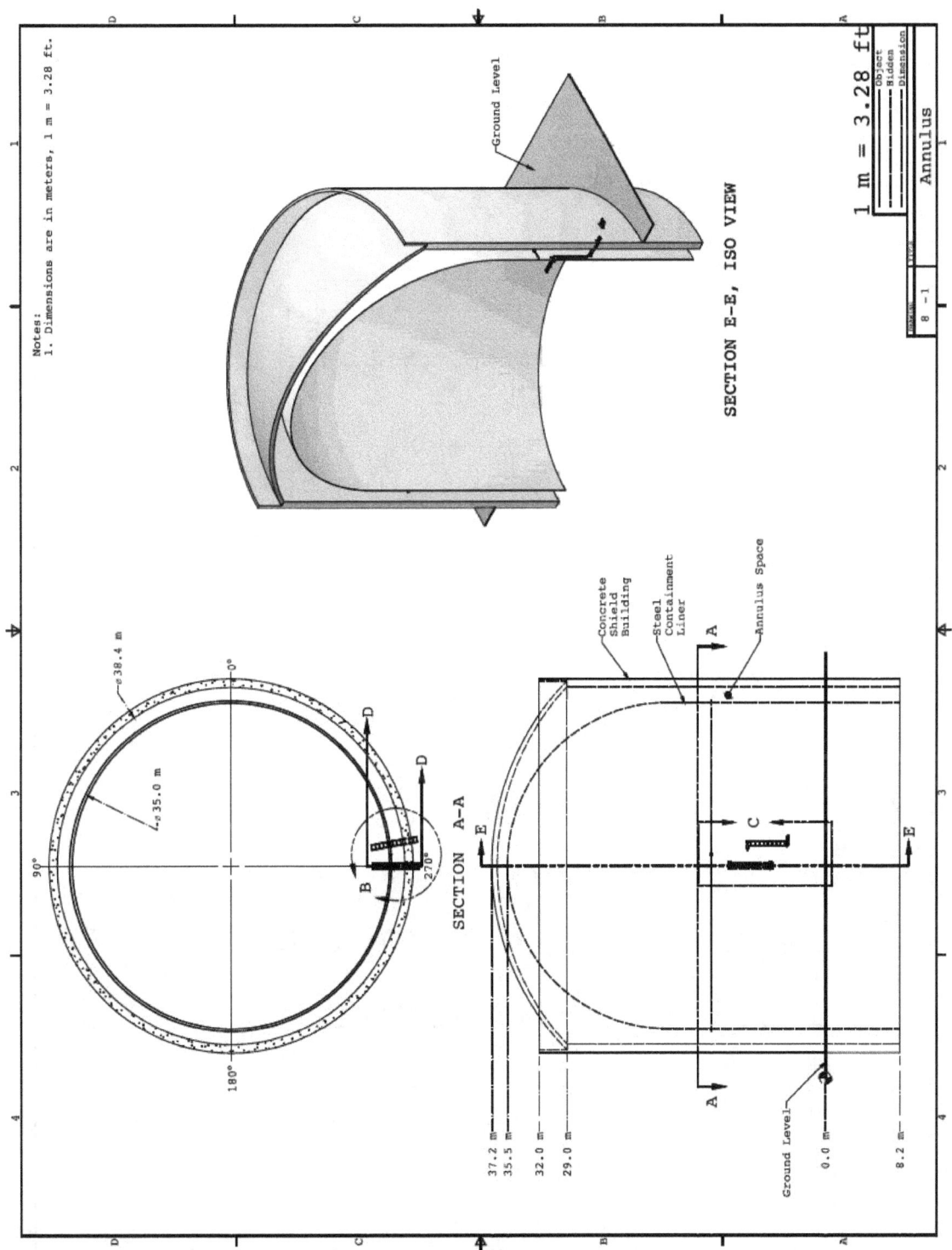

Figure H-1. Geometry of the annulus.

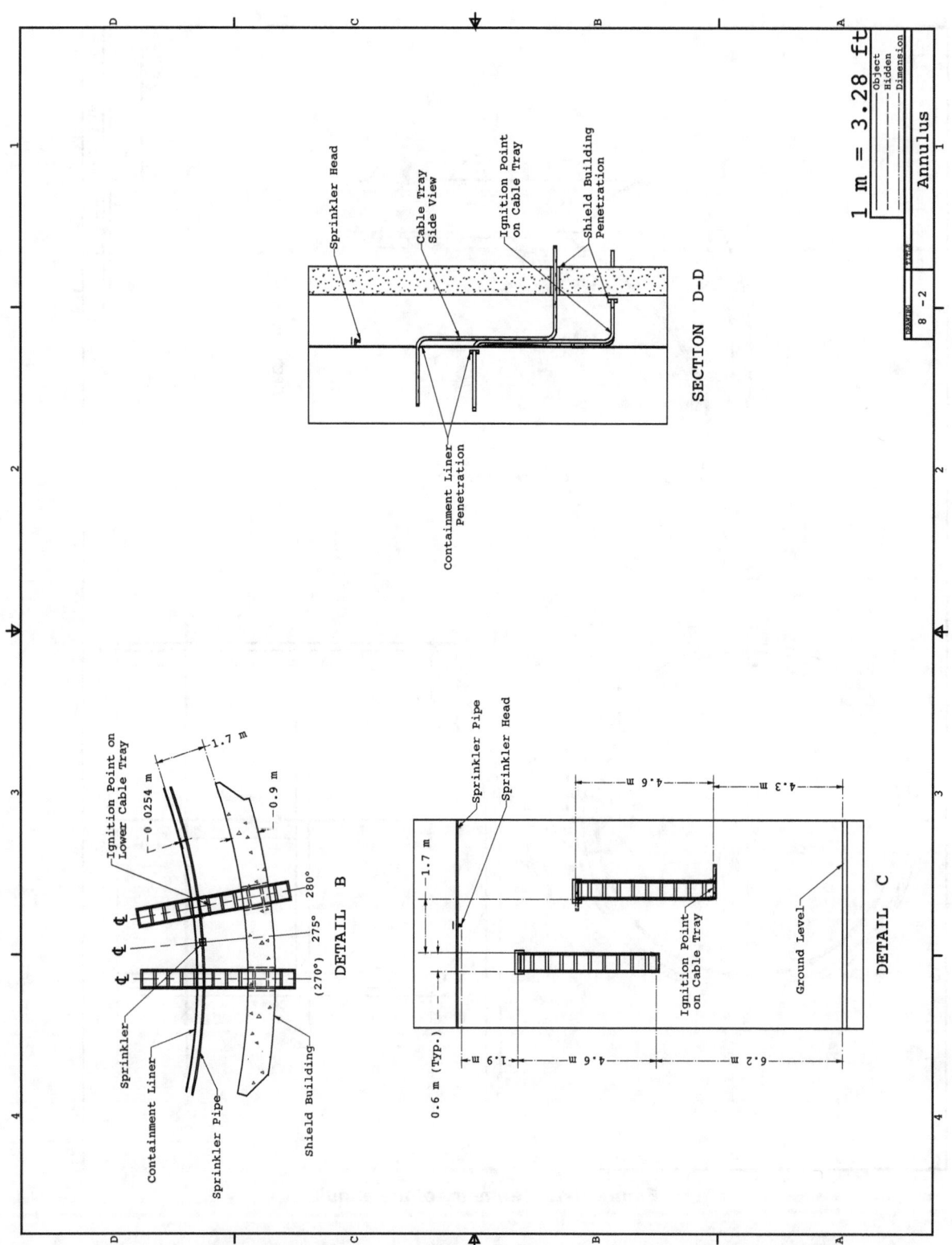

Figure H-2. Geometry details of redundant cable trays located in the annulus.

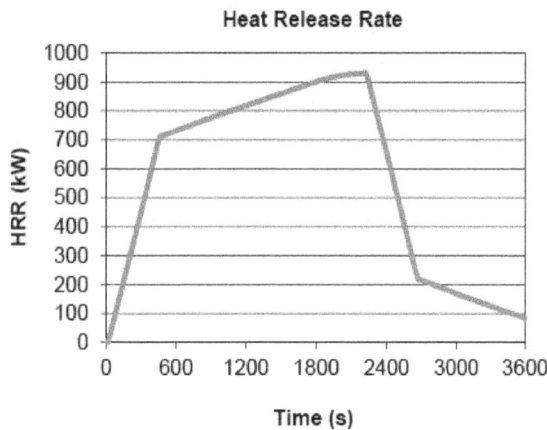

Figure H-3. HRR for a cable fire in the annulus.

Suppression System: Standard response sprinklers are located on the inner wall, as shown in Figure H-2. The sprinklers have a response time index (RTI) of 130 $(\text{m·s})^{1/2}$ and activate at a temperature of 100 °C (212 °F) (NUREG-1805, Chap. 10). Each sprinkler is topped by heat collectors designed to trap heat from a fire. The ambient temperature within the annulus region is typically 35 °C (95 °F).

H.3 Selection and Evaluation of Fire Models

This section describes the overall modeling strategy, the selection of models, and a discussion of the validation exercises justifying the use of these models for this scenario. The discussion separately addresses the prediction of the heat flux to and temperature of the cable targets, and the sprinkler activation.

H.3.1 Damage to Cables

The point source radiation heat flux model included in the Fire Dynamics Tools (FDT^S) and Fire-Induced Vulnerability Evaluation (FIVE-Rev1) provides useful screening estimates for cable damage. A simple extension of the point source heat flux model, in which the sources of heat are distributed at discrete points within the burning tray, provides a refinement of the single point source method. Given the close proximity of the two cable trays, the distributed point source model is expected to provide a more accurate estimate of the heat flux to the adjacent cable tray.

This fire scenario can be categorized as an open rather than a compartment fire. For this reason, zone models are not particularly useful. Computational Fluid Dynamics (CFD) can refine the estimates of the simple heat flux calculations to account for tray geometry and orientation.

H.3.2 Sprinkler Activation

Although the geometry of this scenario is unlike the mostly rectangular compartments found in a nuclear power plant (NPP), it is not particularly difficult to model in the Fire Dynamics Simulator (FDS). In fact, the containment building is so large that the curvature of the walls has little effect on the results of the calculation. For this reason, FDS can provide an estimate of the sprinkler link temperature to determine its potential for activation. Empirical correlations are not reliable because the fire burns along a horizontal and a vertical tray and thus has no well-defined circular base, and the plume abuts the cylindrical wall.

H.3.3 Validation

Only one of the non-dimensional parameters that have been used to characterize the fire scenarios is applicable here (Table H-2), mainly because the other parameters address phenomena unique to compartment fires. The only relevant parameter, r/D, indicates the relative distance separating the fire from the target cables. However, the effective fire diameter, D, is not well-defined when the fire is expected to spread vertically and horizontally within the cable trays. The lateral distance between the burning tray and the target tray is 1.7 m (5.6 ft). The trays are each 0.6 m (2 ft) wide. For International Collaborative Fire Model Project (ICFMP) Benchmark Exercise #3, which is described in NUREG-1824, the relative distance between the fire and the heat flux gauges in some of the tests was similar to this scenario. However, the accuracy of the solid flame and point source algorithms included in the FDT[s] varied considerably in the validation study. The point source heat flux model loses accuracy when the target is relatively close to the fire. The solid flame model is designed to improve the accuracy in the near-field. However, the implementation within the FDT[s] is based on the fire being modeled as a pool fire with a nearly circular base and cylindrical shape. This is not the case in the scenario under consideration. In short, the simple heat flux models included in the FDT[s] have been assessed in NUREG-1824 for experiments of comparable scope to the given scenario, but the models have not been demonstrated to be particularly accurate. For this reason, the models will be used to provide screening estimates, but the predictions will be supplemented by CFD calculations.

The FDS sprinkler activation algorithm was validated using a variety of experimental test series (NIST SP 1018-5). The plume algorithm was also assessed in NUREG-1824. The cable failure algorithm, THIEF, was developed and validated in NUREG/CR-6931 (Vol. 3).

Table H-2. Normalized parameter calculations for the annulus fire scenario.

Quantity	Normalized Parameter Calculation	NUREG-1824 Validation Range	In Range?
Fire Froude Number	N/A – The fire does not conform to classic fire plume theory.	0.4 – 2.4	N/A
Fire Height, $H_f + L_f$, relative to the Ceiling Height, H_c	N/A – The fire does not conform to classic fire plume theory.	0.2 – 1.0	N/A
Ceiling Jet Radial Distance, r_{cj}, relative to the Ceiling Height, H_c	N/A – The ceiling height is essentially infinite.	1.2 – 1.7	N/A
Equivalence Ratio, φ, as an indicator of the Ventilation Rate	N/A – The scenario is outside of a clearly defined compartment.	0.04 – 0.6	N/A
Compartment Aspect Ratio	N/A – The scenario is outside of a clearly defined compartment.	0.6 – 5.7	N/A
Target Distance, r, relative to the Fire Diameter, D	See discussion in Section H.3.3.	2.2 – 5.7	Yes

H.4 Estimation of Fire-Generated Conditions

This section provides specific details on how each model is set up and run.

H.4.1 Algebraic Models

The FDT^s contain several correlations for estimating the heat flux at a fixed distance from a fire:

- 05.1_Heat_Flux_Calculations_Wind_Free.xls (Point Source)

- 05.1_Heat_Flux_Calculations_Wind_Free.xls (Solid Flame 1)

For the point source method, the estimated peak HRR, $\dot{Q}$, is 945 kW, the radiative fraction, χ_r, is 0.49, and the horizontal distance from the center of the burning tray to the edge of the target tray, r, is 2 m (6.6 ft). The calculated heat flux is:

$$\dot{q}''_{ps} = \frac{\chi_r \dot{Q}}{4\pi r^2} = \frac{0.49 \times 945 \text{ kW}}{4\pi \times 2.0^2 \text{ m}^2} \cong 9.2 \text{ kW/m}^2 \tag{H-3}$$

The solid flame calculation is based on the fire having a roughly circular base with which it employs a flame height correlation to estimate the vertical extent of the luminous flame region.

This is not the case for the vertical and horizontal trays filled with burning cables. However, it is possible to use a variation of the point source method to get a refined heat flux estimate that is similar in scope to a solid flame model. Essentially, the fire can be divided into several point sources along the length of the burning cable tray so that the radiative energy is not emanating from just one point. Suppose that the 4.6 m (15 ft) vertical tray segment is divided into four segments, each one 1.15 m (3.8 ft) long and 0.6 m (2 ft) wide. The horizontal tray is the fifth segment, 1.7 m (5.6 ft) long and 0.6 m (2 ft) wide. Each segment generates energy at a rate of 250 kW/m^2. This leads to 172.5 kW for each of the four vertical segments and 255 kW for the horizontal segment. The sum is 945 kW. The distance between the center points of the segments and a point at the intersection of the horizontal and vertical trays nearest to the burning cable trays can be calculated, and the combined heat flux from this set of distributed sources can be calculated:

$$\dot{q}''_{dps} = \frac{\chi_r}{4\pi} \sum_i \frac{\dot{Q}_i}{r_i^2} = \frac{0.49}{4\pi}\left(\frac{255}{2.9^2} + \frac{172.5}{2.4^2} + \frac{172.5}{2.0^2} + \frac{172.5}{2.2^2} + \frac{172.5}{2.9^2}\right)\frac{kW}{m^2} \cong 6.2 \text{ kW/m}^2 \tag{H-4}$$

Much like the solid flame method, the *distributed* point source method provides a refined estimate of the heat flux, based on the fact that the fire is distributed over the two parts of the tray and is not concentrated at a single point. The end result is a lower estimate of the heat flux to a given target point. The fire could be distributed over more than five points, but the answer would not significantly change. Neither the point source nor the distributed point source method accounts for changes in the orientation of the target. The target cables are modeled as having a direct view of the fire with no obstructions, such as the sides of the trays, that can reduce the heat flux to the target cables. The CFD model can provide a more refined estimate.

H.4.2 CFD Model

Geometry: Only the section of the annulus encompassing the cables and relevant targets is included in the computational domain. This volume is 9.6 m (31.5 ft) wide, 2.5 m (8.2 ft) deep, and 12.8 m (42 ft) high. Extra depth is needed to accommodate the slight curvature of the bounding walls. The top, bottom, and sides of the computational domain are specified as open, that is, open to an infinitely large volume. Since the volume of the annulus is very large, neither smoke build-up nor pressure effects would influence the region near the cables. Both the internal and external walls of the annulus are included in the model. Since FDS only allows rectilinear obstructions, a series of obstructions approximately 20 cm (7.9 in) thick approximate the curved walls. The numerical grid conforms to this "stair-stepped" geometry.

Fire: The fire ignites near the base of the vertical portion of the cable tray attached to the inner wall. The spread rates of 258 mm/s (10 in/s) in the vertical direction and 0.9 mm/s (0.035 in/s) in the horizontal are input by using an FDS feature in which a surface is designated as having a fire spread over it at a designated rate. In this case, a surface is specified along the side of the vertical tray and along the top of the horizontal tray with the respective spread rates. The HRR per fire unit area is specified directly and not predicted by the model. A Smokeview rendering of the FDS simulation is shown in Figure H-4.

Cables: One of the objectives of the calculation is to estimate the potential damage to the cables within the redundant train. FDS is limited to only 1-D heat transfer into either a rectangular or cylindrical obstruction. In this simulation, the cables are modeled as 1.5 cm

(5.9 in) cylinders. Following the Thermally-Induced Electrical Failure (THIEF) methodology in NUREG/CR-6931, Vol. 3, electrical functionality is lost when the temperature just inside the 1.5 mm (0.06 in) jacket reaches 205 °C (400 °F). Since the objective of this calculation is to estimate time to failure of the redundant cables, ignition and spread of the fire over the second set of cables is not considered. The in-depth heat penetration calculation is focused on a single cable that is relatively free of its neighbors and that would heat up more rapidly than those buried deeper within the pile.

Sprinkler Activation: FDS uses the conventional RTI concept to predict sprinkler activation. In this scenario, a steel plate has also been added just above the location of the sprinkler to simulate the effect of the actual deflector. Note that the sprinkler itself is just a point in the model, and its activation is determined by the time history of the temperature and the velocity of the hot gases within the numerical grid cell in which the sprinkler is located.

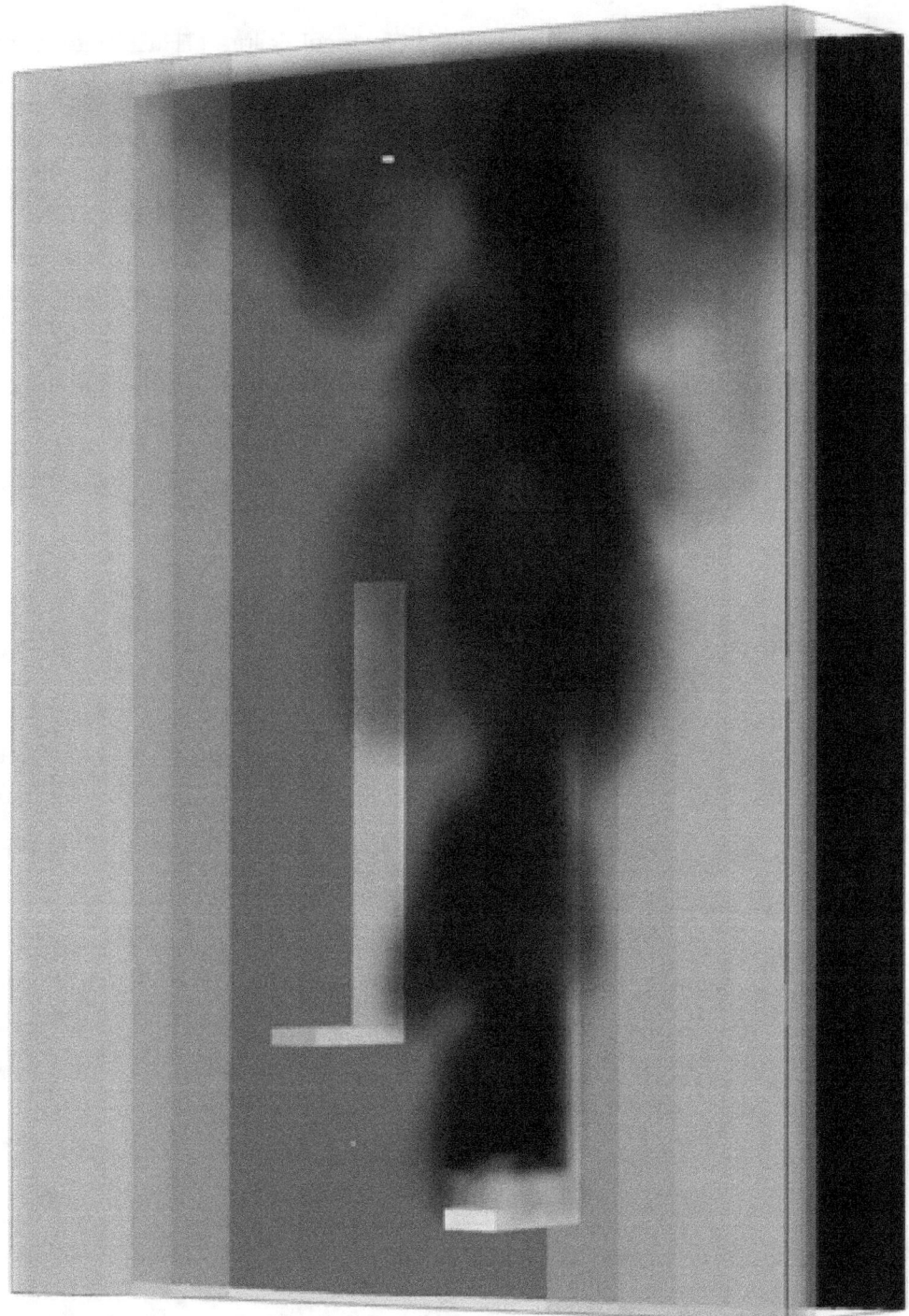

Figure H-4. FDS/Smokeview rendering of the annulus fire scenario.

H.5 Evaluation of Results

The calculations described above assess the potential for damage to redundant safe-shutdown cables due to a fire in an adjacent tray in the annulus region of the containment building. In addition, a CFD calculation is used to determine whether the fire would activate a sprinkler. The results of the calculations are summarized in Table H-3.

Table H-3. Summary of model predictions for the annulus fire scenario.

Model	Bias Factor, δ	Standard Deviation, $\tilde{\sigma}_M$	Target	Predicted Value	Critical Value	Probability of Exceeding
Heat Flux (kW/m²)						
Point Source	1.42	0.55		9.2	6.0	0.553
Distributed Point Source	1.42	0.55	Cables	6.2	6.0	0.248
FDS	1.10	0.17		2.5	6.0	0.000
Target Temperature (°C)						
FDS	1.02	0.13	Cables	120.0	205.0	0.000
Plume Temperature (°C)						
FDS	1.15	0.11	Sprinkler	90.0	100.0	0.001

H.5.1 Heat Flux and Temperature

The simple point source heat flux calculations are used as part of a screening analysis to determine whether more detailed calculations are warranted. Using the peak HRR as a constant input, the point source model estimates the heat flux to be 9.2 kW/m². A slightly more refined estimate using distributed point sources decreases this estimate to 6.2 kW/m². Both of these estimates suggest a potential for damage to the redundant train of cables.

A more detailed calculation is performed with FDS, which yields a predicted peak heat flux of approximately 2.5 kW/m² (see Figure H-5). The reason that the FDS prediction of heat flux is significantly lower than the point source method predictions is that (1) FDS accounts for the side of the tray that blocks much of the thermal radiation, and (2) the radiation transport equation solved by FDS accounts for the orientation of the target tray relative to the fire. In other words, FDS does not model the target cables as directly facing the fire, whereas the point source models implicitly do.

As with the heat flux, the predicted interior cable temperatures predicted by FDS indicate a very low probability for cable damage. However, it should be noted that these low probabilities are based on the particular configuration of the cables within the trays. If some of the target cables were to have a direct view of the fire, the heat flux would more likely be comparable to that predicted by the distributed point source model.

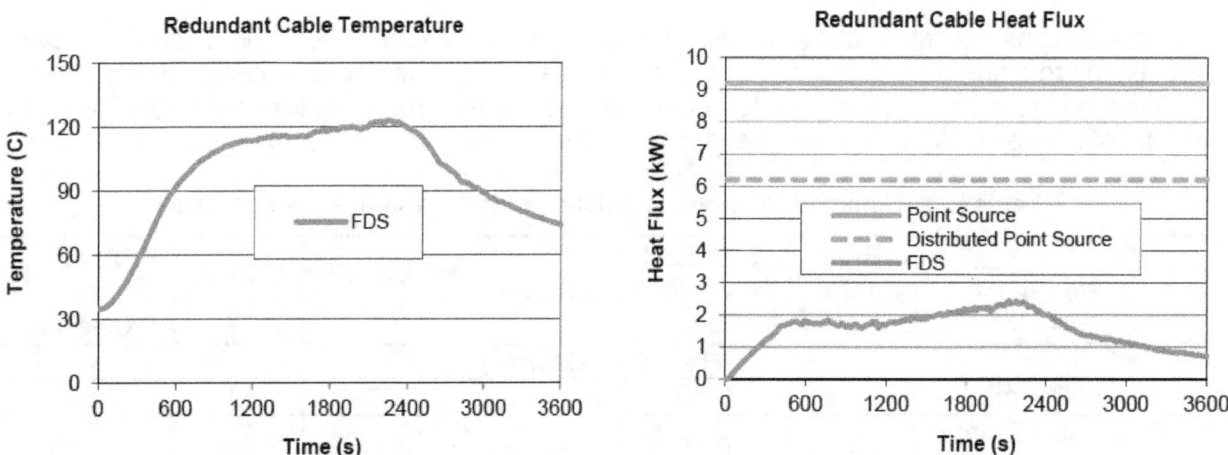

Figure H-5. Summary of simulation results for the annulus.

H.5.2 Fire Protection Systems

The sprinkler link temperature predicted by FDS is shown in Figure H-6. The sprinkler is not predicted to activate in this scenario because the link temperature is predicted to increase to approximately 90 °C (194 °F), less than the activation temperature of 100 °C (212 °F). It should be noted, however, that the sprinkler is located just outside the fire plume. It is expected that for a real fire of this type, the natural air movements within such a large space as the containment annulus would almost certainly bend the plume from the vertical in a way that would be difficult to replicate with a model that is not accounting for the air movements throughout the entire facility. A slight lean in the plume towards the sprinkler could result in activation.

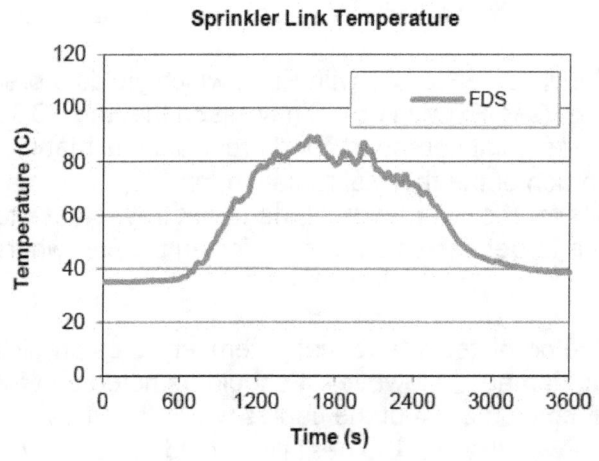

Figure H-6. Predicted sprinkler link temperature for the annulus fire scenario.

It is possible to determine how large a fire is needed to increase the plume temperature by 10 °C. Table 4-3 indicates that the HGL temperature rise is proportional to the HRR to the 2/3 power. The plume temperature behaves similarly. Following the methodology in Section 4.4, in order to increase the predicted plume temperature by 10 °C, the peak HRR, $\dot{Q}$, must increase by approximately:

$$\Delta\dot{Q} = \frac{3}{2}\dot{Q}\frac{\Delta T}{T - T_0} = \frac{3}{2}945 \text{ kW} \times \frac{10\text{ °C}}{90\text{ °C} - 35\text{ °C}} \cong 258 \text{ kW} \tag{H-5}$$

In other words, the peak HRR of the fire would have to be approximately 945 + 258 = 1203 kW to bring the plume temperature into a range to cause sprinkler activation.

H.6 Conclusion

Simple point source heat flux calculations indicate that a fire in one of the cable trays within the annulus region of the containment building might damage the cables in an adjacent tray. However, an additional analysis using FDS indicates that cable damage is unlikely due to the orientation of the target cables and the blockage of thermal radiation by the cable tray itself. This suggests that the details of the cable tray location, orientation, and configuration can significantly impact potential for damage.

FDS predicts that sprinkler activation above the fire is unlikely. However, its prediction is sensitive to the exact location of the sprinkler relative to a fire plume that may be subject to unpredictable air movements throughout the entire facility. Alternative protection strategies, such as shielding between trays or other thermal barriers, should be considered to ensure the protection of the redundant cables.

H.7 References

1. NIST SP 1018-5, *Fire Dynamics Simulator (Version 5), Technical Reference Guide, Volume 3, Experimental Validation*, 2010.
2. NUREG-1805, *Fire Dynamics Tools (FDT^s) Quantitative Fire Hazard Analysis Methods for the U.S. Nuclear Regulatory Commission Fire Protection Inspection Program*, 2004.
3. *NUREG-1824 (EPRI 1011999), Verification and Validation of Selected Fire Models for Nuclear Power Plant Applications*, 2007.
4. NUREG/CR-6850 (EPRI 1011989), *EPRI/NRC-RES Fire PRA Methodology for Nuclear Power Facilities*, 2005.
5. NUREG/CR-6931, *Cable Response to Live Fire (CAROLFIRE), Volume 3: Thermally-Induced Electrical Failure (THIEF) Model*, 2008.
6. *NUREG/CR-7010, Cable Heat Release, Ignition, and Spread In Tray Installations during Fire (CHRISTIFIRE), Volume 1: Horizontal Trays*, 2012.
7. *SFPE Handbook of Fire Protection Engineering*, 4th edition, 2008.

H.8 Attachments (on CD)

1. FDS input file: Annulus.fds

INDEX

INDEX

U.S. NUCLEAR REGULATORY COMMISSION

BIBLIOGRAPHIC DATA SHEET

(See instructions on the reverse)

1. REPORT NUMBER
(Assigned by NRC, Add Vol., Supp., Rev., and Addendum Numbers, if any.)

2. TITLE AND SUBTITLE

3. DATE REPORT PUBLISHED

MONTH	YEAR

4. FIN OR GRANT NUMBER

5. AUTHOR(S)

6. TYPE OF REPORT

7. PERIOD COVERED (Inclusive Dates)

8. PERFORMING ORGANIZATION - NAME AND ADDRESS (If NRC, provide Division, Office or Region, U. S. Nuclear Regulatory Commission, and mailing address; if contractor, provide name and mailing address.)

9. SPONSORING ORGANIZATION - NAME AND ADDRESS (If NRC, type "Same as above", if contractor, provide NRC Division, Office or Region, U. S. Nuclear Regulatory Commission, and mailing address.)

10. SUPPLEMENTARY NOTES

11. ABSTRACT (200 words or less)

12. KEY WORDS/DESCRIPTORS (List words or phrases that will assist researchers in locating the report.)

13. AVAILABILITY STATEMENT
unlimited

14. SECURITY CLASSIFICATION

(This Page)
unclassified

(This Report)
unclassified

15. NUMBER OF PAGES

16. PRICE

Printed
on recycled
paper

Federal Recycling Program

UNITED STATES
NUCLEAR REGULATORY COMMISSION
WASHINGTON, DC 20555-0001

OFFICIAL BUSINESS

NUREG-1934

Nuclear Power Plant Fire Modeling Analysis Guidelines (NPP FIRE MAG)

November 2012

www.ingramcontent.com/pod-product-compliance
Lightning Source LLC
Chambersburg PA
CBHW080236180526
45167CB00006B/2301

* 9 7 8 1 4 9 9 6 2 4 1 0 6 *